COST
MANAGEMENT
OF
CAPITAL
PROJECTS

T0203603

COST ENGINEERING

A Series of Reference Books and Textbooks

Editor

KENNETH K. HUMPHREYS, Ph.D.

Consulting Engineer
Granite Falls, North Carolina

COST MANAGEMENT OF CAPITAL PROJECTS

KURT HEINZE
Heinze & Associates
Toronto, Ontario, Canada

CRC Press
Taylor & Francis Group
Boca Raton London New York

CRC Press is an imprint of the
Taylor & Francis Group, an **informa** business

CRC Press
Taylor & Francis Group
6000 Broken Sound Parkway NW, Suite 300
Boca Raton, FL 33487-2742

First issued in paperback 2019

ISBN-13: 978-0-8247-9783-6 (hbk)
ISBN-13: 978-0-367-40123-8 (pbk)

Library of Congress Cataloging-in-Publication Data

Heinze, Kurt.
  Cost management of capital projects / Kurt Heinze.
      p. cm. (Cost engineering ; 27)
  Includes bibliographical references and index.
  ISBN 0-8247-9783-3 (hardcover : alk. paper)
  1. Engineering economy. I. Title. II. Series: Cost engineering (Marcel Dekker, Inc.) ; 27.
TA177.4.H45   1996
658.15'52—dc20   96-23984

Visit the Taylor & Francis Web site at
http://www.taylorandfrancis.com

and the CRC Press Web site at
http://www.crcpress.com

PREFACE

For a project to be successful in today's cost-driven business environment, the return on investment is an overriding factor. All significant design, procurement and scheduling decisions must be made by *optimizing economic cost*. Only a cost conscious project team can achieve this. Members of the team must develop *the skill of cost management* in addition to their individual technical skills.

This book describes how to manage cost effectively. The focus is on how the project team can accomplish its goals in a disciplined way without compromising the creativity that must be inherent in cost control. Supported by tables and graphs, the text is unique as it contains a mosaic of knowledge modules that together form a comprehensive picture of an integrated cost management system.

Chapter 1 is devoted to elementary business calculations and offers a short introduction to management science. Modern project management requires that the reader have a basic elementary knowledge of business mathematics to appreciate the application of complicated management functions.

Chapter 2 shows how construction and operation fit into the various stages of a capital project. It also discusses the elements of the decision-making process, including the concept of risk management.

Chapter 3 deals with the planning phase of a capital project. It introduces the concept of an integrated project organization and demonstrates how scheduling and cost estimating are used as a base on which to measure performance.

Chapter 4 emphasizes a systematic approach to cost control, how collected costs are compared with the budget, and the effect of changes on project cost.

Chapter 5 deals with cost reporting and describes how the communication of facts can be structured to be conducive to action. It also shows how to identify misleading information that may unintentionally enter the system.

Chapter 6 defines productivity and production. It also describes performance standards and adjustments to a uniform base and outlines a system that measures performance against predetermined standards and explains how to report perform-
,ance.

Chapter 7 outlines the use of method study techniques to increase the effectiveness of systems and to improve the productivity of resources, thereby enhancing the technical aspects of cost management. This topic is seldom found in standard references or textbooks of this nature.

Because of the diverse educational paths of management practitioners, the book intends to *build a bridge* between the academic quantitative world of management science and the more practical world of project cost management.

The text includes some examples and exercises that the novice can use. The examples are in metric units to appeal to an international readership and to meet the need for global standards.

This book serves as an introduction to project cost management for senior level undergraduate students and graduates in business and engineering. It also gives the working professional a better understanding of cost-effective management. In addition, it can be used effectively for professional seminars and training courses.

Kurt Heinze

CONTENTS

Contents

CHAPTER 1
INTRODUCTION TO BASICS

The objectives of this chapter are:

1. To discuss basic prerequisites needed to fully understand the content in the remainder of the text.
2. To provide examples for readers who want to become more proficient with the basics of productivity and cost control.

How often have we read a technical text on a professional subject and had to skip over some mathematical treatment of a problem because

- ◆ We had forgotten basic mathematics from our school days
- ◆ We never did get that far in our education, but we know enough to build on it
- ◆ We are reluctant to make the effort to review related texts in order to recall the basics from our rusty memory?

To help the reader fully understand the material in this book and appreciate the techniques of project cost management, the first chapter is devoted to elementary business calculations and a short introduction to management science.

Introduction

Knowledge is a treasure; practice is the key to it. It is never too late to learn.
Old proverbs

"Knowledge is power" goes an old saying, and that is also true in the field of business management. We often neglect to include "project" management in this category. We seem to believe that project managers come up through the ranks or are promoted into the job because they are excellent engineers or accountants.

While pertinent experience is definitely an asset, it is not the only criterion. Modern project management is knowledge intense. It needs both, theoretical and practical knowledge. The use of productivity software does not turn an individual into a qualified project manager.

There have been students aspiring to become managers, believing that they do not need to know the basics of cost management theory because they will have staff to do this work for them. This is shortsighted because they would not be able to communicate knowledgeably with those who report to them, neither would they be able to make informed decisions. Therefore, a basic elementary knowledge of business mathematics is required to appreciate the application of complicated management functions.

1.1 BASIC PREREQUISITES

This section deals with elementary and basic statistical calculations often used in cost engineering. Some examples are provided.

1.1.1 Elementary Business Calculations

Ratios

The base of a ratio is its unit. The base of the ratio $\frac{5}{9}$ is $\frac{1}{9}$. It has five units.

Ratios can be expressed differently depending on the application. We can say:

> Five out of nine
> Five over nine
> 0.556 or 55.6% etc.

Ratios are applicable when quoting unit rates in contracts, in measuring productivity, accuracy in estimating, trending etc. Ratios cannot be added unless they have a common base (denominator).

Statistical Ratios must have meaningful relationships to merit analysis. (A certain learning institution released their statistics quoting that 33% of all their women students were pregnant. They forgot to mention that out of 100 students only three were women, one of whom was pregnant).

Errors are often made when quoting ratios. An advertising sign for war surplus stated: "We are selling for 200% less than the government paid for it!" Here, the base is 100%. A decrease by 100% means zero.

Hybrid Ratios such as km/h, $/workhour, m^3/day etc. are usually called *rates*. It should be recognized that they are not mathematical ratios. They are actually averages. A work crew installing piping at the rate of 13 m/h on a job will have installed the pipe at various lengths during the installation period. At the end of the job, we divide the total length installed by the hours it took to do the job.

Example:

Q #1.01: A newspaper reported that the number of smokers in the USA has gone down from 45 % (20% women, 25% men) to 35% (15% and 20%) in one year. What is the ratio of smokers to non-smokers for each of the two years?

A #1.01: The newspaper report is wrong! It should not have added the male and female percentages, because male and female populations are almost evenly distributed.

The correct answer is: $(20 + 25)/2 = 22.5\%$ and $35/2 = 17.5\%$ smokers.

Proportions

Proportions express size related ratios such as $4/8 = 2/4$. If one of the numbers is unknown, it can be calculated:

For $4/x = 2/4$, $x = \dfrac{4 \times 4}{2} = 8$ This is often used in making existing mathematical

tables more accurate. It is called *interpolation*.

Q #1.02:
If we have the following mathematical table which shows a listing of logarithm. What is the log of N = 35114? (To solve this problem it is not necessary to be familiar with logarithm; any mathematical table could have been used as an example).

N	1	2	3	4
350	54419	54432	54444	etc.
351	**54543**	**54555**	54568	etc.
352	etc.	etc.	etc.	etc.

A #1.02: The result (mantissa) lies between 54543 and 54555 and the difference between the two numbers is 12. This corresponds to the difference

$$N' = 35120 - 35110 = 10, \text{ and } N'' = 35114 - 35110 = 4$$

by proportion, $\dfrac{x}{4} = \dfrac{12}{10}$ $\therefore x = \dfrac{12 \times 4}{10} = 4.8$; and, the mantissa for 35114 is

$54543 + 4.8 = 545478$. Since the characteristic of a 5-digit number is 4.xxxx,

log 35114 = 4.545478
This can also be repre-
sented graphically:

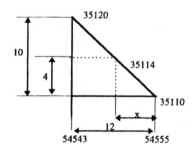

Q #1.03: What is the standard deviation of $Z = 1.016$?

Z	.00	.01	.02	etc.
0.9	.3159	.3186	.3212	etc.
1.0	.3413	**.3438**	**.3461**	etc.
1.1	.3643	.3665	.3686	etc.

A #1.03: The number lies between 0.34<u>38</u> and 0.346<u>1</u>. The difference is 0.00<u>23</u>. This corresponds to 1.010 and 1.020 with a difference of 0.010. Therefore

$$x = \frac{0.0023 \times (1.016 - 1.010)}{0.010} = 0.00138$$

and for $Z = 1.016$: $0.34380 + 0.00138 = 0.34528$

Q #1.04: A project has a net cashflow (expenses and receipts after taxes, but before depreciation) as follows:

$$-1000 \quad +400 \quad +300 \quad +200 \quad +400 \quad +600$$

+400 +300 +200 +400 +600

years   0   1   2   3   4   5

−1000

Find the payout time in months.
Note: The payout time can be found by calculating the location of the break-even point (cumulative zero)!

A #1.04: Cumulative cashflow is: (1000); (600); (300); (100); +300; +900. The break-even point is therefore located before the end of the fourth year.

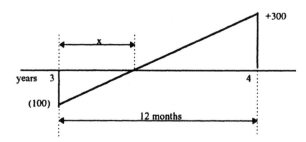

$$x/12 = 100/(100 + 300) \quad \text{and} \quad x = 3 \text{ months.}$$

The payout time is therefore **39** months.

Averages

The common expression "averages" has many different definitions. It can mean "typical" or "usual" (the "average" cost engineer) or it could refer to the result of a very specific process of calculation. There are five types of specific averages:

1) Simple Arithmetic Mean
The most common average is the *simple arithmetic mean* (equally weighted). It is the sum of all items over the number of items

$$\overline{X} = \frac{\sum x}{N} = \frac{25 + 35}{2} = \frac{60}{2} = 30$$

2) Weighted Arithmetic Mean
The *weighted arithmetic mean* uses hybrid ratios to establish its average. Therefore, quantities should be taken into account. It is the sum of the weighted items over the sum of the weights

$$x = \frac{\sum pq}{\sum q}, \quad \text{where q is the weight.}$$

Q #1.05: Assume a contractor paid $120.00/m³ of concrete last month and $140/m³ this month. What is the average price per m³ over the two month period?

A #1.05: It is unlikely to be (120 + 140)/2 = $130/m³ unless the same quantity was bought each month.
Let
$$p = \text{price in \$/unit}$$
$$q = \text{quantity,}$$

then average price $= \dfrac{p_1q_1 + p_2q_2 + \cdots\cdots p_nq_n}{\sum_1^n(q)}$

Assume the quantities were $60m^3$ (q_1) last month and $20m^3$ (q_2) this month. The answer is:

$$\frac{120q_1 + 140q_2}{q_1 + q_2} = \frac{7200 + 2800}{60 + 20} = \frac{10\,000}{80} = \$125/m^3$$

We can use relative weights whereby the weights will total to unity. To do this, we divide each weight by the sum of the weights:

$$60/80 + 20/80 = 3/4 + 1/4$$

and, applying related quantities,

$$\frac{3(120)}{4} + \frac{1(140)}{4} = \frac{500}{4} = \$125/m^3$$

Q #1.06: A foreman monitors the percentage of idle time for various jobs performed by his crew during a week. Total hours worked = 250 Whrs (work hours):

Job #1 = 10% idle time
Job #2 = 4% idle time
Job #3 = 10% idle time
24% idle time in total

He reports average idle time 24/3 = 8% of 250 = 20 Whrs. Is this correct?

A #1.06: No! When individual quantities are not known, a precise answer cannot be given. A better approximation would have been the geometric mean:

$$(10 \times 4 \times 10)^{1/3} = 7.37\% \text{ or } 18.4 \text{ Whrs.}$$

We always should ask X% of what?

Assuming that active time for each job is 70, 120, and 60 Whrs, in that order, then the weighted average idle time is

$$\frac{p_1q_1 + p_2q_2 + \cdots + p_nq_n}{\sum_1^n q} = \frac{10 \times 70 + 4 \times 120 + 10 \times 60}{70 + 120 + 60} = 7.12\% = \mathbf{17.8} \text{ hours,}$$

which is the correct answer.

Q #1.07: A range estimate shows the following degree of confidence for each item estimated:

Item	Optimistic	Most likely	Pessimistic
1	− 10%	50 k$	+ 20%
2	− 5%	100 k$	+ 15%
3	− 15%	200 k$	+ 30%
1 to 3:	− 10%	350 k$	+ 22%

Those totals were shown on the management report. Using a simple arithmetic method, check if this is correct.

A #1.07: It is not! Each item must be calculated:

	45	50	60
	95	100	115
	170	200	260
	310	350	435
=	−11%	350 k$	+24%

Q #1.08: A manager asks the cost engineer to obtain the average unit price on a structural steel erection job. The following figures are available to the engineer:

Description	Unit rate	Quantity
Columns	$100	50 Mg (metric tons)
Girders	$120	100 Mg
Decking	$140	200 Mg

A #1.08: Each weight is divided by the sum of the weights first:

$$\frac{50 + 100 + 200}{350} = \frac{1 + 2 + 4}{7} ; \text{ unit rates are now applied, i.e.}$$

$$\frac{1 \times 100 + 2 \times 120 + 4 \times 140}{7} = \frac{100 + 240 + 560}{7} = \$ 128.57/Mg$$

3) The Geometric Mean
It is an approximation with limited application. It is the Nth root of the product of all numbers larger than zero.

$$G = (\Pi x)^{1/N} ; \quad \log G = \frac{\sum \log x}{N}$$

Q #1.09: A manufacturer decided that the output in a process should triple on the average over the next three years to obtain the desired return on investment. The shop foremen doubled the output in the first year, tripled the output in the second year and quadrupled the output in the third year. Did the manufacturer meet his goal?

A #1.09: No, the stated mean shows that the average output over the three years was

$$(2 \times 3 \times 4)^{1/3} = 2.88 \; ; \; \text{(short of the goal by 4\%)}.$$

It should have been $(3 \times 3 \times 3)^{1/3} = 3.00$ (triple on the average).

To compare, $3 \times 3 \times 3 = 27$, whereby $2 \times 3 \times 4 = 24$

When quantities are not known, a precise answer cannot be given. We must estimate average quantities and state that the answer is an estimate. If only percentages are known, an approximation can be obtained by using the *geometric mean*. This is the Nth root of the product of the numbers.

For example: Assume that the absence of staff over a fixed period in five departments is 1.9%, 1.0%, 0.9%, 10.0%, 6.2%. This gives an average (simple arithmetic mean) of 4%. Without knowing the strength of the departments, this answer is incorrect. Let us assume there were 58 workdays lost from a total 2574. This is 2.25%. By using the geometric mean, we obtain the following result:

The product of $1.9 \times 1.0 \times 0.9 \times 10.0 \times 6.2 = 106.2$; but $\sqrt[5]{106.2} = 2.54$, a much closer approximation.

4) The Harmonic Mean:
Certain types of ratios can be averaged directly by using the *harmonic mean*. It is an approximation which excludes zero. It can be applied when non-mixed ratios are averaged directly.

$$H = \frac{N}{\sum\left(\frac{1}{x}\right)}$$

The harmonic mean is applicable only if the same number of units of the numerator category of the ratio apply to all the items being averaged.

Explanation: Items being averaged: km/hr (hybrid ratio)
Numerator category: km   (equal for all items)
Number of items: N

If the denominator (hr) is chosen as being equal for all items, the simple arithmetic mean applies, not the harmonic mean.

Lets use an every day example. A driver travels at 60 km/h to a point 60 km away. She then makes the return trip at an average speed of 40 km/h. What is the average speed for the 120 km round trip?

It is not (60+40)/ 2 = 50 km/h, it is

$$\frac{2}{1/60 + 1/40} = 48 \text{ km/h}$$

The harmonic mean is a special case of the weighted arithmetic mean. By substituting Σq for N and $\Sigma q/p$ for $\Sigma(1/x)$, we obtain

$$\frac{\Sigma q}{\Sigma q/p}$$

Applying this to our previous example of concrete installation, whereby a contractor bought 60 m^3 concrete at $120.00/m^3 last month and 20 m^3 at $140/m^3 this month:

$$\frac{\Sigma q}{\Sigma q/p} = \frac{60 + 20}{60/120 + 20/140} = \$125 / m^3$$

Q #1.10: Three types of valves are priced at $50.00, $100.00 and $120.00 each. 20 of each type are purchased for a project. Using the shortest way of calculation, what is the average price per valve?

A #1.10: Since the quantity (denominator) is constant for each type and the dollars (numerator) vary, the simple arithmetic mean applies:

$$\overline{X} = (50 + 100 + 120)/3 = \$ 90.00/valve.$$

Q #1.11: The same three types of valves are being purchased with the requirement that the same amount of money be spent on each type of valve. What is the average cost per valve? How many valves of each type can we now purchase?

A #1.11: Since the variable ($/valve) needs to be averaged directly and the cost for each type (numerator) is the same, the harmonic mean can be used:

$$\frac{3}{1/50 + 1/100 + 1/120} = \frac{3}{0.0383} = \$78.26/valve$$

Total amount of purchase order = 20 × 3 × $90.00 = $5400.00
Number of valves purchased = 5400.00/78.26 = 69 valves, and for each type
1800/50 + 1800/100 + 1800/120 = 36 + 18 + 15 = 69 valves.

Q #1.12: To establish an average base unit rate, a company looks at the record of five similar jobs for the daily installation of drywall. The record shows 50 m^2/day, 40 m^2/ day, 45 m^2/day, 35 m^2/day, 55 m^2/day. What is the average base unit rate?

A #1.12:
$$X = (50 + 40 + 45 + 35 + 55)/5 = 45 \ m^2/day$$

Q #1.13: Using the same record for dry wall installation (Q #1.12) but rephrasing the question: The same m^2 of surface area need to be installed. What is the average area per day?

A #1.13:

$$H = \frac{5}{1/50 + 1/40 + 1/45 + 1/35 + 1/55} = \frac{5}{0.114} = 43.86 \text{ m}^2/\text{day}$$

Q #1.14: There are two brands of pencils; one is priced at $0.40 per pencil, the other at $0.60 per pencil. The office supervisor wants to purchase a bulk amount for both brands. Her total budget is $ 100.00. What is the average price for the two brands she purchased? Analyze the question and explain how you derive at an answer.

A #1.14: This question cannot be answered unless it specifies the following:
   a) how many pencils of each brand are to be purchased? or
   b) how much money is to be spent on each brand?
   If the same number of pencils is to be purchased for each brand, then the answer is (denominator is constant):

$$X=(40+60)/2 = \$0.50/\text{pencil or } \$100/0.5 = 200 \text{ pencils}$$

If the same amount of money (numerator) is to be spent on each brand, the harmonic mean will give the correct answer?

$$H = \frac{2}{1/40 + 1/60} = \$0.48/\text{pencil or } \$100/0.48 = 208 \text{ pencils}$$

The number for each brand is $50/0.4 = $50/0.6 = 125 + 83 pencils respectively.

5) The Median
Another of the averages is the *median*. It is the center of a distribution of numbers, i.e. the number of items which precede it is the same as the number of items that follow it. It is mostly used in statistics and applied to range estimating. It is generally defined as the center of a frequency distribution of data grouped in the order of magnitude.

$$Md = N/2$$

It divides the number of items into two equal portions. For ungrouped data,

$$Md = (N+1)/2$$

Example: For the series 1 3 4 6 11, the median is the third item in the series

$$(N + 1)/2 = 6/2 = 3^{rd} \text{ item} = \text{number 4}$$

for those uneven number of digits. If the series has an even number of digits, e.g. 1-3-4-6-11-13, the median is the arithmetic mean of the third and the fourth item, = (4+6)/2 = 5 at the position (N + 1)/2 = 7/2 = 3.5.

The median can only be determined if the numbers are listed in the order of magnitude (array).

For grouped data, the median is at a distance of Md = N/2 item segments (not individual items) from the lower limit of the first class in the frequency distribution. The median divides the area under a frequency curve into two equal parts.

Index Numbers

The *index* is a single ratio which measures the combined change of several variables between two different times, places or situations. It is applied to Consumer Prices, Escalation Tables, Unit Production Changes, Efficiency of Labor Performance and many more.

It is important to remember, that index numbers are ratios, and also averages. Any mathematical manipulation of indexes can only be an approximation.

Example: A department store sold two million dollars of goods in 1990 and three million dollars of goods in 1991. Have they sold 50% more goods?

Only if the prices have not changed. Because of the great variety of goods sold, it is impractical to record each individual unit quantity with the related actual sales price. Therefore, various averaging methods are used to obtain an index number.

Lets assume the price index for the category of goods sold was 150 in 1990 and 210 in 1991. The department store therefore experienced a 210/150 = 1.4 or a 40% increase in price. That means of the 3 M$, 1.4×2= 2.8 M$ would have been needed to buy the same quantity of goods. The rest (or 0.2 M$) is due to an increase in quantity. Using base $:

Year	Value of sales in annual $	Price Index	Real sale in base $
1990	2 M$	150	2/1.50 = 1.333 M$
1991	3 M$	210	3/2.10 = 1.429 M$
sales increase	1 M$		0.096 M$

which is a 7.14% increase due to quantity, i.e. (3/2.1)/(2/1.5)

Q #1.14: A construction project was built in 1993 for 744 k$. The cost index was 238. How much would it cost in 1996 when the projected index is 320?

A #1.14:
The cost changes are

$$C_2 = C_1 \frac{\text{index 2}}{\text{index 1}} \quad \text{or} \quad 744 \frac{320}{238} = 1000 \text{ k\$}$$

Q #1.15: Index numbers are essentially "averages" and "ratios" combined. Name two cases where index numbers are applied in cost estimating and control.

A #1.15: Escalation tables, commodity prices.

Q #1.16: A worker earned $ 20.00/hr in 1990. The price index at that time was 320. For the next four years he increased his wages by 4% p.a. The cost of goods and services (price index) increased by 16% during the same period. What is the price index for 1994? Did the worker gain or lose in real wages, and by how much?

A #1.16:

The 1994 index is	$320 \times 1.16 = 371.2$
Worker's wage increase =	$20 \times 1.044 = \$ 23.40$
Due to inflation =	$20 \times (371.2/320) = \$ 23.20$

Therefore, the worker gained $ 0.20 in 4 years in real wages.

Q #1.17: A published Installed Equipment Cost Index has a 1926 base of 100. In 1985 it was 444, and in 1993 it was 660. A company wants to convert the 1926 base to a more recent 1985 base. What is the 1993 index with the new base?

A #1.17:

$$\frac{660}{444} \times 100 = 149 \text{ for 1993 with 1985} = 100 \text{ (base)}$$

Q #1.18: A company bought 10 pumps for $ 20 000 each in 1982. The price index was 120. Nine pumps were installed and one kept as a spare. The installed pumps had to be replaced in 1993 when the price index was 168. What was the replacement price for each pump in 1993?

A #1.18:

$$x = \frac{20000 \times 168}{120} = \$ 28 000$$

Significant Digits

When computations are performed on a computer, it is tempting to carry results much further than their reliability and significance. This can be misleading. When quoting data, we must not imply more accuracy than exists. The estimated final cost of a project must not be reported as $ 6 549 580 by adding a rough forecast of 1.2 M$ to an accurate present cumulative expenditure of $ 5 349 580. The rough forecast has two significant digits, while actual expenditures are reported with six significant digits. It would be more appropriate to report the final cost of this project to be in the neighborhood of $ 6 550 000 or 6.5 M$.

The numbers 655, 0.0655 and 0.000655 each have three significant digits. But the number 0.06550 has four significant digits, because the final zero is appended only if the accuracy goes to that position. Again, in scientific notation, the expression 2.31×E5 shows three significant digits, but shown like this 2.31000×E5 indicates six significant digits.

Example: A sign which reads "20 miles to the border" was replaced by a kilometer sign "32.18 km to the border". It should have said "32 km to the border." A land surveyor calculates the measured area to be exactly 4.00 hectares. Since there are three significant digits, this should be equated to 9.88 acres, 10 acres would be wrong.

Rounding

Rounding to *nearest* final digit:

       53.845 becomes 53.84 because 4 is an even number.

       16.975 becomes 16.98 because 7 is an uneven number.

This rule is sometimes broken when calculators are used, because either truncation may take place, which changes 16.975 into 16.97, or an automatic rounding occurs, changing 16.975 into 16.98.

However, rounding in calculations should be applied only to the result

Q #1.19: Add the following numbers 322.9, 2500, 2402.8, 2300, 2200.0 and state the result in terms of a rounded number if 2300 has four significant digits.

A #1.19: As a first step mark the last significant digit for each number. Then add and convert the result according to the mark on the total.

                  322.9

                  2500

                  2402.8

                  2300

                  2200.0

                  9725.7 = 9700

Rounding to the *last final* digit:

This applies to intervals. A digit is maintained for a specified interval, then jumps to the next digit at the end of the interval. For example, two weeks before your 30th birthday anniversary you still call yourself as being 29 years of age instead of 29.96.

Rounding to the *next final* digit:

This also applies to intervals, but the digit changes at the beginning of the interval. For example, the post office applies dollar values to letters beginning at a certain weight and maintains the charge to the next interval. As soon as a letter reaches the weight of 50g a new payment applies until it reaches 100g.

Logarithms

Some productivity calculations such as the learning curve may need the application of logarithms to solve a problem. Even though it is fairly simple, some adult students are not familiar with it. The reason may be that many (but not all) applications have been replaced by calculators and computers. For the purpose of explaining the basics, we will review the *origin* of logarithm and use tables for the examples. **A logarithm is simply an exponent of a positive number**

$$\text{If } 10^2 = 100, \text{ then } \log 100 = 2$$

Logarithm to the base 10 are called Common or Briggs Logarithm. There is another type based on the transcendental number

$$e = 2.71828$$

$$\text{If } e^2 = 7.389, \text{ then } \ln 7.389 = 2$$
$$\text{If } e^x = 100, \text{ then } \ln 100 = x$$

For our purpose, we will only use common logarithms. A brief review of elementary exponential notations may be in order:

1) $a^n a^m = a^{n+m}$, where a is any real number and n and m are positive integers.

2) $a^n/a^m = a^{n-m}$ which can also be written as $a^n a^{-m} = a^{n-m}$.

3) There is a special case, when n=m=p, then $\dfrac{a^p}{a^p} = a^{p-p} = a^0 = 1$

4) $a^{1/n} = \sqrt[n]{a}$ and, in general, for any real value of n and m, except zero, and "a" being a positive number, then $a^{m/n} = \sqrt[n]{a^m} = \left(\sqrt[n]{a}\right)^m$

Having said that logarithms are simply exponents which follow the rules of the above exponential notations, we can say

If $X = a^Y$ is a positive number and a is a positive number, then $Y = \log_a X$

Expressed in real numbers $10^4 = 10\,000$ and $\log_{10} 10\,000 = 4$.

How about $10^{3.46538} = ?$ The result must lie between 10^3 and 10^4

To find the result, we will use tables. Those tables will give us the fractional part of a logarithm called the *mantissa* = 46538. (Please refer to Table 1.01a & b.)
We know that

$$3.46538 = 3.00000 + 0.46538 \text{ and } 10^{3.46538} = (10^3)(10^{0.46538})$$

The integral part 3 is called the *characteristic* of the logarithm. It identifies the integral part of the result.

The table shows N = 292 for the mantissa 46538. We know, the number must be between 10^3 and 10^4, that is between 1 000 and 10 000. Therefore,

$$10^{3.46538} = 2\ 920$$

It is a four digit number. We can further deduce
$$10^0 = 1 \quad \text{therefore} \quad \log 1\ = 0$$
$$10^{-1} = 0.1 \quad \text{therefore} \quad \log 0.1 = 0.0 -1\ = -1$$
For ease of calculation, -1 is sometimes written as $9.000 -10$ It follows that

$10^{0.46538-1} = \log 0.292 = 0.46538-1$ or $9.46538-10 = -0.53462$ and

$10^{0.46538-2} = \log 0.0292 = 8.46538-10 = -1.53462$

To summarize:

$\log 1\ 000.000 = 3$
$\qquad \rightarrow \quad \log 292.00000 = 0.46538 +2$
$\log \quad 100.000 = \ 2$
$\qquad \rightarrow \quad \log \quad 29.20000 = 0.46538 +1$
$\log \quad 10.000 = \ \ 1$
$\qquad \rightarrow \quad \log \quad \ 2.92000 = 0.46538 \pm 0$
$\log \quad \ 1.000 = \quad 0$
$\qquad \rightarrow \quad \log \quad \ 0.29200 = 0.46538 -1$
$\log \quad \ 0.100 = \ -1$
$\qquad \rightarrow \quad \log \quad \ 0.02920 = 0.46538 -2$
$\log \quad \ 0.010 = \ -2$
$\qquad \rightarrow \quad \log \quad \ 0.00292 = 0.46538 -3$
$\log \quad \ 0.001 = \ -3$
$\qquad\qquad\qquad \text{etc.}$

The observant student will have noticed by now how the characteristics relate to the logarithms. Some applications follow.

Multiplication:

From $a^n a^m = a^{n+m}$ follows that $\log XYZ.... = \log X + \log Y + \log Z + ...$
We are reducing multiplication to a simple addition.

Example: N = 292 × 36.1 × 0.202 = log 292 + log 36.1 + log 0.202, using the table,

2.46538 + 1.55751 + (9.30535 −10) = 3.32824 ; N = 2129 approx. (The calculated accurate result is N = 2129.3224.)

Division:

From $a^{n/m} = a^{n-m}$ follows $\log \dfrac{x}{y} = \log X - \log Y$

Example: $N = (2920 \times 118)/0.018 = \log 2920 + \log 118 - \log 0.018 =$
$3.46538 + 2.07188 - (8.25527 - 10) = 7.28199$
therefore $N = 19{,}142{,}000$ interpolated.

Powers:

From $\left(a^{n}\right)^{m} = a^{nm}$ follows $\log(x^{m}) = m \log x$

Example:: \$ 400 are invested for 8 years at 9%. From

$$S = P(1+i)^{n} \quad \text{follows} \quad S = 400(1+0.09)^{8}, \text{ and}$$

$\log S = \log 400 + 8(\log 1.09) = 2.60206 + 8(0.03743) = 2.9015 \therefore S = \$\,797$

Roots:

From $a^{m/n} = \sqrt[n]{a^{m}}$ follows that $\log x^{1\,n} = \dfrac{\log x}{n}$

Example: Find the geometric mean of the numbers 384, 310, 360, 355, 375
The result is the fifth root of the product of those numbers.

$$\log N = 1/5(\log 384 + \log 310 + \log 360 + \log 355 + \log 375)$$

$$= 1/5(2.58433 + 2.49136 + 2.55630 + 2.55023 + 2.57403) = 2.55125$$

$$\therefore N = 355.8 \text{ (interpolated)}$$

Another example: Economic studies indicate that prices for three metals in a specific category have changed as follows:

	1994	1995
Copper	100	80
Steel	100	100
Zinc	100	120

What is the *average* 1995 price index for metals in this category?
Using the geometric mean,

$\log G = 1/3(\log 80 + \log 100 + \log 120) = 1/3(1.90309 + 2.00000 + 2.07918)$

$$= 1/3(5.98227) = 1.99409,$$

and taking the antilog the price index for 1995:

$$G = 98.648$$

Table 1-01

Five place logarithms

N	0	1	2	3	4	5	6	7	8	9
0		0.00000	0.30103	0.47712	0.60206	0.69897	0.77815	0.84510	0.90309	0.95424
10	1.00000	1.00432	1.00860	1.01284	1.01703	1.02119	1.02531	1.02938	1.03342	1.03743
11	1.04139	1.04532	1.04922	1.05308	1.05690	1.06070	1.06446	1.06819	1.07188	1.07555
12	1.07918	1.08279	1.08636	1.08991	1.09342	1.09691	1.10037	1.10380	1.10721	1.11059
13	1.11394	1.11727	1.12057	1.12385	1.12710	1.13033	1.13354	1.13672	1.13988	1.14301
14	1.14613	1.14922	1.15229	1.15534	1.15836	1.16137	1.16435	1.16732	1.17026	1.17319
15	1.17609	1.17898	1.18184	1.18469	1.18752	1.19033	1.19312	1.19590	1.19866	1.20140
16	1.20412	1.20683	1.20952	1.21219	1.21484	1.21748	1.22011	1.22272	1.22531	1.22789
17	1.23045	1.23300	1.23553	1.23805	1.24055	1.24304	1.24551	1.24797	1.25042	1.25285
18	1.25527	1.25768	1.26007	1.26245	1.26482	1.26717	1.26951	1.27184	1.27416	1.27646
19	1.27875	1.28103	1.28330	1.28556	1.28780	1.29003	1.29226	1.29447	1.29667	1.29885
20	1.30103	1.30320	1.30535	1.30750	1.30963	1.31175	1.31387	1.31597	1.31806	1.32015
21	1.32222	1.32428	1.32634	1.32838	1.33041	1.33244	1.33445	1.33646	1.33846	1.34044
22	1.34242	1.34439	1.34635	1.34830	1.35025	1.35218	1.35411	1.35603	1.35793	1.35984
23	1.36173	1.36361	1.36549	1.36736	1.36922	1.37107	1.37291	1.37475	1.37658	1.37840
24	1.38021	1.38202	1.38382	1.38561	1.38739	1.38917	1.39094	1.39270	1.39445	1.39620
25	1.39794	1.39967	1.40140	1.40312	1.40483	1.40654	1.40824	1.40993	1.41162	1.41330
26	1.41497	1.41664	1.41830	1.41996	1.42160	1.42325	1.42488	1.42651	1.42813	1.42975
27	1.43136	1.43297	1.43457	1.43616	1.43775	1.43933	1.44091	1.44248	1.44404	1.44560
28	1.44716	1.44871	1.45025	1.45179	1.45332	1.45484	1.45637	1.45788	1.45939	1.46090
29	1.46240	1.46389	1.46538	1.46687	1.46835	1.46982	1.47129	1.47276	1.47422	1.47567
30	1.47712	1.47857	1.48001	1.48144	1.48287	1.48430	1.48572	1.48714	1.48855	1.48996
31	1.49136	1.49276	1.49415	1.49554	1.49693	1.49831	1.49969	1.50106	1.50243	1.50379
32	1.50515	1.50651	1.50786	1.50920	1.51055	1.51188	1.51322	1.51455	1.51587	1.51720
33	1.51851	1.51983	1.52114	1.52244	1.52375	1.52504	1.52634	1.52763	1.52892	1.53020
34	1.53148	1.53275	1.53403	1.53529	1.53656	1.53782	1.53908	1.54033	1.54158	1.54283
35	1.54407	1.54531	1.54654	1.54777	1.54900	1.55023	1.55145	1.55267	1.55388	1.55509
36	1.55630	1.55751	1.55871	1.55991	1.56110	1.56229	1.56348	1.56467	1.56585	1.56703
37	1.56820	1.56937	1.57054	1.57171	1.57287	1.57403	1.57519	1.57634	1.57749	1.57864
38	1.57978	1.58092	1.58206	1.58320	1.58433	1.58546	1.58659	1.58771	1.58883	1.58995
39	1.59106	1.59218	1.59329	1.59439	1.59550	1.59660	1.59770	1.59879	1.59988	1.60097
40	1.60206	1.60314	1.60423	1.60531	1.60638	1.60746	1.60853	1.60959	1.61066	1.61172
41	1.61278	1.61384	1.61490	1.61595	1.61700	1.61805	1.61909	1.62014	1.62118	1.62221
42	1.62325	1.62428	1.62531	1.62634	1.62737	1.62839	1.62941	1.63043	1.63144	1.63246
43	1.63347	1.63448	1.63548	1.63649	1.63749	1.63849	1.63949	1.64048	1.64147	1.64246
44	1.64345	1.64444	1.64542	1.64640	1.64738	1.64836	1.64933	1.65031	1.65128	1.65225
45	1.65321	1.65418	1.65514	1.65610	1.65706	1.65801	1.65896	1.65992	1.66087	1.66181
46	1.66276	1.66370	1.66464	1.66558	1.66652	1.66745	1.66839	1.66932	1.67025	1.67117
47	1.67210	1.67302	1.67394	1.67486	1.67578	1.67669	1.67761	1.67852	1.67943	1.68034
48	1.68124	1.68215	1.68305	1.68395	1.68485	1.68574	1.68664	1.68753	1.68842	1.68931
49	1.69020	1.69108	1.69197	1.69285	1.69373	1.69461	1.69548	1.69636	1.69723	1.69810
50	1.69897	1.69984	1.70070	1.70157	1.70243	1.70329	1.70415	1.70501	1.70586	1.70672
51	1.70757	1.70842	1.70927	1.71012	1.71096	1.71181	1.71265	1.71349	1.71433	1.71517
52	1.71600	1.71684	1.71767	1.71850	1.71933	1.72016	1.72099	1.72181	1.72263	1.72346
53	1.72428	1.72509	1.72591	1.72673	1.72754	1.72835	1.72916	1.72997	1.73078	1.73159
54	1.73239	1.73320	1.73400	1.73480	1.73560	1.73640	1.73719	1.73799	1.73878	1.73957

Table 1-01   (Continued)

N	0	1	2	3	4	5	6	7	8	9
55	1.74036	1.74115	1.74194	1.74273	1.74351	1.74429	1.74507	1.74586	1.74663	1.74741
56	1.74819	1.74896	1.74974	1.75051	1.75128	1.75205	1.75282	1.75358	1.75435	1.75511
57	1.75587	1.75664	1.75740	1.75815	1.75891	1.75967	1.76042	1.76118	1.76193	1.76268
58	1.76343	1.76418	1.76492	1.76567	1.76641	1.76716	1.76790	1.76864	1.76938	1.77012
59	1.77085	1.77159	1.77232	1.77305	1.77379	1.77452	1.77525	1.77597	1.77670	1.77743
60	1.77815	1.77887	1.77960	1.78032	1.78104	1.78176	1.78247	1.78319	1.78390	1.78462
61	1.78533	1.78604	1.78675	1.78746	1.78817	1.78888	1.78958	1.79029	1.79099	1.79169
62	1.79239	1.79309	1.79379	1.79449	1.79518	1.79588	1.79657	1.79727	1.79796	1.79865
63	1.79934	1.80003	1.80072	1.80140	1.80209	1.80277	1.80346	1.80414	1.80482	1.80550
64	1.80618	1.80686	1.80754	1.80821	1.80889	1.80956	1.81023	1.81090	1.81158	1.81224
65	1.81291	1.81358	1.81425	1.81491	1.81558	1.81624	1.81690	1.81757	1.81823	1.81889
66	1.81954	1.82020	1.82086	1.82151	1.82217	1.82282	1.82347	1.82413	1.82478	1.82543
67	1.82607	1.82672	1.82737	1.82802	1.82866	1.82930	1.82995	1.83059	1.83123	1.83187
68	1.83251	1.83315	1.83378	1.83442	1.83506	1.83569	1.83632	1.83696	1.83759	1.83822
69	1.83885	1.83948	1.84011	1.84073	1.84136	1.84198	1.84261	1.84323	1.84386	1.84448
70	1.84510	1.84572	1.84634	1.84696	1.84757	1.84819	1.84880	1.84942	1.85003	1.85065
71	1.85126	1.85187	1.85248	1.85309	1.85370	1.85431	1.85491	1.85552	1.85612	1.85673
72	1.85733	1.85794	1.85854	1.85914	1.85974	1.86034	1.86094	1.86153	1.86213	1.86273
73	1.86332	1.86392	1.86451	1.86510	1.86570	1.86629	1.86688	1.86747	1.86806	1.86864
74	1.86923	1.86982	1.87040	1.87099	1.87157	1.87216	1.87274	1.87332	1.87390	1.87448
75	1.87506	1.87564	1.87622	1.87679	1.87737	1.87795	1.87852	1.87910	1.87967	1.88024
76	1.88081	1.88138	1.88195	1.88252	1.88309	1.88366	1.88423	1.88480	1.88536	1.88593
77	1.88649	1.88705	1.88762	1.88818	1.88874	1.88930	1.88986	1.89042	1.89098	1.89154
78	1.89209	1.89265	1.89321	1.89376	1.89432	1.89487	1.89542	1.89597	1.89653	1.89708
79	1.89763	1.89818	1.89873	1.89927	1.89982	1.90037	1.90091	1.90146	1.90200	1.90255
80	1.90309	1.90363	1.90417	1.90472	1.90526	1.90580	1.90634	1.90687	1.90741	1.90795
81	1.90849	1.90902	1.90956	1.91009	1.91062	1.91116	1.91169	1.91222	1.91275	1.91328
82	1.91381	1.91434	1.91487	1.91540	1.91593	1.91645	1.91698	1.91751	1.91803	1.91855
83	1.91908	1.91960	1.92012	1.92065	1.92117	1.92169	1.92221	1.92273	1.92324	1.92376
84	1.92428	1.92480	1.92531	1.92583	1.92634	1.92686	1.92737	1.92788	1.92840	1.92891
85	1.92942	1.92993	1.93044	1.93095	1.93146	1.93197	1.93247	1.93298	1.93349	1.93399
86	1.93450	1.93500	1.93551	1.93601	1.93651	1.93702	1.93752	1.93802	1.93852	1.93902
87	1.93952	1.94002	1.94052	1.94101	1.94151	1.94201	1.94250	1.94300	1.94349	1.94399
88	1.94448	1.94498	1.94547	1.94596	1.94645	1.94694	1.94743	1.94792	1.94841	1.94890
89	1.94939	1.94988	1.95036	1.95085	1.95134	1.95182	1.95231	1.95279	1.95328	1.95376
90	1.95424	1.95472	1.95521	1.95569	1.95617	1.95665	1.95713	1.95761	1.95809	1.95856
91	1.95904	1.95952	1.95999	1.96047	1.96095	1.96142	1.96190	1.96237	1.96284	1.96332
92	1.96379	1.96426	1.96473	1.96520	1.96567	1.96614	1.96661	1.96708	1.96755	1.96802
93	1.96848	1.96895	1.96942	1.96988	1.97035	1.97081	1.97128	1.97174	1.97220	1.97267
94	1.97313	1.97359	1.97405	1.97451	1.97497	1.97543	1.97589	1.97635	1.97681	1.97727
95	1.97772	1.97818	1.97864	1.97909	1.97955	1.98000	1.98046	1.98091	1.98137	1.98182
96	1.98227	1.98272	1.98318	1.98363	1.98408	1.98453	1.98498	1.98543	1.98588	1.98632
97	1.98677	1.98722	1.98767	1.98811	1.98856	1.98900	1.98945	1.98989	1.99034	1.99078
98	1.99123	1.99167	1.99211	1.99255	1.99300	1.99344	1.99388	1.99432	1.99476	1.99520
99	1.99564	1.99607	1.99651	1.99695	1.99739	1.99782	1.99826	1.99870	1.99913	1.99957

Differentiation

It is not a requirement for the reader of this text to be familiar with calculus. To understand decision-making problems such as optimizing the cost of certain alternatives it would be advantageous though to understand the *concept* of differentiation. Below is a short explanation what it means:

Assume a curve has the function

$$y = x^2 + 3x + 5 \quad \text{..(1)}$$

When we plot this curve for various values of the variable x the curve will look like this:

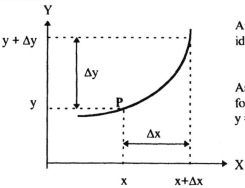

Any point on the curve can be identified by the variables
x and y

Assuming we put x = 4 into the formula (1) above, then
y = 16 + 12 + 5 = 33

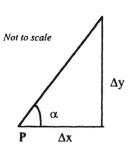

Not to scale

If we move along the curve in a positive direction by a small increment Δx (pronounced delta x) we can calculate Δy. If Δx is very small, it will give us a good approximation of the slope of the curve at point P.

$$\tan\alpha = \frac{\Delta y}{\Delta x}$$

This is called the average rate of change of the function y = f(x) with respect to x or per unit change in x.

By adding that very small change to the equation (1) of the curve, we obtain

$$y + \Delta y = (x + \Delta x)^2 + 3(x + \Delta x) + 5 \quad \text{.............................(2)}$$

Assuming x changes from 4 to 4.1, then

$$\Delta x = 4.1 - 4 = 0.10$$
$$y + \Delta y = 16.81 + 12.30 + 5.00 = 34.11$$
$$\Delta y = 34.11 - 33 = 1.11$$

and $\dfrac{\Delta y}{\Delta x} = \dfrac{1.11}{0.10} = 11.1$ (close to an 85° slope)

Solving the general equation (2) we obtain

$y + \Delta y = x^2 + 2x\Delta x + \Delta x^2 + 3x + 3\Delta x + 5 = \Delta x(\Delta x + 2x + 3) + (x^2 + 3x + 5),$

but $(x^2 + 3x + 5) = y$, therefore $\Delta y = \Delta x(\Delta x + 2x + 3)$ and

$$\dfrac{\Delta y}{\Delta x} = 2x + 3 + \Delta x$$

If Δx is so small that it approaches zero or $\lim_{\Delta x \to 0} \dfrac{\Delta y}{\Delta x}$, it is defined as the *derivative* of the function

$$y = f(x) = \dfrac{\Delta y}{\Delta x} = f'(x) = y' = \lim_{\Delta x \to 0} \dfrac{\Delta y}{\Delta x},\text{ provided the limit to zero exists.}$$

Instead of calling the rate of change average, we call it now the *instantaneous* rate of change, given by $\dfrac{dy}{dx}$

Going back to equation (3) for $\Delta x \to 0$

$$\dfrac{dy}{dx} = \lim_{\Delta x \to 0}(2x + 3 + \Delta x) = 2x + 3 \qquad \text{.........................(4).}$$

This is called the derivative of $y = x^2 + 3x + 5$

Those general differential formulas apply:

$\dfrac{d}{dx}(c) = 0$ where "c" is any constant

$\dfrac{d}{dx}(cx) = c$ and $\dfrac{d}{dx}(x^n) = nx^{n-1}$ and $\dfrac{d}{dx}\left(\dfrac{c}{u}\right) = -\dfrac{c}{u^2}\dfrac{d}{dx}(u)$

specifically in reference to equation (1)

$\dfrac{d}{dx}(5) = 0$; $\dfrac{d}{dx}(3x) = 3$; $\dfrac{d}{dx}(x^2) = 2x$; $\dfrac{d}{dx}(x^{-1}) = -\dfrac{1}{x^2}$

Going back to equation (4), if we set $\dfrac{dy}{dx}$ = zero, then $2x + 3 = 0$

$$\text{and } x = -1.5$$

That means the slope on the curve is zero at $x = -1.5$ and the tangent at that point is horizontal. This can be graphically represented by plotting the curve

$$y = x^2 + 3x + 5$$

x →	4	3	2	1.5	1	0.5	0	−0.5	−1	−1.5	−2	−3	−4
y →	33	23	15	11.75	9	6.75	5	3.75	3	2.75	3	5	9

The curve reaches a minimum at the horizontal line (tangent)

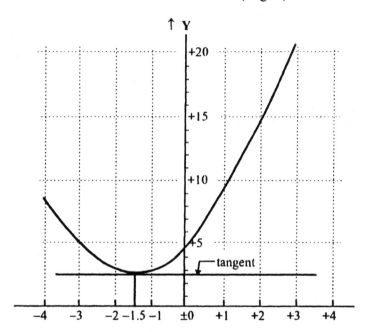

Figure 1-01
Curve at the minimum position.

If we differentiate (4) again, the second derivative is

$$\frac{d^2y}{dx^2} \text{ or } f''(x) \text{ or } y'' = +2$$

This positive value indicates, that the tangent touches the curve at its minimum. (The value of y increases in both directions. A negative value would have indicated a maximum.)

The reader should realize, that the above is a crude example of the meaning of differentiation and to demonstrate how it can be used in the application of minimum and maximum values of a mathematical function. *(For detailed discussions on limits and more complicated functions the reader should consult texts on mathematics.)*

1.1.2 Elementary Statistics

Frequency Distribution

Defining some expressions:

ARRAY   = Listing of numbers in their order of magnitude.
RANGE   = The difference between the largest and smallest.
CLASS INTERVALS (i)   = A further breakdown of the range.
FREQUENCY (f)   = The number of items within each class.

Example: Assume we survey the number of hours worked per week by production employees of N = 100 industries:(4)

Table 1-02
Tabulating statistical data

RAW DATA				ARRAY			
42.3	40.2	38.4	39.4	30.4	38.2	40.2	41.7
41.7	39.7	36.4	40.9	31.7	38.3	40.2	41.8
46.5	37.2	36.8	38.2	34.2	38.3	40.3	41.9
43.0	37.5	38.3	38.9	35.4	38.4	40.3	41.9
30.4	36.7	39.9	39.3	35.4	38.5	40.5	42.1
40.9	38.5	39.3	39.7	35.9	38.9	40.5	42.1
46.0	37.1	35.4	40.9	36.2	38.9	40.6	42.3
46.0	40.8	41.3	42.1	36.2	39.1	40.6	42.3
41.2	40.1	42.8	42.3	36.3	39.1	40.7	42.3
40.3	36.2	42.9	40.6	36.4	39.1	40.8	42.4
41.3	40.0	39.2	39.9	36.4	39.2	40.9	42.8
41.6	37.9	40.5	41.9	36.4	39.2	40.9	42.9
40.3	38.2	41.0	40.6	36.7	39.3	40.9	43.0
42.4	41.6	41.5	41.8	36.7	39.3	40.9	43.0
40.2	44.2	41.0	41.6	36.8	39.4	41.0	43.5
38.2	39.1	40.5	46.4	36.8	39.4	41.0	43.5
43.5	35.4	40.1	45.4	36.9	39.7	41.2	43.5
46.6	36.8	41.5	39.1	37.1	39.7	41.3	44.2
41.4	36.7	43.5	41.9	37.1	39.9	41.3	44.9
37.7	35.9	42.1	43.5	37.2	39.9	41.4	45.4
38.9	34.2	39.2	43.0	37.5	40.0	41.5	46.0
39.4	36.3	40.9	40.7	37.7	40.0	41.5	46.0
39.1	31.7	36.2	36.4	37.9	40.1	41.6	46.4
36.9	37.1	40.0	44.9	38.2	40.1	41.6	46.5
38.3	36.4	40.1	42.3	38.2	40.1	41.6	46.6

The list shows the average number of hours worked for each industry as reported by a survey. The hours are then tabulated in the order of magnitude. This array ranges from 30.4 to 46.6 hours per week. We then assemble them into nominal three-hour class intervals (i). The lower limit for each class interval (L) is the arithmetic average between the stated end of the previous class and the beginning of the subject class, e.g.

$$L = (38.9 + 39.0)/2 = 38.95$$

We now list the number of industries per class (f). This data is plotted in a graph called the frequency polygon or presented as a histogram (Figure 1.02). When class intervals become very small, the polygon resembles a curve. This curve can be symmetric (normal distribution) or asymmetric (skewed).

Hours Worked	Number of Industries	
30.0 - 32.9	2	⎫
33.0 - 35.9	4	⎬ = d_2
36.0 - 38.9	26	⎭
39.0 - 41.9	47	= f
42.0 - 44.9	15	⎫ = d_1
45.0 - 47.9	6	⎭

The *average hours worked* is the arithmetic mean of the frequency distribution. The simplest but not the shortest method is to add the total hours listed divided by the number of industries, i.e.

$$\overline{X} = 4006/100 = 40.06 \text{ hours}$$

To establish the *median* (center of distribution), we are using the three hour interval for the hours worked 39.0 – 41.9 (Table 1-02).

Number of Industries

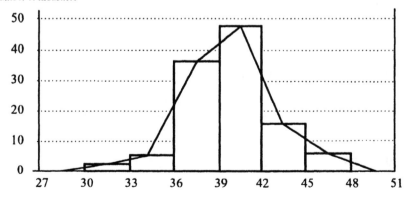

Figure 1-02
Frequency polygon.

The number of industries within this interval is $f = 47$. The number of industries reporting less than 39.0 hours/week is $d_2 = (26+4+2) = 32$.
The formula to calculate the median is

$$Md = L + i(N/2 - d2)/f \quad \text{or} \quad 38.95 + 3(50 - 32)/47 = 40.1$$

The point of the greatest density in a frequency distribution is called the mode. This is also the most likely event when making predictions, such as risk assessment when estimating. In our example the mode is 40.14. At this point, the slope on the curve is horizontal. An exact calculation is complicated, but an approximation is accurate enough for most applications:

$$M_o = L + i\frac{d_1}{d_1 + d_2} = 38.95 + 3\frac{15 + 6}{21 + 32} = 40.14$$

d1 is the number of industries with >42 hours/week.

The difference between the median and the mode is very small in our example. This means that the curve has very little skewness (close to normal distribution).

Q #1.20: For 100 similar jobs, statistics were obtained to establish the time required to install 100 m^2 of floor tile. The results were tabulated in the order of magnitude:

Hours worked:	Frequency:	
20.46 to 21.85	4	Using a class interval i = 2.0 hours and
22.55 to 23.75	10	starting with 20.0 hours rounded to the
24.00 to 25.85	26	nearest tenth of an hour, find
26.05 to 27.90	48	
28.18 to 29.79	7	a) the median of the distribution
30.09 to 31.88	5	b) the approx. mode of the distribution

A #1.20:

a)

Class Interval:	Frequency:	Median:
20.0 - 21.9	4	
22.0 - 23.9	10	$L + i(N/2 - d_2)/f =$
24.0 - 25.9	26	$25.95 + 2(50 - 40)/48 =$
26.0 - 27.9	48	$25.95 + 0.42 = 26.37$ Whrs.
28.0 - 29.9	7	
30.0 - 31.9	5	

b) Mode:

$$M_o = L + i\frac{d_1}{d_1 + d_2} = 25.95 + \frac{2 \times 12}{12 + 40} = 26.41 \text{ Whrs}.$$

The Estimating Line

One application of simple statistical correlation is the *regression line* or sometimes called the *estimating line*.

When related data is widely scattered over a period of time, a trend is hard to visualize. Plotting the scattered data on a time series diagram, it is obvious, that the curve has to be "fitted" to establish an "average" relationship.

There are several methods to fit this estimating line to the dots. One solution is the method of maximum likelihood. This method is elaborate and will not be discussed here.

We could use an analog gadget (37) as shown by "Scientific American", page 18, Oct. 1985: Data points are plotted on a wood surface and a nail is driven partly into the wood at each point (Figure 1.03). A number of uniform rubber bands are then slipped onto a rod, one band for each nail. The rod assumes an equilibrium position which minimizes the total energy of the system. (We do not recommend this method of a physical model for the project office.)

The most popular and easily understood method is the *method of least squares*. This means the squares of the distance from the dots to the line are made as small as possible rather than the absolute distances. This would come close to the physical energy equilibrium experienced in the model below.

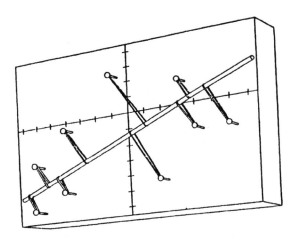

Figure 1-03
Analog gadget.

The functional relationship between two variables represented by a straight line is:

$$Y = a + bX$$

where the abscissa X is the accumulation of intervals and the ordinate Y is the value per interval. From this we can derive at the least squares requirement for two simultaneous equations:

1) $\Sigma Y = Na + b \Sigma X$
2) $\Sigma XY = a\Sigma X + b \Sigma X^2$

where N is the number of dots on the diagram. Listing the measured X and Y data is shown in the example below:

measured data		calculated values		
X	Y	XY	X^2	Y^2
1	8	8	1	64
2	9	18	4	81
3	15	45	9	225
4	11	44	16	121
5	17	85	25	289
15	60	200	55	780

Average = $\overline{X}$ = 60/5 = 12
Entering the totals into the equations:

$$60 = 5a + 15b$$
$$200 = 15a + 55b$$

Solving for a and b yields a = 6 and b = 2, therefore, the resulting equation is

$$Y = 6 + 2X$$

from which the estimating line can be plotted by using different values of X (see Fig. 1-04):

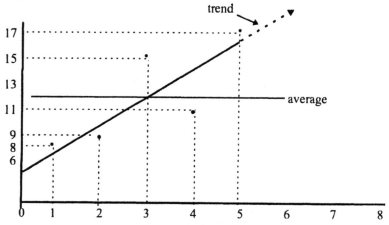

Figure 1-04
The estimating line.

Q #1.21: A department spends the following monthly amounts on computer training over a 12 months period: k$15, 14, 6, 7, 12, 9, 11, 8, 4, 5, 6, 6.
What are the abscissae (vertical distances on the time scale) over the 12 months period for the estimating line? Use the method of least squares. Analyze the trend. What is the projected cost of the training program?

A #1.21:

Months(X)	k$(Y)	XY	X^2
1	15	15	1
2	14	28	4
3	6	18	9
4	7	28	16
5	12	60	25
6	9	54	36
7	11	77	49
8	8	64	64
9	4	36	81
10	5	50	100
11	6	66	121
12	6	72	144
78	103	568	650

$\Sigma Y = aN + b\Sigma X \quad Y = a + bX$
$\Sigma XY = a\Sigma X + b\Sigma X^2$

$103 = 12a + 78b \quad (\times 6.5)$
$568 = 78a + 650b \quad \text{subtract}$

$669.5 = 78a + 507b$
$\underline{568.0 = -78a - 650b}$
$+101.5 = \quad 0 - 143b \quad b = -0.71$

$12a = 103 + 55.36 \quad a = +13.20$

Y = 13.20 − 0.71X

This equation yields

X =	1	2	3	4	5	6	7	8	9	10	11	12
Y =	12.5	11.8	11.1	10.4	9.6	8.9	8.2	7.5	6.8	6.1	5.4	4.7

from which a graph can be produced.

Analysis:

A negative b indicates a downward trend. The equation Y = a + bX indicates a straight line. Therefore, only the two end points need to be plotted:

For X = 0 ; y = 13.20 $\qquad$ For X = 12; y = 4.68

The arithmetic mean is $\quad$ Ma = ΣY/N = 103/12 = $ 8 580.

The correctness of the estimating line can be checked by averaging Y1 and Y12, i.e.

$$(12.49 + 4.68)/2 = \$ \ 8 \ 580$$

If the trend continues, it is likely that the training will continue for another x months.

For Y = 0, x = 13.20/0.71 = 18.6 or approximately 7 months more. Expenditures remaining:

for X = 13; y = 3.97 ; $\quad$ for X = 18; y = 0.42 (intervals reduced by 0.71)

28

Chapter 1

therefore,

$$\left(\frac{3.97 + 0.42}{2}\right) \times 6 = \textbf{13.17 k\$}$$

total projected cost = 103.00 + 13.17 = **116.17 k\$**

Q #1.22: In some cases it is desirable to fit the scattered data with a curve rather than a line. This would require a quadratic function in the form

$$Y = a + bX + X^2$$

Three normal equations are obtained when applying the least squares criterion:

$$\Sigma Y = aN \quad + b\Sigma X \quad + c\Sigma X^2$$
$$\Sigma XY = a\Sigma X \quad + b\Sigma X^2 + c\Sigma X^3$$
$$\Sigma X^2Y = a\Sigma X^2 + b\Sigma X^3 + c\Sigma X^4$$

For the following scattered data

X	3	5	10	14	16	20
Y	15	12	6	5	4	4

    a)   find the coordinates for the straight estimating line
    b)   find the fitted curve using the quadratic function.

(Calculations can easily be performed on calculators with statistical keys or by using available computer programs.)

Below is a step-by-step calculation for the uninitiated student:

A #1.22:

X	Y	XY	X^2	X^2Y	X^3	X^4
3	15	45	9	135	27	81
5	12	60	25	300	125	625
10	6	60	100	600	1 000	10 000
14	5	70	196	980	2 744	38 416
16	4	64	256	1 024	4 096	65 536
20	4	80	400	1 600	8 000	160 000
68	46	379	986	4 639	15 992	274 658

a) <u>Estimating line:</u>   46 = 6a + 68b   ($\times$ 11.33...)
                              379 = 68a + 986b
                              <u>521 = 68a + 771b</u>
                              −142 =   215b   b = − 0.66
                              6a = 46 + 44.91   a = 15.15

$$Y = 15.15 - 0.66X$$

For X = 0 , y = 15.15 ; for X = 20 , y = 1.95 ; which establishes the slope of the line.

b) For a second-degree curve, we need to calculate a, b, c.

1) $46 = 6a + 68b + 986c$
2) $379 = 68a + 986b + 15\,992c$
3) $4\,639 = 986a + 15\,992b + 274\,658c$

We can now solve for a, b, and c by using determinants or by conventional methods which will give the following results:

$$a = 20.13 \qquad c = 0.0557 \qquad b = -1.9074$$

Finally, $\qquad Y = 20.13 - 1.9074X + 0.0557X^2$

Y can now be calculated for the various values of X:

X	0	2	4	6	8	10	12	14	16	18	20	22
Y	18.3.	16.5	13.4	10.7	8.4	6.6	5.3	4.3	3.9	3.8	4.2	5.1

The surprising fact here is, that the linear trend approaches zero very fast, while the curve indicates a slight recovery (Figure 1-05). The decision maker will have to resort to other indicators to determine a most likely trend.

There are additional non-linear trend curves in use such as exponential smoothing, seasonal adjustment or rolling averages.

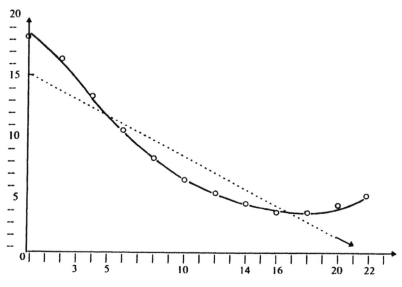

Figure 1-05
Curvilinear trend.

1.2. THE TIME VALUE OF MONEY

We all know, that time only flows in one direction. Time never stands still. By reading this sentence, 30 seconds have already passed. Those 30 seconds have gone. They cannot be retrieved. No wonder that time is a valuable commodity.

There is a convention in cost management whereby time is displayed as a horizontal line, counted from left to right. The time units "t" can be hours, days, weeks, months or years, depending on the size of the project or the project element that is depicted.

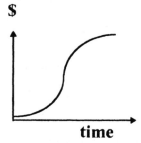

1.2.1 Simple Interest Calculations

We all are involved with interest payments in our daily lives and are probably familiar with routine interest calculations. Project management and staff must be continuously conscious what monetary effect any deviation from the schedule commitment can have on total project cost. Any time delay during construction will cause a rise in interest cost. Money can be saved by being on time. Below is a short review of the mathematics needed to understand the basics of interest calculations:

Suppose a present amount P earns at the rate i per year (expressed as a decimal), and let S_n be the future amount "n" years later, then, at the end of the first year

$$S_1 = P + iP = P(1+i)$$

At the end of the second year, P becomes P(1+i), therefore,

$$S_2 = P(1+i)+iP(1+i)=P(1+i)^2$$

And for any future year $\qquad S_n = P(1+i)^n$

or future worth = (present worth)(compound interest factor)

And for derived present worth $\qquad P = S(1+i)^{-n}$
This leads to the following important statement:

> **Future Worth moves with the calendar, →**
> **Present Worth moves against the calendar. ←**

Uniform Payment Series

Consider a uniform end-of-year annual amount R for a period of n years (Jelen calls R the unacost)(5)

Present worth: $P = R\dfrac{(1+i)^n - 1}{i(1+i)^n}$, but $P = S(1+i)^{-n}$

therefore, $S = R\dfrac{(1+i)^n - 1}{i}$ = compounded uniform payment

from this, $R = S\dfrac{i}{(1+i)^n - 1}$ = unacost sinking fund payment

The reverse of unacost is *capital recovery*. This factor converts a single zero time cost to an equivalent uniform end-of-year annual cost:

$$R = P\frac{i(1+i)^n}{(1+i)^n - 1}, \text{ from this, } P = R\frac{1-(1+i)^{-n}}{i}$$

There are numerous other relationships in formulas such as blended equal and beginning-of-year payments, sinking fund and annuities with their various applications. Most applications in project management deal with the concept of the time value of money. The bottom line is, when money moves in the direction of the calendar, it grows by earning interest, it is *compounded*. Money is *discounted* when moving against the calendar. Discounting to a present worth value is especially useful when comparing several project options by bringing them onto the same level of time.

For other than annual factors, the formula is $S = P\left(1 + \dfrac{i}{p}\right)^{np}$

where p is the number of periods per year.

The monthly compounding for three years would therefore be

$$S = P\left(1 + \frac{i}{12}\right)^{36}$$

When the number of periods approach infinity $p \rightarrow \infty$, the compounding for one year is

$\left(1 + \dfrac{i}{p}\right)^{p}$; since the limit of $\left(1 + \dfrac{i}{p}\right)^{p}$ is $e = 2.71828$ (Naperian constant)

the continuous compounding factor is
$$\left[\left(1+\frac{i}{p}\right)^{\frac{p}{i}}\right]^{i} = e^{i}$$

and $\quad S = Pe^{in} \ ; \quad P = Se^{-in}$

Table 1-03

Compound interest tables

	Single Payment		Uniform Payment Series				
	Compound-Amount Factor	Present-Worth Factor	Compound-Amount Factor	Sinking-Fund Factor	Present-Worth Factor	Capital-Recovery Factor	
	Given P to find S $(1+i)^{n}$	Given S to find P $\dfrac{1}{(1+i)^{n}}$	Given R to find S $\dfrac{(1+i)^{n}-1}{i}$	Given S to find R $\dfrac{i}{(1+i)^{n}-1}$	Given R to find P $\dfrac{(1+i)^{n}-1}{i(1+i)^{n}}$	Given P to find R $\dfrac{i(1+i)^{n}}{(1+i)^{n}-1}$	
n			Number of Years				*n*
1	1.1000	0.9091	1.0000	1.00000	0.9091	1.10000	1
2	1.2100	0.8264	2.1000	0.47619	1.7355	0.57619	2
3	1.3310	0.7513	3.3100	0.30211	2.4869	0.40211	3
4	1.4641	0.6830	4.6410	0.21547	3.1699	0.31547	4
5	1.6105	0.6209	6.1051	0.16380	3.7908	0.26380	5
6	1.7716	0.5645	7.7156	0.12961	4.3553	0.22961	6
7	1.9487	0.5132	9.4872	0.10541	4.8684	0.20541	7
8	2.1436	0.4665	11.4359	0.08744	5.3349	0.18744	8
9	2.3579	0.4241	13.5795	0.07364	5.7590	0.17364	9
10	2.5937	0.3855	15.9374	0.06275	6.1446	0.16275	10
11	2.8531	0.3505	18.5312	0.05396	6.4951	0.15396	11
12	3.1384	0.3186	21.3843	0.04676	6.8137	0.14676	12
13	3.4523	0.2897	24.5227	0.04078	7.1034	0.14078	13
14	3.7975	0.2633	27.9750	0.03575	7.3667	0.13575	14
15	4.1772	0.2394	31.7725	0.03147	7.6061	0.13147	15
16	4.5950	0.2176	35.9497	0.02782	7.8237	0.12782	16
17	5.0545	0.1978	40.5447	0.02466	8.0216	0.12466	17
18	5.5599	0.1799	45.5992	0.02193	8.2014	0.12193	18
19	6.1159	0.1635	51.1591	0.01955	8.3649	0.11955	19
20	6.7275	0.1486	57.2750	0.01746	8.5136	0.11746	20
21	7.4002	0.1351	64.0025	0.01562	8.6487	0.11562	21
22	8.1403	0.1228	71.4027	0.01401	8.7715	0.11401	22
23	8.9543	0.1117	79.5430	0.01257	8.8832	0.11257	23
24	9.8497	0.1015	88.4973	0.01130	8.9847	0.11130	24
25	10.8347	0.0923	98.3471	0.01017	9.0770	0.11017	25

PRESENT-WORTH FACTORS FOR INTEREST RATES FROM i = 6% TO i = 30%

n	6%	8%	10%	12%	15%	20%	25%	30%	n
				Given S, Find P					
1	0.9434	0.9259	0.9091	0.8929	0.8696	0.8333	0.8000	0.7692	1
2	0.8900	0.8573	0.8264	0.7972	0.7561	0.6944	0.6400	0.5917	2
3	0.8396	0.7938	0.7513	0.7118	0.6575	0.5787	0.5120	0.4552	3
4	0.7921	0.7350	0.6830	0.6355	0.5718	0.4823	0.4096	0.3501	4
5	0.7473	0.6806	0.6209	0.5674	0.4972	0.4019	0.3277	0.2693	5
6	0.7050	0.6302	0.5645	0.5066	0.4323	0.3349	0.2621	0.2072	6
7	0.6651	0.5835	0.5132	0.4523	0.3759	0.2791	0.2097	0.1594	7
8	0.6274	0.5403	0.4665	0.4039	0.3269	0.2326	0.1678	0.1226	8
9	0.5919	0.5002	0.4241	0.3606	0.2843	0.1938	0.1342	0.0943	9
10	0.5584	0.4632	0.3855	0.3220	0.2472	0.1615	0.1074	0.0725	10
11	0.5268	0.4289	0.3505	0.2875	0.2149	0.1346	0.0859	0.0558	11
12	0.4970	0.3971	0.3186	0.2567	0.1869	0.1122	0.0687	0.0429	12
13	0.4688	0.3677	0.2897	0.2292	0.1625	0.0935	0.0550	0.0330	13
14	0.4423	0.3405	0.2633	0.2046	0.1413	0.0779	0.0440	0.0254	14
15	0.4173	0.3152	0.2394	0.1827	0.1229	0.0649	0.0352	0.0195	15
16	0.3936	0.2919	0.2176	0.1631	0.1069	0.0541	0.0281	0.0150	16
17	0.3714	0.2703	0.1978	0.1456	0.0929	0.0451	0.0225	0.0116	17
18	0.3503	0.2502	0.1799	0.1300	0.0808	0.0376	0.0180	0.0089	18
19	0.3305	0.2317	0.1635	0.1161	0.0703	0.0313	0.0144	0.0068	19
20	0.3118	0.2145	0.1486	0.1037	0.0611	0.0261	0.0115	0.0053	20

CAPITAL- RECOVERY FACTORS FOR INTEREST RATES FROM i = 6% TO i = 30%

n	6%	8%	10%	12%	15%	20%	25%	30%	n
				Given S, Find P					
1	1.06000	1.08000	1.10000	1.12000	1.15000	1.20000	1.25000	1.30000	1
2	0.54544	0.56077	0.57619	0.59170	0.61512	0.65455	0.69444	0.73478	2
3	0.37411	0.38803	0.40211	0.41635	0.43798	0.47473	0.51230	0.55063	3
4	0.28859	0.30192	0.31547	0.32923	0.35027	0.38629	0.42344	0.46163	4
5	0.23740	0.25046	0.26380	0.27741	0.29832	0.33438	0.37185	0.41058	5
6	0.20336	0.21632	0.22961	0.24323	0.26424	0.30071	0.33882	0.37839	6
7	0.17914	0.19207	0.20541	0.21912	0.24036	0.27742	0.31634	0.35687	7
8	0.16104	0.17401	0.18744	0.20130	0.22285	0.26061	0.30040	0.34192	8
9	0.14702	0.16008	0.17364	0.18768	0.20957	0.24808	0.28876	0.33124	9
10	0.13587	0.14903	0.16275	0.17698	0.19925	0.23852	0.28007	0.32346	10
11	0.12679	0.14008	0.15396	0.16842	0.19107	0.23110	0.27349	0.31773	11
12	0.11928	0.13270	0.14676	0.16144	0.18448	0.22526	0.26845	0.31345	12
13	0.11296	0.12652	0.14078	0.15568	0.17911	0.22062	0.26454	0.31024	13
14	0.10758	0.12130	0.13575	0.15087	0.17469	0.21689	0.26150	0.30782	14
15	0.10296	0.11683	0.13147	0.14682	0.17102	0.21388	0.25912	0.30598	15
16	0.09895	0.11298	0.12782	0.14339	0.16795	0.21144	0.25724	0.30458	16
17	0.09544	0.10963	0.12466	0.14046	0.16537	0.20944	0.25576	0.30351	17
18	0.09236	0.10670	0.12193	0.13794	0.16319	0.20781	0.25459	0.30269	18
19	0.08962	0.10413	0.11955	0.13576	0.16134	0.20646	0.25366	0.30207	19
20	0.08718	0.10185	0.11746	0.13388	0.15976	0.20536	0.25292	0.30159	20

Chapter 1

Using basic formulas and the interest tables 1-03, answer the questions below:

Q #1.23: Find the net present value for the following annual cashflow, based on a 10% discount factor: -1000, +400, +300, +300, +400.

A #1.23: The net present value is the sum of all positive and negative values:

$$
\begin{array}{ccccc}
+400 & +300 & +300 & +400 & \\
(0.9091) & (0.8264) & (0.7513) & (0.6830) & \text{(p.w. factors)} \\
364 & 248 & 225 & 273 &
\end{array}
$$

```
    0*      1       2     3      4
    |       |       |     |      |
 -1000
```

* The asterisk denotes the present value (reference).

The net present value is $-1000 + 364 + 248 + 225 + 273 = \textbf{\$ 110}$

Q #1.24: Find the net present value for the following annual cash flow based on a 10% discount factor if the present value is hinged to the end of the third year (focal point).

$$-2\,000, \; -100, \; +400, \; +800, \; +500, \; +1\,000$$

A #1.24:

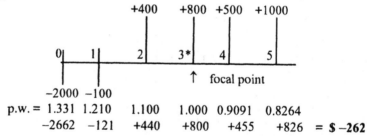

Q #1.25: A facility will cost one million dollars if purchased five years from now. How much has to be invested annually (end-of-year) at 10% to have those fund available?

A #1.25: Sinking fund factor is (0.1638), therefore **\$ 163 800** annually.

Had the question been asked: If I invest \$ 163 800 annually at 10%, what is the amount after five years, the answer is

Compound interest factor is $6.105 \times 163\,800 =$ one million dollars.

Q #1.26: There are linear relationships between various simple interest factors. Demonstrate these relationships by substituting the compound factor $(1+i)^n$ with an equivalent "C".

A #1.26: If $C = (1+i)^n$, then

Single payment present worth:	$P = 1/C$
Unacost compound factor:	$S = (C-1)/i$
Sinking fund:	$R = i/(C-1)$
Present worth:	$P = (C-1)/iC$ or $[1 - (1/C)]/i$
Capital recovery:	$R = iC/(C-1)$ or $i/[1-(1/C)]$

Those relationships can be used to find various methods of repayment of invested capital.

Q #1.27: Why is the use of the capital recovery factor for calculating repayment of invested capital equivalent to a bulk repayment of a loan or to the sinking fund provision of repayment? Show the three methods of capital recovery, assuming a capital investment of $10 000 at 10% during a period of five years:

A #1.27: Repayment of invested capital:
  1) Borrow $ 10 000, pay interest at 10% for five years ($ 5 000) and then repay the $ 10 000.

```
+10 000
  |                            3.170  0.6209 (p.w.)
 0|     1|    2|    3|    4|    5|
  |      |     |     |     |     |           0.6209 × 11 000
  |    1000  1000  1000  1000  11 000       +3.1700 ×   1000
  |                                        = 6830+3170 =1M$
```

  2) Repay the $ 10 000 in equal payments of $ 2 000 plus interest on money owing ($ 3 000).

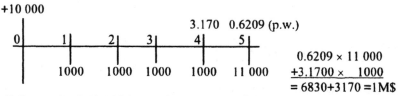

```
+10 000                                              2 728
  |   0.9091  0.8264 0.7513  0.6830  0.6209 (p.w.)   2 314
 0|     1|    2|    3|    4|    5|                    1 953
  |      |     |     |     |     |                    1 639
  |                                                   1 366
  |    2000  2000  2000  2000  2000 (repay)          10 000
       1000   800   600   400   200 (interest)
```

  3) Pay equal amounts of capital into a sinking fund to have the total investment of $ 10 000 available when due in 5 years.

```
                                      (3.791)×2638 = 10 000
  0    1|    2|    3|    4|    5|
        |     |     |     |     |
      1638  1638  1638  1638  1638  (sinking fund)
      1000  1000  1000  1000  1000  (interest)
      2638  2638  2638  2638  2638
```

Q #1.28: A project needs 4 M$ each year for a duration of four years. To finance the program, the owner invests money a year ahead of time at 10%, increasing the deposit in increments of 10% each year. What is the initial deposit the owner has to make? For comparison with other proposals, the focal point is at the end of the second year.

A #1.28: Contribution to the fund must equal withdrawals.

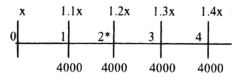

$x(1.21)+(1.1x)(1.1)+(1.2x)+(1.3x)(0.9091)+(1.4x)(0.8264)$ (investment) equals
$4000(1.1 + 1 + 1.7355)$ (project budget)
$x(1.21+1.21+1.2+1.182+1.157) = 15\ 342$ $\therefore$ x = **$2575** (initial deposit)

Economic Comparison

Q #1.29: Two projects with different starting times have the following end-of-year cash flows:

Year	1986	1987	1988	1989	1990	1991	1992	1993	Total
Project #1 (k$)	500	200	300 ,	100	200	--	--	--	1300
Project #2 (k$)	--	--	--	500	400	300	200	100	1500

Assuming a 10% discount rate, which project is less expensive to build and by how much? Use 1990 dollars for comparison.

A #1.29:

		Project #1		Project #2	
Year	p.w.	k$	p.w.$	k$	p.w.$
1986	1.464	500	732	--	--
1987	1.331	200	266	--	--
1988	1.210	300	363	--	--
1989	1.100	100	110	500	550
1990	1.000	200	200	400	400
1991	0.9091	--	--	300	273
1992	0.8264	--	--	200	165
1993	0.7513	--	--	100	75
		1300	**1671**	1500	**1463**

Project #2 is less expensive by 208 k$

Q #1.30: A contractor is in need of a new bulldozer. Bulldozers **A** and **B** are suitable for the job. The following cost information is available:

	A	B
Initial cost	$ 50000	$ 70 000
Average annual maintenance	$ 4 000	$ 2 000
Expected service life	15 years	15 years
Salvage value at disposal	$ 3 000	$ 5 000

Which purchase is more economical if money has to be borrowed at 10% p.a. with end-of-year repayment?

A #1.30:

Bulldozer A:

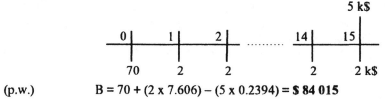

(p.w.)   A = 50 + (4 x 7.606) – (3 x 0.2394) = **$ 79 706**

Bulldozer B:

(p.w.)   B = 70 + (2 x 7.606) – (5 x 0.2394) = **$ 84 015**

The purchase of bulldozer A is more economical.

In the above example it is assumed that maintenance and repair costs are evenly distributed when in fact those costs increase with age. The purchase price of equipment is not necessarily related to the discount rate only. Technical improvements such as better leveling through laser application and computerization can increase the price of replacement equipment with time.

A contractor may opt to review the estimated service life earlier or even each year to decide whether to keep the equipment or sell and replace it.

If the business is very slow, he may have to calculate to rent equipment when needed. Depreciation and related tax considerations also enter the picture.

Q #1.31: If bulldozer A in the above example (Q.#1.30) has a service life of 10 years and bulldozer B a service life of 20 years, which purchase is more economical? Money has to be borrowed at 10%.

A #1.31: Because of uneven service lives, comparisons can now be made only on the basis of a common denominator of their service lives, which is 20 years.
Present worth at time of installation

Bulldozer A: 50 + (4 × 6.144) – (3 × 0.3855) = $ 73 420
Bulldozer B: 70 + (2 × 8.514) – (5 × 0.1486) = $ 86 285

```
A    0|          10|          20|
        |            |            |
     73 420       73 420
```

p.w. = 73 420 (1 + 0.3855) = **$ 101 723**

```
B    0|                        20|
        |                          |
```

$ 86 285

The purchase of bulldozer B is more economical by approx. 15%.

Interest During Construction

Q #1.32: $ 960 000 has been spent on a construction project during the last three years, excluding interest. Total estimated installation cost (excluding. interest) is $ 1 920 000, scheduled to be completed 2½ years later. 12% interest, compounded annually, is charged to pay for the funds which are released quarterly to the project manager, who must pay the interest on a quarterly basis. Calculate the financial cost. The quarterly released funds are:
k$ 10, 20, 30, 40, 50, 60, 70, 80, 100, 140, 160, 200, 200, 200, 180, 120, 100, 80, 40, 20, 10, 10.

A #1.32:
A calculator or spread sheet can be used to calculate interest cost:

# OF PERIODS	3 MONTH PERIODS	CUMULATIVE CASH FLOW	COST OF MONEY 3% PER PERIOD
1	10	10	0.30
2	20	30	0.90
3	30	60	1.80
4	40	100	3.00
		→+	6.00 = 106.00
5	106.00 + 50 =	156	4.68
6	60	216	6.48
7	70	286	8.58
8	80	366	10.98
		→+	30.72 = 396.72
9	396.72 +100 =	496.72	14.90
10	140	636.72	19.10
11	160	796.72	23.90
12	200	996.72	29.90
	960	→+	87.80 = 1084.52

Estimates:

# OF PERIODS	3 MONTH PERIODS	CUMULATIVE CASH FLOW	COST OF MONEY 3% PER PERIOD
13	1084.52 + 200 =	1284.52	38.54
14	200	1484.52	44.54
15	180	1664.52	49.94
16	120	1784.52	53.54
		└───→+	186.56 = 1971.08
17	1971.08 + 100 =	2071.08	62.13
18	80	2151.08	64.53
19	40	2191.08	65.73
20	20	2211.08	66.33
		└───→+	258.72 = 2469.80
21	2469.80 + 10 =	2479.80	74.39
22	10	2489.80	74.69
	1920	└───→+	149.08 = 2638.88

Therefore, total estimated cost of money = 2638.88 - 1920.00 = **k\$ 719**

1.3 PROBABILITIES

Before discussing probabilities in detail, we should have a look at all solutions that are *possible* under given circumstances. The easiest and most quoted example is the throwing of a die. It can readily be visualized and experimented with. When we throw a die, it is possible that a 1, or 2, or 3, or 4, or 5, or 6 turn up at the first throw. Therefore, the *possibilities* of outcome are *certain*. Obviously, it is not possible to throw a 7 or a 10. With three dice, we can throw a total of 7 or 10 dots, but not 1 or 2.

It is not certain, however, which of the six numbers will come first. However, it will give us combinations for which we can *predict probable outcomes.*

1.3.1 Combinations Under Certainty

If we call the six numbers on the die n = 6 and the one die (any number) under consideration r = 1, then the combination is

$$\frac{6!}{(1!)(5!)} = 6$$

Generally speaking, for a set of n things taken r at a time the combination is

$$C_r(n) = \frac{n!}{(r!)(n-r)!} = \binom{n}{r}$$

n! is the factorial of n, n! = $1 \times 2 \times 3 \times \ldots\ldots \times (n-1) \times n$

Assuming we throw *two* dice with the stipulation *not* to allow duplicate numbers such as 1,1 or 2,2 etc. then the combinations without duplications are

          1,6
          1,5   2,6
          1,4   2,5   3,6
          1,3   2,4   3,5   4,6
          1,2   2,3   3,4   4,5   5,6   giving us 15 possibilities.

Calculated $C_r(n) = \binom{6}{2} = \frac{6!}{(2!)(2!)} = \frac{1 \times 2 \times 3 \times 4 \times 5 \times 6}{1 \times 2(1 \times 2 \times 3 \times 4)} = 15$

For three dice, $C_3(6) = 20$ and so on. $\binom{n}{r}$ is also called the binomial coefficient

and is defined as $\binom{n}{r} = \frac{n(n-1)(n-2) \ldots\ldots (n-r+1)}{r!}$, whereby

$$\binom{n}{1} = n \; ; \; \binom{n}{n} = 1 \; ; \; \binom{n}{0} = 1 \; ; \; \binom{n}{r} = 0 \text{ if } r > n \text{ and } 0! = 1$$

We may remember from highschool that the series

$$(1+x)^n = 1 + \binom{n}{1}x + \binom{n}{2}x^2 + \ldots\ldots \binom{n}{r}x^r + \ldots\ldots \text{ is the binomial series.}$$

Let us look at our dice example again. If we include duplications we would add the following combinations to our two-dice problem: 1,1 2,2 3,3 4,4 5,5 6,6
In total:

          1,6
          1,5 2,6
          1,4 2,5 3,6
          1,3 2,4 3,5 4,6
          1,2 2,3 3,4 4,5 5,6
          1,1 2,2 3,3 4,4 5,5 6,6   thus giving us 21 possibilities.

The formula for this is

$$C'_r(n) = \binom{n+r-1}{r} \text{ which in our case would be } \binom{7}{2} = 21$$

Under the same conditions, the combination for three dice with repetition, i.e. 1,1,1; 2,2,2 etc. added, would be

$$C'_3(6) = \binom{8}{3} = 56$$

We must not forget, the conditions were that a simple exchange of dice *do not* form a new combination. Throwing the numbers

      is the same as

If the two dice have different colors, say red (R) and blue (B) and it matters which color is assigned to a number, i.e. "Red-1" plus "Blue-2" is a different combination as "Blue-1" plus "Red-2". Positions are considered:

      is <u>not</u> the same as

but if we have *no repetitions* such as 1,1 2,2 etc., then the number of variations of n elements, taken r at a time

$$V_r(n) = \binom{n}{r} \times r! \text{ and with 2 dice, } \binom{6}{2} \times 2! = 30 \text{ variations.}$$

and with three dice, $V_3(6) = \binom{6}{2} \times 3! = 120$ variations

Following this, if we take the same number of dice as there are numbers on the dice, then

$$n = r \text{ and } V_n(n) = n! = 720$$

this is called the *permutation* of "n" elements, also written as

$$P(n) = n!$$

Since this is the borderline of variations, permutations do not have repetitions.

Table 1-04

Two-dice frequency distribution

Range $\Sigma_d \rightarrow$	2	3	4	5	6	7	8	9	10	11	12
1						6,1					
3					5,1	5,2	6,2				
5				4,1	4,2	4,3	5,3	6,3			
7			3,1	3,2	3,3	3,4	4,4	5,4	6,4		
9		2,1	2,2	2,3	2,4	2,5	3,5	4,5	5,5	6,5	
11	1,1	1,2	1,3	1,4	1,5	1,6	2,6	3,6	4,6	5,6	6,6
fa	1	2	3	4	5	6	5	4	3	2	1

If we include repetitions (1,1 2,2 etc.) into the variations, the total number would be

$$V'_r(n) = n^r$$

there are $6^2 = 36$ variations in the case of our two dice.

The sum of the numbers on the dice which we will call Σ_d, range from
$$1,1 = 2 \text{ to } 6,6 = 12.$$
The distribution of all throws would look like shown in Table 1-04.

"$f_{6,2}$" is the discrete distribution of all possibilities for a set of six taken two at a time. The sum of the dots on the pair of dice ranges from $\Sigma_d = 2$ to $\Sigma_d = 12$. In case of three dice, the number of variations are $6^3 = 216$

For $f_{6,3}$ the discrete distribution is
$$1+3+6+10+15+21+25+27+27+25+21+15+10+6+3+1 = 216$$
and counting along the vertical scale
$$1\times16+2\times14+3\times12+4\times10+5\times8+6\times6+4\times4+2\times2 = 216$$
That means there are 10 possible ways (fa=10) that the 3 dice add up to a total of
$$15 \ (\Sigma_d = 15)$$
and only 3 ways that they can add up to 17 etc., as shown in Table 1-05.

Here is another example, where the elements of combinations are treated separately. Lets assume there are six courses, a to f, in an educational program. There are no restrictions as to the sequence of the courses to be taken. At the time of a survey, some students have taken only one course, others more, to a maximum of six. The instructor wants to know what the maximum possible combination of courses taken is at this time. The solution could be arrived at in a very elaborate way by counting all the elements. It is much easier to use the binomial series

$$(1 + x)^n = 1 + \binom{n}{1} x + \binom{n}{2} x^2 + \ldots \binom{n}{r} x^r + \ldots$$

Table 1-05

Frequency distribution of 3 dice

fa →	1	3	6	10	15	21	25	27	27	25	21	15	10	6	3	1
2 ←								631	641							
2 ←								622	632							
4 ←							621	613	623	651						
4 ←							612	541	614	642						
4 ←							531	532	551	633						
4 ←							522	523	542	624						
6 ←						611	513	514	533	615	661					
6 ←						521	441	451	524	561	652					
6 ←						512	432	442	515	552	643					
6 ←						431	423	433	461	543	634					
6 ←						422	414	424	452	534	625					
6 ←						413	351	415	443	525	616					
8 ←					511	341	342	361	434	516	562	662				
8 ←					421	332	333	352	425	462	553	653				
8 ←					412	323	324	343	416	453	544	644				
8 ←					331	314	315	334	362	444	535	635				
8 ←					322	251	261	325	353	435	526	626				
10 ←				411	313	242	252	316	344	426	463	563	663			
10 ←				321	241	233	243	262	335	363	454	554	654			
10 ←				312	232	224	234	253	326	354	445	545	645			
10 ←				231	223	215	225	244	263	345	436	536	636			
12 ←			311	222	214	161	216	235	254	336	364	464	564	664		
12 ←			221	213	151	152	162	226	245	264	355	455	555	655		
12 ←			212	141	142	143	153	163	236	255	346	446	546	646		
14 ←		211	131	132	133	134	144	154	164	246	265	365	465	565	665	
14 ←		121	122	123	124	125	135	145	155	165	256	356	456	556	656	
16 ←	111	112	113	114	115	116	126	136	146	156	166	266	366	466	566	666
Σd =	3	4	5	6	7	8	9	10	11	12	13	14	15	16	17	18

The sum of the coefficients equals the number of possible combinations of courses taken, which is n=6

$$\Sigma C_r(n) = \binom{6}{1} + \binom{6}{2} + \binom{6}{3} + \binom{6}{4} + \binom{6}{5} + \binom{6}{6} = 6 + 15 + 20 + 15 + 6 + 1 = 63$$

1

2 → 1 = **3**
   ↓

3 → 3 → 1 = **7**
   ↓   ↓

4 → 6 → 4 → 1 = **15**
   ↓   ↓   ↓

5 →10→10→ 5 →1 = **31**
   ↓   ↓   ↓   ↓

6 →15→20→15→ 6→ 1 = **63**
   ↓   ↓   ↓   ↓   ↓

7 →21→35→35→21→7 → 1 = **127**
            etc.

A binomial table can be constructed that depicts the coefficients for various exponents of the series.

The additions to the next row are

a → + b
   ↓
   =
   c

In the above example it did not matter how many courses were taken and in which order. If we ask the question: "What possible combinations were there for all six courses?" This would be the *permutation* of n = 6.

Mathematically, for n = 6, it is simply $P(n) = n! = 6! = 720$

But if we combine some of the courses for which the frequency does not matter, say b,c,d = A and e,f = B, then

$$P_{AB}(n) = \frac{n!}{a!\, A!\, B!} = \frac{720}{1 \times 6 \times 2} = 60 \text{ possibilities}$$

There are other variations that apply to give the decision-maker an idea of the number of possible solutions. For example, if we made it a prerequisite that course "a" must be the *first* course to be taken, the permutations would be greatly reduced, i.e.

$$P(n) = (n - 1)! = 120$$

1.3.2 Probability Distributions

Within a certain environment, the solution to a problem is known. In such an environment alternative courses of action can be predicted (deterministic model). For example, if we choose alternative # 3, the outcome will be X, for alternative # 5 the outcome will be Y etc. Choosing the best alternative is usually straight forward.

When we specify all *possible* outcomes in a real project environment, we may not be certain *which* outcome will happen.

Under those conditions we are using the common expressions *"the chances are..."* or *"the odds are..."* or *"it is likely that..."*

Throwing one die, we do not know which number will turn up first. We can predict, however, that the probability of a six (or any given number) is 1/6 or 0.167 (stochastic model). This prediction is not the absolute truth. We may be unfortunate enough to throw the die twelve times before the six turns up or we may have three sixes within the first six throws. The more often we repeat the experiment the closer we will converge toward the outcome of 0.167.

If I bet even money that you would throw at least one six in four throws, my chances of making money are good because there is a fifty-fifty chance (0.5) that a six will occur with one die in three throws. In four throws it will be 4/6 = 0.67 or 67%. In spite of this mathematical prediction, we may still end up a loser.

Probability calculations yield only a frequency distribution limited by an infinite random series of events.

Generally, the probability that event "a" will occur among all possible events "m" is

$$P(a) = \frac{\text{the number of ways "a" can occur}}{\text{the number of ways any outcome of "m" can occur}} = \frac{a}{m}$$

where "a" is the sum of the numbers on the range of possibilities (frequency of occurrence), and "m" equals the total number of possible combinations.

With two dice the *possibility* to throw a sum of 5 would be (1,4 and 2,3) = 2; to throw a sum of six is (1,5 and 2,4 and 3,3) = 3.

The total number of possible combinations (previous section) for two dice of equal color is

$$\text{from } C'_r(n) = \binom{n + r - 1}{r} ; C_2(6) = \binom{6 + 2 - 1}{2} = 21$$

Therefore, the *probability* to throw a sum of 5 is 2/21 = 9.5% and for a 6 it is 3/21 = 14%.

We also know from the previous section for colored dice, that the position of the dice matter, and that the total number of possible events is the variation

$$m = V'_r(n) = n^r, \text{ and for two dice: } m = 6^2 = 36$$

In this case, the number of ways "a" (to throw the sum of 5) increases to 4 and the probability is

$$P(a) = a/m = fa/\sum fa = 4/36 = 0.11$$

Many practical problems that involve uncertainties are based on the above criteria, the last of which n^r is the most common.

If we want to tabulate the probabilities for various values of "a" for our dice problem, the probabilities for each frequency are

For one die:

d	1	2	3	4	5	6
ƒa	1	1	1	1	1	1
P(a)	0.67	0.67	0.67	0.67	0.67	0.67

For two dice:

d	2	3	4	5	6	7	8	9	10	11	12
ƒa	1	2	3	4	5	6	5	4	3	2	1
P(a)	.03	.06	.08	.11	.14	.17	.14	.11	.08	.06	.03

For three dice:

∑d = 3	4	5	6	7	8	9	10	11	12	13	14	15	16	17	18
ƒa→ 1	3	6	10	15	21	25	27	27	25	21	15	10	6	3	1
d 3	4	5	6	7	8	9	10	11	12	13	14	15	16	17	18
P(a).005	.014	.028	.046	.069	.097	.116	.125	.125	.116	.097	.069	.046	.028	.014	.005

where d is the number of dots on the 3 dice per throw and ƒa = the frequency at which this sum occurs.

Shown below are the probabilities graphically represented on a frequency diagram for the three-dice example (Figure 1-06).

The statistical mean of this distribution is interpreted as the sum of the likely value of event "a" over the total of equally likely events "m".

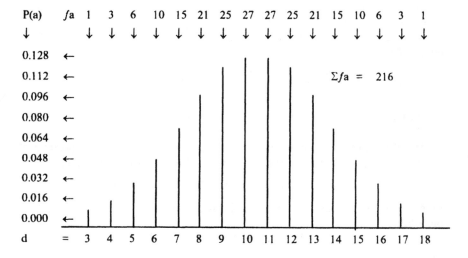

Figure 1-06
Normal distribution of 3-dice.

For our dice example

$$\overline{X} = \frac{\sum\{d(f_a)\}}{\sum f_a}$$

$\overline{X} = \{3(1)+4(3)+5(6)+6(10)+....16(6)+17(3)+18(1)\}/(216) = \textbf{10.5}$

but P(a) was $f_a/\sum f_a$, therefore, in terms of probabilities

$\overline{X} = 3(0.005)+4(0.014)+5(0.028)+...17(0.014)+18(0.005) = \textbf{10.5}$

In case of three dice the mean is 10.5 per throw.

This statistical average is called the *expected value* E of "a"

$$E(a) = \sum a_i P(a_i)$$

In other words, the expected value E(a) is the sum of the products

a_i = the ith outcome of an alternative for the event "'a" measured in specific units such as dots on dice, Whrs or $, and

$P(a_i)$ = the probability ratio of the outcome a_i

Further, the probability P(a) of an event a_i occurring among the number of all possible events is a ratio between zero and one:

$$0 \leq P(a) \leq 1.0 \quad \text{therefore} \quad \sum P(a) = 1.0$$

Example:

If somebody will pay you $ 30.00 every time your three dice will show a sum of five, and you will have to pay the bank $1.00 if this is not the case, would you play the game?

The probability distribution is (see distribution diagram for three dice):

a_i	$P(a_i)$
$ 30.00	0.028
$ (1.00)	0.972
	1.000

The expected value is:

$$E(a) = (\$30)(0.028) + (-1)(0.972) = 0.84 - 0.97 = -\textbf{\$ 0.13}$$

The result shows that the expected value is negative. You would over the long run register an average *loss* of $ 0.13. Once in a while you may win $ 30.00, but quite often you will loose $1.00. Would you play? This depends on your *attitude toward risk*. The alternative not to play at all yields zero gain, which is a better alternative than the loss of $ 0.13.

We have assumed of course that the dice are truly balanced. You may find that the cheap dice you bought at a corner store do not seem to roll easily.

So you have them tested. The test results show that the two dots on each die turn up much more frequently than they should be.

Be a Winner!

Loaded dice

The "two" is on the lighter side, i.e. the dice are "loaded" in favor of throwing the "two" 70% more often than normal. This means that the heavy side of the dice (five dots) will only show up 70% *less* often than normal or 30% of the time. We now do not have a normal distribution any more.

How does this affect the outcome and the distribution? The "skewed" f_a is now $6 \times 1.7 = 10.2$ and
$$P(a) = 10.2/216 = 0.047,$$
therefore $E(a) = \$30(0.047) + (-\$1)(0.953) = +\$0.46$
The expected value is now positive. It would have paid to play the game.

Comparing the normal distribution (balanced dice) with the skewed distribution (loaded dice) we should count how often the "twos" and "fives" appear, apply +70% and −70% accordingly to obtain the skewed $f(a)$.

Σd	twos	fives	+0.7twos	−0.7fives	normal $f(a)$	skewed $f(a)$	P(a)
3	0	0	0	0	1	1	0.0046
4	3	0	+2.1	0	3	5.1	0.0236
5	6	0	+4.2	0	6	10.2	0.0472
6	9	0	+6.3	0	10	16.3	0.0755
7	12	3	+8.4	−2.1	15	21.3	0.0986
8	15	6	+10.5	−4.2	21	27.3	0.1264
9	18	9	+12.6	−6.3	25	31.3	0.1449
10	15	12	+10.5	−8.4	27	29.1	0.1347
11	12	15	+8.4	−10.5	27	24.9	0.1153
12	9	18	+6.3	−12.6	25	18.7	0.0866
13	6	15	+4.2	−10.5	21	14.7	0.0681
14	3	12	+2.1	−8.4	15	8.7	0.0403
15	0	9	0	−6.3	10	3.7	0.0171
16	0	6	0	−4.2	6	1.8	0.0083
17	0	3	0	−2.1	3	0.9	0.0042
18	0	0	0	0	1	1	0.0046
168	108	108	+75.6	−75.6	216	216.0	1.0000

The peak of the distribution or the mode Mo has now shifted to the left, to d = 9 (Mo *would be 9.15 on a continuous distribution*). Throwing a total of 9 points with three dice is happening more often than throwing any other sum.

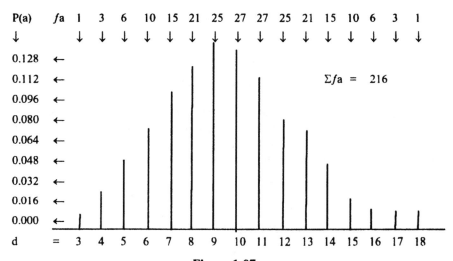

Figure 1-07
Distribution with loaded dice.

It has also increased in magnitude because P(a) = 0.1449 instead of 0.125 for the normal distribution. This is called a *leptokurtic* distribution (Figure 1-07).
The statistical mean is

$$\overline{X} = \{3(1)+4(5.1)+5(10.2)+.....+16(1.8)+17(0.9)+18(1)\}/(216) = \mathbf{9.45}$$

In regard to probabilities, the average of occurrences is the expected value

$$E(a) = \Sigma\{dP(a)\} = 3(0.0046)+4(0.0236)+...+18(0.0046) = \mathbf{9.45}$$

which is identical to the statistical mean. Each frequency of occurrence under a normal distribution deviates from the mean symmetrically. The true dice with a mean of 10.5 would have "distances" from the mean

fa	$\overline{X}$	$fa - \overline{X}$
1	−10.5	= −9.5
3	−10.5	= −7.5
6	−10.5	= −4.5
10	−10.5	= −0.5
15	−10.5	= + 4.5
21	−10.5	= +10.5
25	−10.5	= +14.5
27	−10.5	= +16.5
	.....etc.......	
3	−10.5	= −7.5
1	−10.5	= −9.5
	± 0.0	for a total of $\Sigma(fa - \overline{X}) = 0.0$

In order to obtain an *average deviation*, we are forced to ignore the signs and deal with *absolute* values. This will now give us an average deviation of

$$\Sigma\,|\,\mathit{f}a - \overline{X}\,|\,/N = 136/16 = 8.5,$$

called the average deviation from the mean. It measures the dispersion of grouped data.

Because we are using absolute values, the method is nonalgebraic. This can be a handicap when working out sampling distributions. For this reason, another measure of dispersion, the *variance* is quite common. Here, the deviation from the mean is squared to eliminate negative values. The variance is defined as

$$\sigma^2 \;=\; \frac{\Sigma\left(\mathit{f}_a - \overline{X}\right)^2}{\Sigma \mathit{f}_a - 1}\quad \text{and } \sigma \text{ is the standard deviation.}$$

$$\sigma^2 = 2(90.25 + 56.25 + 20.25 + 0.25 + 20.25 + 110.25 + 210.25 + 272.25)/215$$

$$\sigma^2 = 7.256 \text{ and } \sigma = 2.69$$

The standard deviation represents an average departure above or below a central position or, in our case, the mean.

The smaller the standard deviation, the closer the items are grouped about the arithmetic mean.

One, two, and three standard deviations will give us

$$\overline{X} + 1\sigma = 13.19 \qquad \overline{X} - 1\sigma = 7.81$$
$$\overline{X} + 2\sigma = 15.89 \qquad \overline{X} - 2\sigma = 5.12$$
$$\overline{X} + 2\sigma = 18.57 \qquad \overline{X} - 2\sigma = 2.43$$

The range of our 16 frequencies go from 3 to 18. Why then are the values for three standard deviations 2.43 and 18.57? The area under the distribution extends beyond the end points.

So far, we have been counting individual frequencies. Statistics rely mainly on continuous frequency distributions, where an unlimited number of intervals are considered. This would result in a smooth curve.

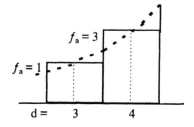

For a normal distribution, the curve is split into equal areas and may look like the one shown in Figure 1-08.

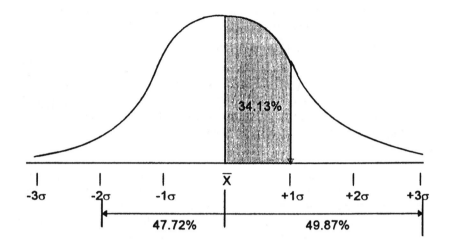

Figure 1-08
Normal distribution curve.

The total area under the curve is 100%. One standard deviation from the mean will then represent 68.27% of the total area or total frequencies under the curve. Two standard deviations represent 95.45% and 3σ represent 99.74% of the area under the curve.

This has profound implication in the application of probabilities!

If an estimate has a range that falls under a normal distribution, the probability that it will not differ by more than one standard deviation is 0.68, i.e. 1σ in both directions from the point estimate. It will fall 19 out of 20 times (95.45%) within two standard deviations from the mean. It is almost certain not to exceed 3 standard deviations because only 0.13% of the area lie above

$$\overline{X} + 3\sigma \text{ and below } \overline{X} - 3\sigma$$

When we throw the three honest dice (normal distribution) we will throw either 8, 9 ,10 ,11 ,12 ,13 two-thirds (68%) of the time because d8 to d13 has the frequencies 21, 25, 27, 27, 25 ,21 with the sum of 146 (area), which is 68% of the total Σf_a=216 or one standard deviation.

Standard deviations are usually expressed as

$$x/\sigma = 1.96 \text{ (standard units)}, \quad \text{where } x = X - \overline{X}$$

In making use of the normal curve, we need to use proportions of areas (frequencies) included *between* almost any two points on the abscissa.

There are tables available that show fractional multiples of the standard deviation. A condensed version is shown below:

STANDARD NORMAL DISTRIBUTION

x/σ	.0	.1	.2	.3	.4	.5	.6	.7	.8	.9	.96
0.	0.00	3.98	7.93	11.79	15.55	19.15	22.58	25.80	28.81	31.59	
1.	34.13	36.43	38.49	40.32	41.92	43.32	44.52	45.54	46.41	47.13	47.5
2.	47.72	48.21	48.61	48.93	49.18	49.38	49.53	49.65	49.74	49.81	
3.	49.87	49.90	49.93	49.95	49.97	49.98	49.98	49.99	49.99	49.996	
4.	49.9968										

The above is based on a total area = 100%. Fractional parts of the total area under the normal curve between the mean and a vertical projection at various numbers of standard deviations from the mean are shown as percentages.

The standard unit 1.96 is approximately 47.5 which means that 47.5% of the area under the normal curve lies between the mean and the point at a distance of 1.96 standard deviations above the mean. The "tails" at each end of the curve contain 2.50% of the total area.

What is the standard deviation of *the loaded dice*? We do not have to go back to do manual calculations again which was considered necessary so far in order to better understand the principles involved. There are calculators and computer programs available that do all kinds of statistical calculations.

We must be aware, however, that the calculations are based on samples of large frequency distributions. Some calculators show $x\sigma_{n-1}$ as the sample standard deviation (small sample, grouped data) and $x\sigma_n$ as the population standard deviation (large sample, grouped data).

The loaded dice with the mean 9.45 would theoretically have the following standard deviations from the mean ($\sigma = 3.045$):

$$\overline{X} + 1\sigma = 12.50 \qquad \overline{X} - 1\sigma = 6.45$$
$$\overline{X} + 2\sigma = 15.55 \qquad \overline{X} - 2\sigma = 3.35$$
$$\overline{X} + 3\sigma = 18.59 \qquad \overline{X} - 3\sigma = 0.31$$

The distribution has the approximate range of $-2\sigma < \overline{X} < +3\sigma$ (see Figure 1-09). The degree of skewness can be expressed by the ratio

$$Sk = \frac{\overline{X} - Mo}{\sigma}$$

The skewness can be positive or negative, depending if the resultant curve is leaning to the right or to the left.

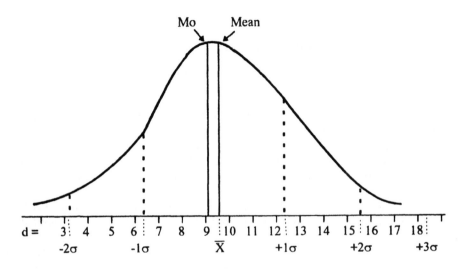

Figure 1-09
Skewed distribution curve.

Figure 1-10 depicts the calculation of the mode.

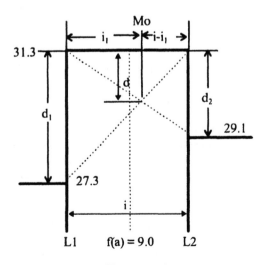

Figure 1-10
Calculating the mode

The mode Mo for a continuous distribution is based on class limits. If we consider $f(a)=9.0$ to be the center of a class with limits $L1 = 8.5$ and $L2 = 9.5$, by proportion

$$\frac{i_1 d_2}{i} = \frac{d_1(i - i_1)}{i}$$

and $\quad d_1 i = d_1 i_1 + d_2 i_1;$

$$\therefore \quad i_1 = \frac{d_1}{d_1 + d_2}$$

$$Mo = L1 + i\frac{d_1}{d_1 + d_2}$$

and solving our example

$$Mo = 8.5 + \frac{31.3 - 27.3}{4.0 + (31.3 - 29.1)} = 8.5 + \frac{4.0}{6.2} = \mathbf{9.15}$$

finally, the degree of skewness $$Sk = \frac{9.45 - 9.15}{3.045} = \mathbf{+0.1}$$

The positive result indicates a skewness to the right, i.e. the median is at the "right" of the mode on the horizontal scale.

The standard distribution gives us a measure of the "flatness" of the curve, the skewness tells us where frequencies are concentrated away from the mean. Two (or more) distributions can have the same frequency (area), the same arithmetic mean, the same standard deviation and skewness and still differ in shape if some of their frequencies have a higher or lower concentration near the common mean, compensated for by corresponding lower or higher concentration of frequencies at other locations under the curve.

This property is called the *kurtosis* of the distribution. It is measured by calculating "moments" about the arithmetic mean. It is somewhat elaborate and will not be discussed here except that the fourth moment

$$M_4 = \Sigma fd^4 \div N$$

is entered into the calculation to measure K1 < 3.00 < K2. 3.00 is the value for a normal distribution. K1 would indicate a more flat-topped and K2 a more peaked curve than normal.

Sampling

One of the basic requirements when taking statistical samples is that *it must be based on probabilities.*

The sample of N items taken from the statistical "universe" (or population) is considered a *simple random sample* only if any combination of N items has an *equal chance* of being selected. The sampling theory is based on the mathematics of an infinite universe. In practice, the samples taken come from a finite universe. When we draw two cards, one at a time, from a well-shuffled deck, we took N = 2 items as a simple random sample from the universe of 52 cards, knowing, that any other two cards had the same chance (probability) of being selected.

We can select a combination of $\binom{52}{2}$ = 1326 samples of two cards from the

deck. If all 1326 combinations (without repetition) were used and the arithmetic mean of each sample were calculated, a distribution of 1326 numbers would be obtained. Each of those sample distributions then have an arithmetic mean and a standard deviation of their own.

Central Limit Theorem

If random samples of size **N** are taken from a universe distribution which has the arithmetic mean **m** and standard deviation σ, the sampling distribution will very closely approximate a normal distribution with the same arithmetic mean and standard deviation $\dfrac{\sigma}{\sqrt{N}}$, provided N is relatively large.

The standard deviation of the sampling distribution is also called the standard error of the mean.

$$\varepsilon = \frac{\sigma}{\sqrt{N}}$$

The larger the sample size N, the smaller is the "error" in predicting the frequency distribution (incl. the mean) of the universe.

For example, a box contains 1000 marbles. 700 are black, 230 are red and 70 are blue. We now shake the box to mix them all up. With eyes closed we pick a marble, record the color, throw it back into the box, shake again, pick another marble, and so on. We do this 100 times. We may have recorded 50 black, 40 red and 10 blue marbles.

We will go through the same process again, picking additional 200 marbles. The new record with 300 marbles may now show 170, 94 and 36. We then try 500. In terms of percentages:

	black	red	blue
Universe	70%	23%	7%
100 samples	50%	40%	10%
300 samples	57%	31%	12%
500 samples	64%	28%	8%

Obviously, the more samples we take, the closer we come to the true ratio. In many practical applications the universe (total mixture of the colored marbles) is not known. We are therefore *inferring* that the sample has a very high probability of representing the universe. The sampling error is an average error, it is the difference between *any* sample mean and the universe mean.

To determine probabilities in "either/or" situations such as the head or tail of a coin, success or failure, work or not work etc., we use the binomial formula

$$P(a) = \binom{N}{a} p^a (1-p)^{N-a} \text{ with the variance } \sigma^2 = Np(1-p)$$

If the mean of the distribution $Np > 5$, the standard error of the distribution is

$$\sigma_p = \sqrt{\frac{p(1-p)}{N}}$$

The standard error of the binomial distribution has many practical applications in quality control, value engineering and work sampling.

Example: The historical record for a shipment of valves over a period of time shows that, on the average, 6% of the valves have been defective in the past. The supplier was changed recently. The new supplier indicates that he is capable of a 'similar performance. The buyer's inspector samples 100 valves at random from the inventory of the manufacturer and finds, that 3 are defective. How representative is the sample?

$$P(a) = \frac{100!}{3!\,97!}(0.02)^3(0.98)^{97} = 0.0864$$

There is only a 9% chance for 3 defective valves in a sample of 100 provided the defect rate is 6%. From the number of samples taken, the following sampling error is indicated

$$\sigma_p = \sqrt{\frac{0.06 \times 0.94}{100}} = 0.024 \text{ or a} \pm 2\% \text{ error.}$$

There are some applications for which we would rather prefer a more nearly continuous distribution. When the number of random events are small as compared to the potential number that could occur, we may want to use the *Poisson distribution*

$$P(x) = \frac{\lambda^x e^{-\lambda}}{x!} \text{ where } \lambda \text{ is both, variance and mean.}$$

When we plot the distribution curves for various values of λ we can directly read off the values for the chance that various events x will occur.

For example: The desk at a construction warehouse has been observed to normally handle 450 material requests per hour. The maximum the staff can handle without serious disruption in service are 10 requests per minute. What are the chances that disruptions may happen ?

$$\lambda = \frac{450}{60} = 7 \text{ per minute (mean),} \quad x = 10$$

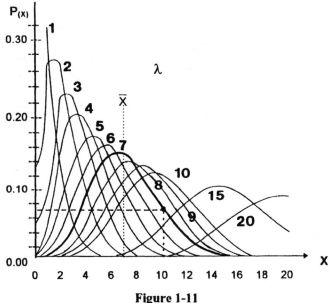

Figure 1-11
Poisson distribution.

We could calculate P(x), use charts or a graph. Looking at the graph (Figure 1-11) where the vertical line at x = 10 touches the Poisson distribution curve λ = 7, the reading is P(x) = 0.07 (calculation yields 0.071).

That means there is a 7% chance that a disruption in service may take place. It is now up to the manager to decide if the risk is acceptable or if some action has to be taken.

1.3.3 Other Probabilities

When we did our dice examples, we were dealing with specific data. We did not have to guess how many dots there were on a die. The calculations were based on existing or prior knowledge and logic. Those calculations are called *à priori probabilities*. If we call "a" all possible outcomes that can happen to an event A and "b" all possible outcomes that are unfavorable to the occurrence of event A then the probability of event A occurring is

$$P(A) = \frac{a}{a+b} \qquad \qquad ......(1)$$

The chances to throw an odd number with one die is   3/6 = 0.5. The probability that A does **not** happen is

$$1 - P(A) \text{ or } 1 - 0.5 = 0.5$$

What is the probability that we throw an even number larger than 3? A Venn diagram may help to visualize this situation:

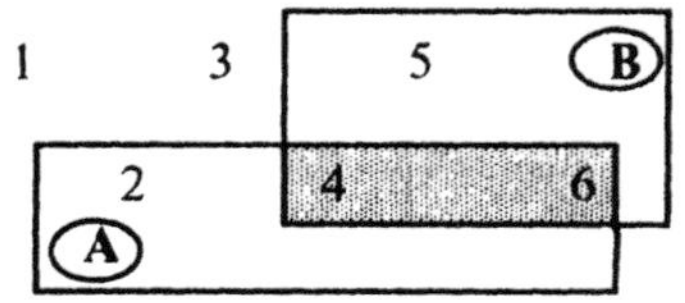

A represents even numbers and B numbers larger than 3. The probability of both, A and B occurring is shown in the shaded area. This is expressed as

$P(A \cap B) = 2/6 = 0.33$, where $\cap$ means "and".
The probability to throw an odd number >3 is

$$P(A \cap B) = 1/6 = 0.17$$

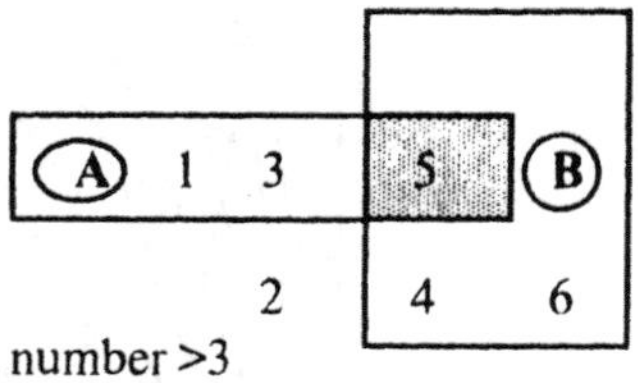

The probability that we throw an odd number or a number >3 or both, (A or B) or (A and B), where $\cup$ means either/or or both, is

$$P(A \cup B) = P(A) + P(B) - P(A \cap B) = 0.5 + 0.5 - 0.17 = 0.83$$

Conditional Probability

The probability that event A will occur, *given that event B has already occurred* is a conditional probability.

This may be demonstrated by using the example of a box with colored marbles. Assuming we have six black and six white marbles mixed up in a box. The first draw will reduce the number of marbles in the box to 11. Depending on the color we pick first, it will affect the probabilities of the second draw (we are not returning the marble).

If we picked black, we only have 5 black and 6 white left in the box. Obviously, the chance to pick another black marble is less than picking a white marble, i.e. 5/11 = 0.45. Because we did not put the black marble back, the chance of picking either black or white on the second draw is *conditional* on the outcome of the first draw. This is expressed as

$$P(A \mid B) = \frac{P(A \cap B)}{P(B)} \qquad \text{...... (2)}$$

The symbol | means "given" that B has occurred.

Regarding our dice example, the probability for an odd number to occur, given that a number >3 has already occurred, is

$$P(A \mid B) = \frac{1/6}{3/6} = 0.33$$

Conditional probabilities are widely used in the decision making process where prior information was based on subjective assessments and new additional information causes a need for reassessment. To better visualize the calculations involved, here is a simple *example*:

Consider three identical jars, J1, J2 and J3. Each jar contains two marbles as follows:

J1 = two green marbles .... .(G;G)
J2 = one green, one white ..(G;W)
J3 = two white ....(W;W)

Assuming the jars are placed in a dark room, what are the chances to touch jar J1 (or any other specified)?

$$P(J1) = 1/3 = 0.33$$

If we are required to pick a marble from any of the three jars (we do not know which one – they are in the dark), what are the chances to pick a green one ?

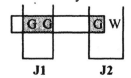 $P(G) = 3/6 = 0.50$

J1   J2   J3

What is the probability to select a green marble given that jar "Ji" was selected previously?

$$P(G \mid Ji) = \frac{P(G \cap Ji)}{P(Ji)} ; \qquad \dots\dots\dots\dots\dots(3)$$

For Jar J1 it would be $P(G \mid J1) = \dfrac{\text{two green out of six}}{\text{one jar in three}} = \dfrac{2/6}{1/3} = 1.00$

For jar J2 and J3 the probabilities would be

$$P(G \mid J2) = \frac{\text{one green out of six}}{\text{one jar in three}} = \frac{1/6}{1/3} = 0.50$$

$$P(G \mid J3) = \frac{\text{no green out of six}}{\text{one jar in three}} = \frac{0/6}{1/3} = 0.00$$

What is the probability that the jar which was originally selected was the one with the two green marbles? Here, the prior probability is "green". Formula (3) applies with G and Ji interchanged:

$$P(Ji \mid G) = \frac{P(Ji \cap G)}{P(G)} \qquad \dots\dots\dots\dots\dots(4)$$

Specifically, $P(J1 \mid G) = \dfrac{2/6}{3/6} = 0.67$ and $P(J2 \mid G) = \dfrac{1/6}{3/6} = 0.33$

$$P(J3 \mid G) = \frac{0/6}{3/6} = 0.00$$

We can summarize all of the above in the form of a table:

Table 1-06

Conditional probabilities

Ji	P(Ji)	P(G∣Ji)	P(Ji∩G)	P(Ji∣G)
J1	1/3	1	2/6	2/3
J2	1/3	½	1/6	1/3
J3	1/3	0	0	0
Total	1		P(G) = ½	

Bayes's Rule

The formula $P(G \mid Ji) = \dfrac{P(G \cap Ji)}{P(Ji)}$ (equation 3) can be expressed as

$$P(G \cap Ji) = P(Ji)P(G \mid Ji)$$

and $P(G) = \Sigma P(G \cap Ji) = \Sigma[P(Ji)P(G \mid Ji)]$, because

$$P(G) = P(J1 \cap G) + P(J2 \cap G) + P(J3 \cap G) + \dots \text{etc.} \dots\dots\dots(5)$$

and specifically for jar J1, using the above substitute for P(G), and equation (4)

$$P(J1 \mid G) = \frac{P(J1)P(G \mid J1)}{\Sigma\left[P(Ji)P(G \mid Ji)\right]} \quad\dots\dots\dots\dots(6)$$

This equation is known as Bayes's rule. Here is a simple *example* how Bayes's rule can be applied:

It was estimated by the head office purchasing department that 70% of all valves that were purchased for a particular construction project were supplied by manufacturer M1 and 30% by manufacturer M2. Inspection sampling indicates that 4% of valves produced by M1 are defective, compared to 6% for M2.

Unfortunately, the site warehouse does not stock valves separated by supplier, they are all intermingled on the shelves. The installation of a defective valve is costly. The Project Manager needs backup to present a claim to the manufacturers. He needs to know

what is the probability that a valve was delivered by manufacturer M1...
what is the probability that a valve was delivered by manufacturer M2...
given that the valve was defective, or P(Mi∣D) =?
Using equations (3) and (4)

$$P(M1 \mid D) = \frac{P(M1)P(D \mid M1)}{P(D)} = \frac{(0.70)(0.04)}{P(D)}$$

$$P(M2 \mid D) = \frac{P(M2)P(D \mid M2)}{P(D)} = \frac{(0.30)(0.06)}{P(D)}$$

Using equation (5), we can now solve for P(D),

$$P(D) = \Sigma[P(Mi)P(D \mid Mi)] = (0.70)(0.04) + (0.30)(0.06) = 0.046$$

and now applying Bayes's rule

$$P(M1 \mid D) = \frac{(0.70)(0.04)}{0.046} = 0.61 \text{ and } P(M2 \mid D) = \frac{(0.30)(0.06)}{0.046} = 0.39$$

The manager can now assume that there is a 60% chance that a defective valve comes from the manufacturer M1.

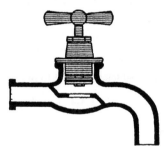

The original estimate of 70% supply by M1 and 30% supply by M2 is called the *prior probability*, i.e. this condition exists prior to the acquisition of additional information. The sampling adds to the given information and combines both probabilities by using Bayes's rule. This new result is called the *posterior probability*.

1.4 MANAGEMENT SCIENCE

A serious student who wants to advance further in a management profession should study management science techniques (also referred to as operations research). This includes

- ♦ Linear programming
- ♦ Dynamic programming
- ♦ Queuing models
- ♦ Decision analysis
- ♦ Simulation
- ♦ Forecasting models
- ♦ Inventory techniques

The application of management science is a decision-making tool that helps a manager to control project costs more effectively.

Detailed treatment of the above is beyond the scope of this text but inspiring project managers should at least have an *appreciation* what management science is all about. There are books on the market that deal adequately with this subject. Among them are

Introduction to Management Science by Davis F. Groebner & Patric W. Shannon, Macmillan Publishing Co., N.Y.,1992 and Jelen's (5) *Cost and Optimization Engineering* as sponsored by the AACE.

Knowledge of management science techniques and their application will give any business manager a distinct advantage over others who are not familiar with those techniques.

Project control means *acting* on available *information*. Taking *action* in turn means making *decisions;* and to make good *decisions,* the available *information* must be in a form conducive to *action*. Management science techniques can give managers greater confidence in making better informed decisions.

This section will give a short overview of some of those important management science techniques:

1.4.1 Linear Programming

What is linear programming? It is a mathematical tool used over a broad range of applications in decision making, such as determining the optimum solution to a product mix. When formulating a mathematical model, the term "linear" implies that variables are raised to the first power. The general formula is

$$a_1x_1 + a_2x_2 + ... a_vx_v + ... a_nx_n = B1 \text{ , where x is + or 0 (zero).}$$

The basic system model INPUT$\Rightarrow$ $\boxed{\text{PROCESS}}$ $\Rightarrow$ OUTPUT can lead to the linear programming model (LP model):

$$
\begin{array}{ccc}
a = \text{fixed rates} & & \text{objective} \\
\text{INPUT} \Longrightarrow & \boxed{\text{CONSTRAINT}} & \Longrightarrow \text{OUTPUT} \\
x = \text{variable activity} & & \text{function Z}
\end{array}
$$

The variable activity x is called the *decision variable*.
Calling the objective function(output) Z and a related cost coefficient C, then

$$Z = C_1x_1 + C_2x_2 + ...C_vx_v + ...C_{n-1}x_{n-1} + ...+ C_nx_n \text{ or } Z = \sum_{v=1}^{n} C_v x_v$$

(Z derives from the German word "Zielfunktion" or target function).
The cost coefficient C can be expressed as a ratio in dollars per quantity, such as

$$\text{\$/m, \$/kg, \$/kW ... etc.,}$$

and the variable x in quantity per hour, such as m/h, kg/h, kW/h ... etc., combining both, the objective function has the unit \$/h

Q #1.33: Assuming Z = 3x and the coefficient C = \$/h, what are the units for x if Z is expressed in \$/kg?

A #1.33: x must be expressed in h/kg!
We know from basic algebra that we need at least as many independent equations as there are variables to solve a problem accurately, e.g.

$$5x_1 + 2x_2 = 20$$
$$x_1 - x_2 = -3 \qquad \text{There is no problem solving for } x_1 \text{ and } x_2.$$

Using determinants

$$x_1 \begin{vmatrix} 5 & 2 \\ 1 & -1 \end{vmatrix} = \begin{vmatrix} 20 & 2 \\ -3 & -1 \end{vmatrix} \quad \text{and} \quad x_2 \begin{vmatrix} 5 & 2 \\ 1 & -1 \end{vmatrix} = \begin{vmatrix} 5 & 20 \\ 1 & -3 \end{vmatrix}$$

$$x_1 = \dfrac{\begin{vmatrix} 20 & 2 \\ -3 & -1 \end{vmatrix}}{\begin{vmatrix} 5 & 2 \\ 1 & -1 \end{vmatrix}} = \dfrac{-20 + 6}{-5 - 2} = 2 \qquad \text{and} \qquad x_2 = 5$$

(a minus b)

For "n" variables there are n^2 elements and n! terms.

Having only one equation with two unknowns, the number of solutions can be infinite:

$5x_1 + 2x_2 = 20$ yields many results, some of which are

$$x_1 = 1, \text{ then } x_2 = 7.5$$
$$x_1 = 2, \text{ then } x_2 = 5.0$$
$$x_1 = 4, \text{ then } x_2 = 0.0 \text{ etc.}$$

We can see that there is a relationship between x_1 and x_2. x_1 is dependent on x_2.

In general, if n denotes the number of variables and m the number of equations, then for n = m in the matrix, we are able to solve for all variables. However, if n>m, there is a huge number of solutions. In this case, what we need is a process that will give the optimum value to the objective function.

Below is a matrix of equations, where m = the number of equations and n = the number of unknowns:

$$a_{1,1}x_1 + a_{1,2}x_2 + \dots a_{1,v}x_v + \dots a_{1,n}x_n = B_1$$
$$a_{2,1}x_1 + a_{2,2}x_2 + \dots a_{2,v}x_v + \dots a_{2,n}x_n = B_2$$
$$\text{etc.}$$
$$a_{u,1}x_1 + a_{u,2}x_2 + \dots a_{u,v}x_v + \dots a_{u,n}x_n = B_u$$
$$\text{etc.}$$
$$a_{m,1}x_1 + a_{m,2}x_2 + \dots a_{m,v}x_v + \dots a_{m,n}x_n = B_m$$

yields the following general expression, which is called the *"constraint"* of the linear programming model:

$$\sum_{v=1}^{n} a_{u,v}x_v = B_u ; \quad \text{where } u = 1, 2, 3, \dots m$$

Q #1.34: Assuming there are n = 8 variables with m = 5 equations. If we set n - m variables equal to zero, then we can solve for those remaining five "basic variables". This will give us a "basic feasible" solution. What are the possible combinations? Since we considered only 5 out of 8, how many possible ways are there to obtain an optimal solution?

A #1.34:
For a set of "n" things taken "m" at a time, there are

$$\binom{n}{m} = \frac{n!}{m!(n-m)!} = \frac{8!}{5!\,3!} = 56 \text{ combinations possible.}$$

56 solutions will have to be tabulated and all non-feasible solutions eliminated.
Assuming a variable is expressed in the form of inequality, for *example* $x_1 < 20$.

This is *not* an equation and must be converted by introducing a *"slack"* variable:

$$x_1 + x_s = 20$$

This expression can now be entered into the matrix above.

Another example of information that needs to be converted into a suitable LP form is the *"quality constraint."* If the quality of a product is restricted to reach a certain level by mixing or blending certain components, then

$$Q = \frac{\sum q_{u,v} x_v}{\sum x_v}$$

Q #1.35: Three liquids are mixed such that they do not exceed 40% alcohol by volume. The alcohol content of each liquid is 60%, 30% and 50% respectively. What is the variable activity?

A #1.35:
$$\frac{60x_1 + 30x_2 + 50x_3}{x_1 + x_2 + x_3} = \leq 40 \text{ or } 60x_1 + 30x_2 + 50x_3 \leq 40(x_1 + x_2 + x_3)$$
and with a slack variable,
$$20x_1 - 10x_2 + 10x_3 + x_4 = 0$$

which has turned into a standard equation that fits the matrix.

There are several methods to obtain optimum solutions, some of which are
♦ the graphical solution
♦ the algebraic method
♦ the simplex method

and there are three basic steps in formulating a problem:

♦ Define the decision variable
♦ Specify the objective function
♦ Determine the constraints

We will now go through simple linear programming examples by using the graphical solution:

Q #1.36: Two products, A and B, are produced in a manufacturing plant. Two groups of workers, one group preparing the products (cut, bend), the other finishing (weld, grind, paint) are processing the products. Group #1 requires 2 hours to prepare product A and 1 hour for product B. Group #2 needs 2 hours to finish product A and 6 hours for product B. Group #1 can put in a maximum of 400 h/week and group #2 a maximum of 900 h/week. The estimated profit for product A is $4.00/unit and product B $6.00/unit.

To *maximize* profit, what is the *optimum* weekly production?

A #1.36: The fixed rates are

GROUP	Product A	Product B	Max. h/week
#1	2 h/unit	1 h/unit	400
#2	2 h/unit	6 h/unit	900
Profit	$4.00	$ 6.00 per unit	

We now define the decision variables: x_1 and x_2 in units/week, (x_1 for group #1 and x_2 for group #2). Converted to LP equations:

$$2x_1 + x_2 = 400 \text{ or } 2x_1 + x_2 + x_3 = 400 \text{ h/week} \quad\quad\quad ....(1)$$
$$2x_1 + 6x_2 = 900 \text{ or } 2x_1 + 6x_2 + x_4 = 900 \text{ h/week} \quad\quad ....(2)$$

We now specify the objective function:

$$Z = 4x_1 + 6x_2 \quad\quad\quad ....(3)$$

To determine the constraints, we will look at both, the algebraic and the graphic solution. For equation (1) setting the slack variable at zero,

$$2x_1 + x_2 = 400$$

By doing this, we now have one equation with two variables, resulting graphically in a straight line (Figure 1-12).

$$\text{if } x_1 = 0 \quad \text{then } x_2 = 400 \quad ....(A)$$
$$\text{if } x_2 = 0 \quad \text{then } x_1 = 200 \quad ....(B)$$

The area within the triangle OABO satisfies the constraints $X_1 \times 10^2$ of the first equation. (Remember, x_v must not be negative!)

For equation (2) see Figure 1-13. Setting the slack variable at zero again,

$$2x_1 + 6x_2 = 900$$

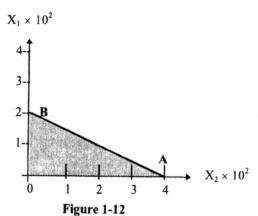

Figure 1-12
Production problem - LP graph 1.

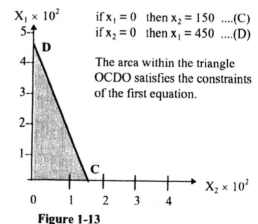

if $x_1 = 0$ then $x_2 = 150$ ....(C)
if $x_2 = 0$ then $x_1 = 450$ ....(D)

The area within the triangle OCEDO satisfies the constraints of the first equation.

Figure 1-13
Production problem - LP graph 2.

Those two equations combined are depicted in Figure 1-14. The shaded area OCEBO represents the *basic feasible region*.

The optimal solution can be found within the shaded area by setting various values for Z through corner points B, C, and E. The optimum line is found on the graph by calculating the intersection of the constraint lines:

(2) .... $2x_1 + 6x_2 = 900$
(1) .... $\underline{2x_1 + x_2 = 400}$ subtracted
                      $5x_2 = 500$ i.e., $x_2 = 100$; $x_1 = 150$

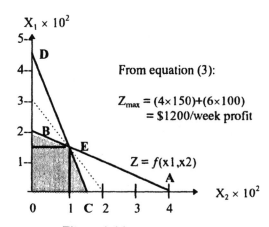

From equation (3):

$$Z_{max} = (4 \times 150) + (6 \times 100)$$
$$= \$1200/week\ profit$$

$$Z = f(x1, x2)$$

Figure 1-14
Production problem - LP graph 3.

The production is   150 units/week of product A and
                                     100 units/week of product B

The slope of the objective function can be found by using

$$Z = 4x_1 + 6x_2$$

and isolating x_2

$$X_2 = \frac{Z}{6} - \frac{2}{3} x_1$$

The slope of the objective function is therefore – 2/3 (negative slope).
Further

$$Z_B = f(x_1, 0) = 6(2/3) \times 200 = \$800$$
$$Z_C = f(0, x_2) = 4(3/2) \times 150 = \$900$$

It means that the line through E yields the maximum profit with given constraints.
This is shown on Figure 1-15.
The slack variables were set to zero. x_1 and x_2 are called the *basic variables* x_3 and
x_4 are *non-basic variables*.
We considered two out of four variables. We know, that there are six combina-
tions to solve the problem or $\dfrac{4!}{2!(4-2)!} = 6$

Using the following previous equations

$$2x_1 + x_2 + x_3 = 400\ h/week..............(1)$$
$$2x_1 + 6x_2 + x_4 = 900\ h/week..............(2)$$
$$Z = 4x_1 + 6x_2 \quad ....(3)$$

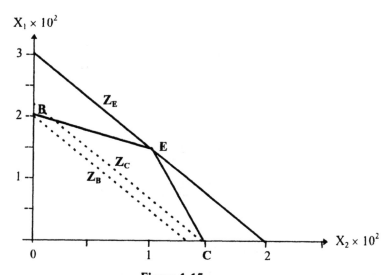

Figure 1-15
Production problem - LP graph 4.

Table 1-07 tabulates the results.

In the example we took the data for the two products A and B to be correct. We did not question the hours needed to finish the product. We have also assumed that the profit is known with certainty. Any figures that are projected into the future are uncertain to some degree. For example, the estimated profit for product A is $ 4.00/unit. Will this be the case next year or the year after? We have not introduced probabilities into our deterministic linear programming model.

Each factor in the equations has a different impact on the outcome. Some factors are more *"sensitive"* than others in affecting feasible solutions. Of all feasible solutions Z_E is the optimal solution.

Table 1-07
Production problem - feasible results

x_1	x_2	x_3	x_4	Z	Graph
0	0	400	900	zero	O
200	0	0	500	800	B
150	100	0	0	**1200**	**E**
0	400	0	−1500	neg.	nonfeasible
450	0	−500	0	neg.	nonfeasible
0	150	250	0	900	C

To allow the decision maker to include uncertainties when setting up and reviewing linear programming models, so-called *sensitivity analysis* is applied. What we first want to find out, is how much variation can we *allow* before the solution moves *beyond the optimum point*. This is also called "the dual" to a linear programming model.

We now look at the sensitivity of the x_1 and x_2 coefficients.
From equation (1)

$$x_2 = 400 - 2x_1 \qquad \text{the slope is } -2$$

From equation (2)

$$x_2 = 900/6 - 1/3x_1 \quad \text{the slope is } -1/3$$

The lines of those two constraints intersect at the optimal corner point E. This corner point will remain optimal for the objective function as long as its slope stays within the boundary of -2 to $-1/3$, or

$$-2 \leq \text{slope of objective function} \leq -1/3$$

Since the slope of the objective function was previously calculated to be $-2/3$, the above statement holds true.

Similarly, it is also true for the sensitivity of the x_2 coefficient, from equation (1)

$$x_1 = 400/2 - 1/2x_2 \quad \text{the slope is } -1/2$$

from equation (2)

$$x_1 = 900/2 - 3x_2 \qquad \text{the slope is } -3$$

The slope of the objective function is within the boundary of

$$-3 \leq -2/3 \leq -1/2$$

Let's now determine the effect that comparable percentage changes in the parameter values (profit/unit) have on the performance measure. If we assume that the estimate of the parameter values have a 20% uncertainty, then the slopes change as shown in Table 1-08.

Table 1-08
Production problem - sensitivity

Change in Parameter	Performance Z = $/week	Change in Profit	Slope
4 + 20% = 4.8	1320	+ 10%	– 0.80
4 – 20% = 3.2	1080	– 10%	– 0.53
6 + 20% = 7.2	1320	+ 10%	– 1.80
6 – 20% = 4.8	1080	– 10%	– 1.20

There is a linear relationship between parameter changes and changes in profit in our particular example. All slopes are within allowed boundaries (tolerance limits). In practice, a linear relationship is rare. For example, if we are testing the sensitivity of a rate of return (performance measure) to 10% changes in sales volume (A), selling prices (B), operating cost (C), etc., the result may look as follows:

Change in Parameter	Change in Rate of Return
A + 10%	+ 2%
A − 10%	− 3%
B + 10%	+ 20%
B − 10%	− 10%
C ...etc.	

Obviously, the rate of return is much more sensitive to selling prices than to the sales volume. B is the critical parameter which should be afforded a greater amount of attention than A.

The next step would be to look at the right-hand-side of the equations. If the maximum hours per week are uncertain for preparing and finishing the products A and B, a review may indicate that weekly hours for group #1 could be raised to 450 and group #2 hours could be reduced to 780. How does this affect the weekly profit?

With the slack variables set to zero, equations (1) and (2) become

$$2x_1 + x_2 = 450 \qquad\qquad 2x_1 + 6x_2 = 780$$
$$x_2 = 450 \ (A') \qquad\qquad\quad x_2 = 130 \ (C')$$
$$x_1 = 225 \ (B') \qquad\qquad\quad x_2 = 390 \ (D')$$

For the intersection E'
$$2x_1 + 6x_2 = 780$$
$$2x_1 + \ x_2 = 450$$
$$5x_2 = 330 \quad \therefore \ x_2 = 66 \text{ and } x_1 = 192$$

and

$$Z' = 4(192) + 6(66) = \$1164/\text{week}$$

The weekly profit is not drastically affected. We must realize, that any change in the right-hand side value of a binding constraint (setting slack variables to zero) will change the optimal solution because the corner points of the feasible region will be changed. If we now include performance-measured sensitivity we obtain the results tabulated in Table 1.09.

The decision-maker can now

a) select the base case best action
b) pick one of the other actions
c) gather additional information (which may lead to an expansion of the matrix).

Table 1-09
Production problem - performance sensitivity

Change in Parameter	Performance Z = $/week	Change in Profit	The change in profit
$4 + 20\% = 4.8$	1318	+ 13.5%	is more pronounced
$4 - 20\% = 3.2$	1010	- 13.4%	with changes in
$6 + 20\% = 7.2$	1259	+ 8.5%	group #1 than in
$6 - 20\% = 4.8$	1101	- 5.4%	group #2.

Q #1.37:

The ABC Cement Company produces two types of pre-mixed dry concrete, which it sells to building supply stores in 40 kg bags. There are two types of mixes:

Type #1 is a rich mix consisting of 16 kg coarse aggregate (pea size gravel) and 15 kg fine aggregate (sand).

Type #2 is a lean mix of 24 kg coarse and 12 kg fine aggregate.

It is estimated, that ABC's sales contract allows for a profit of $1.40 per bag of type #1 mix and $1.50 per bag of type #2 mix. The retailers require at least

1600 bags of type #1 $\geq$ 49.6 Mg/month and
800 bags of type #2 $\geq$ 28.8 Mg/month

The suppliers of aggregates are able to ship a maximum of

80 Mg/month that make up type #1 and
60 Mg/month of type #2.

The cement company adds two ingredients to the aggregates before filling the bags:

Rich mix: One bag cement to four bags aggregates.
One bag lime to 20 bags of aggregates.
Lean mix: One bag cement to five bags of aggregates.
One bag lime to 30 bags of aggregates.

The production capacity for cement is 50 Mg/month and that for lime is 6 Mg/month. For maximum profit find the optimum production. Use the graphic method.

A #1.37:

The *decision variables* are:

$$x_1 = \text{type \#1 aggregate}$$
$$x_2 = \text{type \#2 aggregate}$$

The *objective function* is:

$$Z = 1.40x_1 + 1.50x_2, \text{ where } Z_{max} \text{ is in \$/month.}$$

Constraints are

a) No negative kg of mix is produced. Therefore, $1x_1 + 0x_2 > 0$
$$0x_1 + 1x_2 > 0$$

Supply and delivery constraints are

$$\begin{aligned}
&\text{b)} && 1x_1 + 0x_2 > 49.6 \\
&\text{c)} && 0x_1 + 1x_2 > 28.8 \\
&\text{d)} && 1x_1 + 0x_2 < 80.0 \\
&\text{e)} && 0x_1 + 1x_2 < 60.0
\end{aligned}$$

The production capacity for cement is

$$\text{f) } (1/4)x_1 + (1/5)x_2 \le 50 \text{ Mg/month}$$

Availability of lime is restricted to

$$\text{g) } (1/20)x_1 + (1/30)x_2 \le 6 \text{ Mg/month}$$

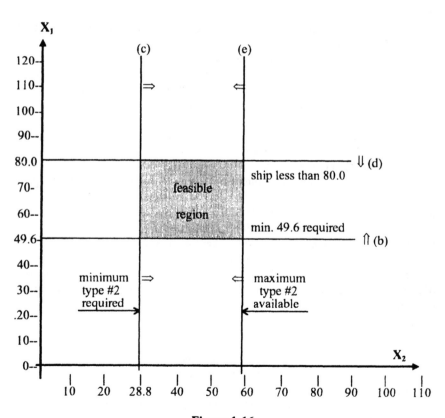

Figure 1-16
Cement company - LP graph 1.

This can be summarized for computer input as

(a) x_1 $x_2 \geq$ 0
(b) x_1 $\geq$ 49.6
(c) $x_2 \geq$ 28.8
(d) x_1 $\leq$ 80.0
(e) $x_2 \leq$ 60.0
(f) $0.25\ x_1 + 0.20\ x_2 \leq 25.0$
(g) $0.050x_1 + 0.033x_2 \leq$ 6.0

To demonstrate this graphically, we need to establish feasible regions by plotting (b), (c), (d) and (e) (Figure 1-16).

We still need to plot (f) and (g):

For (f), $x_1 = 100$ and $x_2 = 125$
For (g), $x_1 = 120$ and $x_2 = 180$

Points A or B seem to be the optimal corner points (see Figure 1-17):

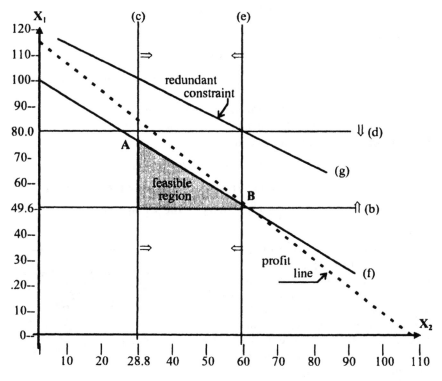

Figure 1-17
Cement company - LP graph 2.

(c) and (f) intersect at A

$$x_2 = 28.8 \text{ Mg/month}; x_1 = 77 \text{ Mg/month, because } x_1 + 0.2x_2 = 25$$

(e) and (f) intersect at B

$$x_2 = 60 \text{ Mg/month}; x_1 = 52 \text{ Mg/month because } x_1 + 0.2x_2 = 25$$

This will give us the slope of the profit line through the origin of the coordinates. The corner point of the feasible region with the greatest distance from the line through the origin is the optimal corner point. *(For simple problems, that corner point can be found through trial and error.)*

We can now find the line of maximum profit by calculating the maximum feasible distance from the line.

$$Z_{min} = 1.40x_1 + 1.50x_2 = \text{zero}$$

For intersection at A:

$$Z_{min} = [1.40(77) + 1.50(28.8)]/0.040 = \textbf{\$3775/month}$$

For intersection at B:

$$Z_{min} = [1.40(52) + 1.50(60)]/0.040 = \textbf{\$4070/month}$$

which yields the higher profit.

And the optimum production is

Type #1: 52/0.040 = 1300 bags/month
Type #2: 60/0.040 = 1500 bags/month

The slope of the profit line though point B is

$$1.50x_2 = Z - 1.40x_1 = 4070 - 1.40x_1$$

and for $x_2 = 0$, $x_1 = 4070/1.4 = 2907 \times 0.040 = \textbf{116 Mg/month}$

and $x_2 = 2713 \times 0.040 = \textbf{109 Mg/month}$.

Q #1.38:

A farmer can use up to 100 ha to plant corn and potatoes. His budget for this is $11 000. To plant and harvest the corn crop costs $200/ha, which is twice as much as that for potatoes. He can only spend up to 160 hours, of which one hour per hectare is allowed for potatoes and 4 hours/ha for corn. The yield to the farmer is an equivalent $400/ha for potatoes and $1200/ha for corn. How much should he plant for maximum profit? See Figure 1-18

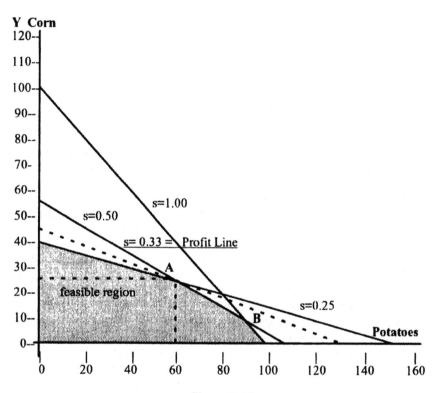

Figure 1-18
LP planting problem.

A #1.38:

	Potatoes(x)	Corn(y)	Available
Outlay	$100/ha	$200/ha	$11 000
Workhours	1 h/ha	4 h/ha	160 h
Income	$400/ha	$1200/ha	Maximum

Objective Function:

$$Z_{max} = 400x + 1200y$$

Decision Variables:

$100x + 200y \leq 11\ 000$	for x = 0	for y = 0	slope:
$x + \quad 2y \leq \quad 110$	y = 55	x = 110	0.50
$x + \quad 4y \leq \quad 160$	y = 40	x = 160	0.25
$x + \quad \quad y \leq \quad 100$	y = 100	x = 100	1.00

For $Z = 0$: $400x + 1200y = 0$, the slope is   0.33 which cuts through point A.

This affects the following lines

$x + 2y = 110$		$y = 25$ ha corn
$\underline{x + 4y = 160}$		$x = 60$ ha potatoes
$2y = 50$	and	$100\text{-}x\text{-}y = 15$ ha *no crop*

This should be the most profitable mix for planting and harvesting

$$Z_{max} = 24\ 000 + 30\ 000 = \textbf{\$54 000} \text{ maximum profit.}$$

The project manager will encounter considerably more complex problems in practice, but the above examples will give the reader an understanding of the meaning of LP.

There are usually more than two variables envolved in a decision-making process. It is impractical, if not impossible to produce a graph with more than two variables. This calls for a more complex mathematical solution.

The most common method is the *simplex method*, first developed by Dantzig (Princeton 1963). This is a mathematical technique which uses algebra to solve a number of simultaneous equations to optimize the objective function. It is beyond the scope of this text to explain the simplex method other than to say that

♦   all constraints are non-negative
♦   the matrix includes slack variables, which also must be non-negative
♦   all constraints are "tested" by moving in an orderly manner from one basic feasible solution to the next (simplex tableau)
♦   the final tableau provides information about the optimal solution.

The computations, using the simplex method, are tedious. This method, therefore, lends itself very much to computer application. However, the computer can never set up the equations, neither can it interpret the results. Human judgment is needed here.

To summarize, linear programming models are used to help solving many practical decision making situations. The algebraic method and simplex procedure lend themselves to effective computer application. Sensitivity analysis is a tool to determine the responsiveness of the conclusions of an analysis to changes in the parameter of the problem.

1.4.2 Dynamic Programming

This is basically linear programming under uncertain decision steps. Instead of having a definite objective function, the outcome of previous decisions, i.e. the objective function, changes with the solution of interrelated subproblems. The solution is obtained in stages usually starting from the last stage and working backward (recursive). Probability analysis is used for uncertain conditions.

$$P_m = \binom{n}{m} p^m (1-p)^{n-m}$$

where
n = sample size (production run)
m = number of outcomes (defective product)
p = probability (% failure expressed as ratio)

While LP is rigidly structured and requires the formulation of the entire problem before the model can be used, dynamic programming uses a more flexible approach by evaluating each step separately. The output of a previous stage becomes the input of the next stage, thereby drastically reducing the number of calculations required.

In LP, a 6-stage problem with 3 decisions at each stage would require $3^6 = 729$ calculations to list all possible alternatives whereby $3 \times 6 = 18$ one-stage optimization problems need only to be addressed in dynamic programming.

There are, however, disadvantages. To identify the sequences of decisions to be made requires insight and experience. Because it is not strictly a mathematical model, it requires intuition to reach the optimal solution.

While computer application is relatively inexpensive for LP, only specialized software seems to exist at this time for dynamic programming applications.

A dynamic programming model may be depicted as shown below:

$$X_1 \qquad X_2 \qquad\qquad X_n$$

Input 0 → $\boxed{1}$ Input 1→ $\boxed{2}$ Input 2→ .... Input n → $\boxed{n+1}$ Input (n+1) etc.

1.4.3 Queuing Models

Simply speaking, queuing theory is a study of waiting lines.

Without even considering to join the long line-up at a restaurant you go to the next eating place. This is called *balking*. When you arrive at the next eating place, you may find that the line is shorter and you decide to eat there. Then you look at the menu which is attached to the door and decide that the food is not to your liking and you leave. This "truncating of the queue" is called *reneging* (Figure 1-19).

It is relatively easy to understand waiting-line systems because we can personally relate to them, but the mathematics for some systems can be extremely complicated. *It is a closed system.*

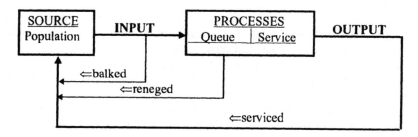

Figure 1-19
The queuing system.

In manufacturing, the queue can be an assembly line. Units arrive at the service at certain intervals and it takes time to serve them.

Services are also provided to dispense materials at a construction site. Here, the arrival rate is not as steady as it is at an assembly line. Even though the service capacity may exceed total demand, at times a queue may build up because of a short term increase in demand. Those times may or may not be predictable. For example, the guard checking cars entering a construction site will handle traffic easily except during the start of work hours (rush hour). This is *predictable*. When a personnel officer who usually interviews applicants for 15 minutes on the average spends 30 minutes with one individual he/she will cause a waiting line. This can be *unpredictable*. If a service system is constantly under capacity, this in theory will cause a waiting line to build up *infinitely*. This situation calls for a review of the system. If there are very few customers entering a store, there may be an excess service function. Here, services are idle most of the time.

Arrivals are usually characterized as "units." This includes customers (people), vehicles, telephone calls, materials and many other items of demand. The number of units arriving are usually counted hourly.

The *average* arrival rate at a doctor's office is controlled by appointments, such as six patients per hour. The "units" are *individuals*. We can also have *groups* arriving such as families at a restaurant. Arrivals can be *dependent* on each other or *independent*. If a scanner detects a faulty unit at a production line, this unit is rejected independently from any other unit that has arrived or will arrive later on. If the conveying device is defective, causing the units to bump into each other, thereby causing faulty readings by the scanner, a unit's arrival time is dependent on other units (Figure 1-20).

Arrivals can be serviced on a *first-come-first-served* basis or *last-come-first-served basis* (stacked units) or on a *random* order basis (lottery draw) or by a *priority* scheme (emergency cases in hospitals). A single waiting line is called a *single channel*.

If the units to be served have a constant arrival time, and the service duration is also constant, we call this

**Single queue with constant arrivals, constant
service times, and first-come-first-served service.**

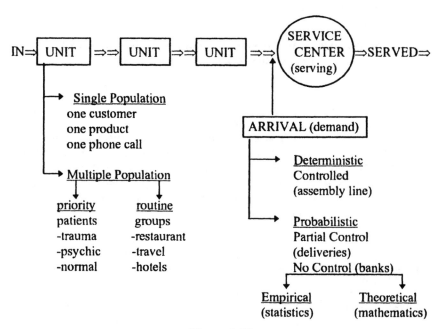

Figure 1-20
Queuing sources and arrivals.

If the time in-service is equal to the processing time, there is no *idle time*. But consider the processor taking some rest before the next unit arrives, then there is idle time in the service facility.

If the individual units arrive faster than they can be serviced, i.e. the *arrival rate* has increased, a *queue buildup* will occur. For each time period, the queue buildup is

Arrival Rate minus Service Capacity

If we assume an infinite population and arrivals come at equally spaced intervals (units per hour) and require equal service time (units per hour), we have the following situation:

Constant arrival	Constant service	Units or time	Situation
A = 30/h	S = 30/h	A – S = zero	No queue, no idle time
A = 20/h	S = 30/h	1 – A/S = 33%	No queue, positive idle
A = 40/h	S = 30/h	A – S = 10/h	Queue build-up, no idle

Most waiting lines do not have known constant arrival or service times. The probability of occurrence must be estimated.

The *Poisson probability distribution* is used for those probabilistic arrivals. This is called

Single queue with Poisson-distributed arrivals, exponential service, and first-come-first-served service.

The Poisson distribution is a ratio mathematically determined by

$$P(x) = \frac{\lambda^x e^{-\lambda}}{x!} \qquad \dots\dots\dots\dots\dots(1)$$

where λ (lambda) is the estimated average (mean) arrival rate, usually per hour, and x other possible arrivals during the same period. There are tables available, giving P(x) for various λ and x values (see also 1.3.2 probability distributions).

For example, it is estimated that $\lambda = 6$ units/hour is the mean arrival rate. What are the chances, that no units will arrive (x=0), or that 14 units will arrive (x=14)? The table below shows values for $\lambda = 6$.

x	0	1	2	3	4	5
P(x)	.0025	.0149	.0446	.0892	.1339	.1606
x	6	7	8	9	10	11
P(x)	.1606	.1377	.1033	.0688	.0413	.0225
x	12	13	**14**	15	16	17
P(x)	.0113	.0052	**.0022**	.0009	.0003	.0001

Calculations or using *Poisson* ratio tables will tell us, that there is a small probability of 0.25% that no units will arrive and a still smaller probability that 14 units will arrive. This knowledge is important when making decisions on the layout of a facility.

A different distribution is used for *service rates*. It is called the exponential probability distribution. This expresses the probability that a unit can be served within some time "t", where the average (mean) number of units the facility can serve during the same time period (usually hours) is μ and if the applicable service time is $\leq t$, then

$$P_t = 1 - e^{-\mu t} \qquad\qquad ....(2)$$

Q #1.39:, It will take a personnel officer usually 15 minutes to interview an applicant, i.e. $\mu = 4$ per hour. What is the probability for an interview to take up to 10 minutes or up to 20 minutes?

A #1.39: Here, the service rate of the interviewer is not constant. The time spent by the personnel officer depends on the complexity of the case and many other factors.

$$P{\leq}10 = 1 - e^{-4(1/6)} = 1 - e^{-0.67} = 1 - 0.513 = 49\%$$
$$P{\leq}30 = 1 - e^{-4(1/3)} = 1 - e^{-1.33} = 1 - 0.264 = 74\%$$

We can determine *waiting-line characteristics*, using μ and λ.

One prerequisite is that the average arrival rate does not exceed the service capacity, or $\mu > \lambda$.

Obviously, if only 8 units per hour ($\mu = 8$) can be served, there is no need to consider more than 8 units to arrive during the same time period.

Below are some waiting line characteristics for an average arrival time of 4 units per hour and service capacity of 6 units per hour:

Probability of idle time at the service facility:

$$P_i = 1 - \frac{\lambda}{\mu} \qquad\qquad ....(3)$$

therefore $Pi = 1 - (4/6) = 33\ \%\ $ or 20 min/hour

Average expected waiting time in the queue:

$$T_w = \frac{\lambda/\mu}{\mu - \lambda} \qquad\qquad ....(4)$$

therefore $T_w = \dfrac{4/6}{6-4} = 1/3 = 20$ minutes queue build-up.

Average time in the system:

Combining the waiting time and the service time for a unit, expected total time is

$$Ts = \frac{1}{\mu - \lambda} \qquad \qquad \text{.................... (5)}$$

in our example, $Ts = 1/(6 - 4) = 1/2 = 30$ minutes

Average number of units waiting for service:

The average length of the waiting line will determine the number of units waiting for service, which is average waiting time multiplied by the arrival rate or

$$U_w = \lambda T_w = \frac{\lambda^2/\mu}{\mu - \lambda} \qquad Uw = \frac{16/6}{6-4} = 1.33 \text{ units} \qquad \text{.................... (6)}$$

Average number of units in the system:

The number of units both, waiting and being served at any one time is

$$U_s = \frac{\lambda}{\mu - \lambda} = 4/(6-4) = 2 \text{ units.} \qquad \text{....................(7)}$$

Probability of having units in the waiting line:

We found from equation (6) that the average waiting line length is 1.33 units. Average means that it could be more or less at different times. To find the probability of "n" units waiting, we use this equation :

$$P_w = (1 - \lambda/\mu)(\lambda/\mu)^n \qquad \text{....................(8)}$$

The probability for three units waiting is

$$P_w = [1 - (4/6)](4/6)^3 = 10\%$$

Q #1.40: If the service facility can only serve a *maximum* of six units, what is the probability that extra customers have to be turned back (balk the queue)?

A #1.40: We can tabulate the units waiting (n) and the probability P_w and summarize all probabilities larger than six units (remember $\Sigma P_w = 1.0000$):

,n	0	1	2	3	4	5	6	7	8	9	10	>10
P_w	.3333	.2222	.1481	.0988	.0658	.0439	.0293	.0195	.0130	.0087	.0058	..0116

$$\Sigma = \quad 0.0586$$

Therefore, the chance that this may happen is approx. 6%.

The two systems discussed so far represent *single channel waiting line* models, such as a single waiting line at a bank with several tellers to serve the public.

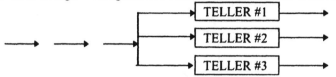

If a separate line-up forms for each teller, this represents a *multiple parallel channel:*

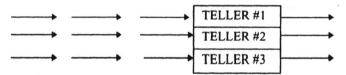

These are

**Multiple parallel channels with Poisson-distributed arrivals,
exponential service, and first-come, first-served service.**

With a multiple queue and with different serving times, one line may move faster than another and customers may jockey between waiting lines in the believe that they may reduce their waiting time. In this case, we do *not* have a first-come, first-served situation.

Equations for multiple-channel waiting-line systems are more complex than for single-channel systems. It is not the purpose of this chapter to get involved in those detailed calculations. The serious reader should gain an appreciation of queuing systems and be curious enough to study the subject further on his or her own.

Cost Issues of Queuing

We all know that time is money. The waiting time and the service time both have a dollar value. Any idle time or waiting delays will have an impact on the efficiency of operation and thereby on the economy of the enterprise.

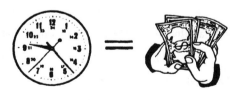

To reduce waiting time, we need to increase service. However, with a higher service rate we may incur more idle time which increases the cost of service. The question is, how much waiting time are we willing to allow?

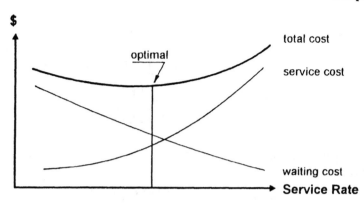

Figure 1-21
Minimum queuing cost.

If a welder on a construction site must walk a long distance to the warehouse and then stand in line for a considerable time to pick up a few welding rods, the waiting cost can be considerable. *Those* internal waiting cost are easy to determine.

External waiting cost are more difficult to formulate or even to identify. Customers who encounter long waiting periods may balk or renege the queue and take their business elsewhere. To assign cost here is often guesswork. Service cost and waiting cost move in opposite directions. There is an economic trade-off (Figure 1-21).

To determine the optimal point on the total cost curve, which will give us minimum cost, some calculations are required.

For a single channel waiting-line model, the total queuing cost are

$$C_T = \mu C_s + C_w U_s \qquad \dots\dots\dots\dots(9)$$

where
- ♦ C_s is the marginal cost of service
- ♦ C_w is the waiting cost per time period
- ♦ U_s is the average number of units in the system.

We had already established (equation 7), that

$$U_s = \lambda/(\mu - \lambda), \qquad \text{where } \mu \text{ is the service capacity.}$$

The total queuing cost is now

$$C_T = \mu C_s + C_w[\lambda/(\mu - \lambda)] \qquad \dots\dots\dots\dots(10)$$

In order to minimize, we take the first derivative of C_T with respect to μ

$$\frac{dC_T}{d\mu} = 0 \text{ or } \mu_{opt} = \lambda + [(C_w/C_s)\lambda]^{\frac{1}{2}} \qquad \dots\dots\dots\dots (11)$$

where μ_{opt} is the optimal service rate, and finally

$$C_{T(opt)} = \mu_{opt}C_s + C_wU_s \qquad \dots\dots\dots\dots\dots(12)$$

Example: Assuming units arrive at an average of four per hour and can be serviced at a rate of six per hour, then

$$\mu_{opt} = 4 + [4(C_w/C_s)]^{\frac{1}{3}}$$

We now have to evaluate C_w and C_s. It is sometimes easier to compare the two with each other, i.e. we may judge that the service cost are four times that of the waiting cost. Therefore,

$$C_w/C_s = 0.25 \text{ and } \mu_{opt} = 4 + 1 = 5 \text{ units/hour}$$

Assuming it costs $80.00 to serve a unit, then the waiting cost is $20.00 per unit. The combined optimum cost is

$$CT(opt) = 5 \times 80 + 20 [4/(6 - 4)] = \$ \textbf{440.00}$$

There are software packages available that have queuing analysis capability, including sensitivity (what if) studies.

Q #1.41:
The equipment at a site which places 20 Mg (tonnes or metric tons) stone blocks for shore protection has the capacity (μ) to handle four floats with 5 blocks each per hour. The arrival rate (λ) of the floats is estimated to be three per hour. If both rates are constant, what is the percentage of idle time?

A #1.41:
$$P_i = 1 - (\lambda / \mu) = 1 - 3/4 = \textbf{25\%}$$

Q #1.42: What are the minimum total queuing cost for Q #1.41 above if it costs $100 to install one block? The idle cost for the next float to arrive is $50 per hour.

A #1.42: The average number of units in the system is

$$U_s = \frac{15}{20 - 15} = 3 \text{ blocks/hour}$$

There are five blocks arriving per hour. The waiting cost is therefore $10/h.

$$\therefore \mu_{opt} = 15 + \sqrt{15 \times 0.1} = \sqrt{16.5} = 4.06$$

and

$$C_{T(opt)} = 4.06(100) + 10(3) = \textbf{\$436}$$

1.4.4 Inventory Techniques

Inventories are physical assets of monetary value.

In management science terms they are *items in a queue* waiting to be serviced (idle resources) (5). Even though the concept of inventory management is simple, the installation and maintenance of an effective inventory control system can be quite complicated.

Managing inventory is a very important function of cost management. Every project manager should have a good understanding how the various inventory models work.

Basic Model

The simplest model assumes that

a) demand is known and constant
b) all new items are stocked at the time the inventory is reduced to zero
c) the time between placing an order and receiving it (leadtime) is certain
d) the costs are accurately known
e) no shortages are allowed
f) all items are received at one time
g) unit costs are independent of quantities ordered
h) ordering costs do not vary with quantities ordered
i) no quantity discounts are given

The main question to be asked is:

How many "units" or "items" do we need to order or produce to meet the demand and when should the order take place?

On a construction site, materials and tools must be available when needed. Material that is not available when needed will be the cause of costly delays (stockout cost). See also 4.62 - Storage and transport of materials.

Stocking too much material will require additional storage space and material handling (holding cost). It will also consume funds earlier than needed (cost of capital). What every manager tries to achieve is a healthy balance between the number of items in inventory (availability) and the demand which draws those items from inventory; in other words the *most economical* inventory management.

Performance can be measured by monitoring inventory and related costs, which includes

Unit Cost
Holding Cost
Order- and Delivery Cost
Stockout Cost

Management science uses mathematical *models* to deal with inventory problems. There are *deterministic* models dealing with information that is assumed to be certain. For example, the quantity of items withdrawn is assumed to be the actual demand.

There are also *probabilistic* models where information are considered uncertain and are described by a probability distribution.

To illustrate the basic model, also called the economic order quantity model, we will use a simple example:

Example: Electricians at a construction project are using 240 meter of electric cable per day which costs $ 2.00 per meter. A four-day supply of cable is picked up and delivered for a total cost of $ 60.00. Holding the cable in inventory is calculated to be 10% of the unit cost. All assumptions of the basic inventory model apply (Table 1-10 and formulas (1) to (4)).

What is the inventory level at any time t_n?

The slope or items per period is $U_T = \dfrac{Q \text{ Items}}{T_D \text{ Sum of Periods}}$ and

$$Qt = \frac{Q(T_D - t_n)}{T_D} = \frac{QT_D}{T_D} - \frac{Qt_n}{T_D} = Q - \frac{Q}{T_D}(t_n) \quad \text{.................... (5)}$$

Table 1-10

Inventory example - 1

Equ.#	Description	Nomenclature	Result
	Unit Cost	C_U	$2.00/m
	Holding Cost Factor	F_H	10% of CU
	Ordering Cost	C_O	$60.00
	Demand Rate (Usage)	U_T	240 m/day
	Monthly Demand	D	(240)(21.67)=5200 m/mo.
	Depletion Time	T_D	4 days
(1)	Order Quantity	$Q=T_D U_T$	4(240)=960 m
	Average Inventory	Q/2	480 m
	Inventory at any time t_n	Q_T	see formula (17)
	Unit Holding Cost	$H=F_H C_U$	0.1(2.00)=$0.20
(2)	Aver.Invent. Holding Cost	(Q/2)H	0.2(480)=$96.00
	# of Orders/month	D/Q	5200/960=5.42/month
(3)	Monthly Ordering Cost	(D/Q)C_O	5.42(60)=$325.20/month
(4)	Total Inventory Cost	TC=(2)+(3)	96+325.20=$421.20/mo.

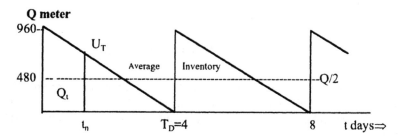

Figure 1-22
Inventory example - 1.

or

$$Q_t = Q - U_t t_n \qquad\qquad .... (6)$$

The basic model assumes no shortage of supplies. Our example calls for a four-day ordering plan with a total inventory buildup to 960 meters of cable (Figure 1-22). We have not yet determined what the impact on total inventory cost is.

If the ordering cost remain constant, the Table 1-11 indicates that a seven-day depletion period would result in lowest total inventory cost. Using the table, the result can be plotted on graph paper (Figure 1-23). It will show the minimum total cost as the horizontal tangent on the resultant total cost curve. It would be impractical to plot curves for hundreds of items of inventory. Minimum cost can easily be obtained by calculation. From equation (4), Table 1-10:

$$TC = (Q/2)H + (D/Q)C_o$$

Table 1-11
Inventory Example - 2

T_D	Q	(Q/2)H	(D/Q)C_o	TC
1	240	24	1300	1324
2	480	48	650	698
3	720	72	433	505
4	960	96	325	421
5	1200	120	260	380
6	1440	144	217	361
→7	**1680**	**168**	**186**	**354←**
8	1920	192	163	355
9	2160	216	144	360
10	2400	240	130	370
11	2640	264	118	382
12	2880	288	108	396
.	.	.	.	etc.

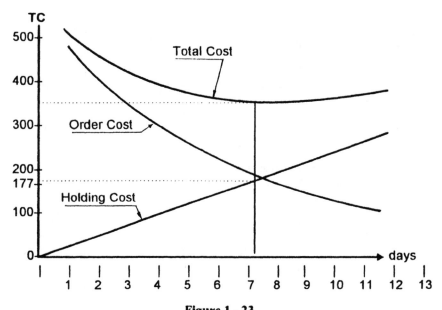

Figure 1 - 23
Inventory example - 2.

Taking the first derivative with respect to Q

$$\frac{dTC}{dQ} = H/2 - (D/Q^2)C_O \qquad \text{setting to zero and solving for Q}$$

$$HQ^2 = 2DC_O \text{ and } Q_{min} = [(2DC_O)/H]^{1/2} \qquad \dots\dots\dots\dots\dots(7)$$

Since the second derivative is positive, we are satisfied that Q is a minimum. For our example

$$Q_{min} = \sqrt{\frac{(2)(5200)(60)}{(0.20)}} = \textbf{1766 m}$$

which is the optimal quantity to be ordered. With this quantity, the monthly inventory holding cost is (Eq. 2)

$$(1766/2)(0.20) = \$ \ 176.60$$

the monthly ordering cost is (Eq. 3)

$$(5200/1766)(60) = \$ \ 176.60$$

Holding cost and ordering cost will always be the same for the basic economic model (crossing point on the graph Figure 1-23).

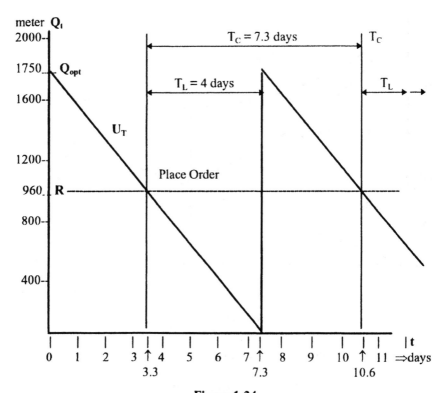

Figure 1-24
Inventory cycle time.

Inventory Cycle Time

We have previously stated, that the lead time is certain. That means the time be-
tween placing an order and receiving the shipment is known with certainty.
Suppliers usually stipulate that larger quantities be ordered in multiples of 50 m.
We will therefore change the optimum quantity from 1766 m to Q_{opt} = 1750 m.
At what point will we need to reorder more electrical cable (figure 1-24)?

The daily demand is

$$U_T = 240 \text{ m/day}$$

If we assume a lead time (T_L) identical to the originally proposed depletion period, then the reorder point is

$$R = U_T T_L = (240)(4) = 960 \text{ m} \qquad ....(8)$$

When the inventory quantity Q_t is reduced to 960 m, another 1750 m of cable will be ordered. The interval between orders is called the "cycle time" T_C.

The cycle time can be calculated by multiplying the number of working days per month W by the economic inventory level Q_{opt} and dividing the result by the monthly demand:

$$TC = \frac{WQ_{opt}}{D} = \frac{21.67 \times 1750}{5200} = 7.3 \text{ working days} \qquad ......(9)$$

Planned Backorder

The basic model portrays an ideal situation, i.e. *always* have goods available when needed. There are occasions when specific items do not lend themselves to be stored in inventory. A retailer may not want to stock all styles or too many colors of an article a customer desires.

The retailer may actually plan for stockouts on those items that are not standard, expecting the customer to wait for a backorder which may have to be placed with the manufacturer. This is also the case when an item has a low demand or is very expensive to tie up in inventory The manager would rather pay a penalty than having specialty items on demand. In this case, we are effectively removing the requirements b) and e) of the basic model. In fact, we are introducing *negative* units into the inventory system. This *stockout model* looks schematically as shown in Figure 1-25.

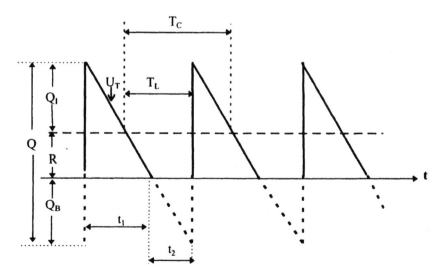

Figure 1 - 25
Inventory stockout.

Table 1-12

Inventory backorder

Description	Nomenclature	Result
Unit cost	C_U	$2.00/m
Holding cost factor	F_H	10% of C_U
Ordering cost	C_O	$60.00
Demand rate (usage)	U_T	240 m/day
Monthly demand	D	$(240)(21.67)=5200$ m/month
Depletion time	T_D	4 days
Backorder cost factor	F_B	50% of CU

We are adding some more expressions:
- ♦ Q_I = Maximum units in inventory at the time of material receipt.
- ♦ Q_B = Maximum amount of backorder at the time of material receipt.
- ♦ F_B = Backorder cost factor in decimal or percent of unit cost.
- ♦ t_1 = Time period (duration) while drawing from inventory.
- ♦ t_2 = Duration during backorder.

Backorder cost are now added to the total inventory cost.

Backorders on inventory items are usually not applicable to the construction industry, but let us stretch our example and say that the electricians at the construction project have backorders with a penalty cost factor of 50% (Table 1-12).

Similar to the basic model, we now need to find the minimum total inventory cost. In addition to determining the optimum order quantity we also need to determine the optimum stockout level. Using differential calculus, the *optimum quantities* which minimize cost are a modification of equation (7):

$$Q_{opt} = [(2DC_O/H)(F_H + F_B)/F_B]^{1/2} \quad \text{...............}(10)$$
$$= \{[(120)(5200)/(0.20)][(0.6)/(0.5)]\}^{1/2} = \textbf{1935 m}$$

Optimal stockout level:

$$Q_{B(opt)} = Q_{opt}[F_H/(F_H + F_B)]^{1/2} \quad \text{...............}(11)$$
$$= 1935 [(0.1)/(0.6)]^{1/2} = \textbf{833 m}$$

Because $Q_I = Q - Q_B$, equation (11) changes to

$$Q_{I(opt)} = Q_{opt} - Q_{B(opt)} = 1935 - 833 = \textbf{1102 m} \quad \text{...............}(12)$$

After having taken care of filling the backorder for 833 m of cable at the time of delivery there is an inventory level of 1102 m remaining. The reorder point of equation (1.4.4.8) is reduced by $Q_{B(opt)}$

$$R = U_T T_L - Q_{B(opt)} = (240)(4) - 833 = \textbf{127 m} \quad \text{...............}(13)$$

After the inventory reaches a level of 127 m, a new order is placed. If R is negative, shortages have occurred before the new order is placed. Using the slope U_T (demand rate)

$$t_1 = Q_I/U_T = 1102/240 = \textbf{4.6 days} \qquad ....(14)$$

These are the days that are spent drawing from inventory. The backordering takes

$$t_2 = Q_B/U_T = 833/240 = \textbf{3.5 days} \qquad ....(15)$$

for an inventory cycle of

$$T_C = t_1 + t_2 = \textbf{8.1 days} \qquad ....(16)$$

The minimum total inventory cost can be calculated by using the optimum quantities above. To establish the *holding cost*, we need to multiply the unit holding cost $H = F_H C_U$ of equation (2) by the average inventory during the cycle period T_C. The holding costs are

$$\{(H)[(Q_I/2)t_1 + 0t_2]\}/(t_1 + t_2) = [(0.20)(0.5)(1102)] / (4.6 + 3.5) = \textbf{\$ 313.78}$$

or $ 62.76 per month.

Average inventory can also shown to be $\dfrac{Q_I^2}{2Q} = \dfrac{1102^2}{3870} = 113.80$

Ordering Cost is calculated using equation (3)

$$(D/Q)C_O = 5.42(60) = \textbf{\$325.20/month}$$

Backorder Cost $= (FB_C)(Q_B^2/2Q) = (0.5)(2)(833^2)/3870 = \textbf{\$179.30/month}$

Total Cost = $\qquad TC = 62.76 + 325.20 + 179.30 = \textbf{\$567.26/month}$

Making sure that no shortages occur will reduce the holding cost, but increase the total cost by $146/month mainly because of a high backorder penalty.

Finally, to find the inventory level *at any time* t_n

$$Q_T = Q - U_T t_n - Q_B = Q_I - U_t t_n \qquad ....(17)$$

Safety Stock

If the project manager wants to be on the safe side by adding more items ($+Q_S$) than required to the inventory, the inventory at any time t_n is

$$Q_T + Q_S - U_T t_n$$

As shown on the diagram below, the optimum order for the lowest cost will not be affected by the safety stock (see equation 7).

$$Q_{opt} = [(2DC_O)/H]^{1/2}$$

However, the optimum period
cost will be increased because
of additional holding cost.

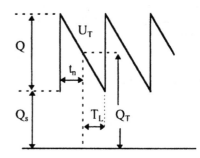

$$TC_{min} = (Q/2)H + (D/Q)C_O + Q_sH$$

expressed in $/period. And similarly,

$$R = U_TT_L + Q_s \quad \text{items}$$

Gradual Inventory Buildup

A finite rate delivery of items is mainly applicable in manufacturing where parts
for a finished product are produced in-house.

Suppliers are accumulating units before shipments are made to the buyer. This
continuous buildup of goods needs a _production_ type model, which is quite different from the basic order quantity model discussed so far.

A manufacturer may, among other parts, produce brake linings that are made
in the fabrication shop. When the production process is set up, a large quantity
(Q), called a "lot" or a "batch" is produced. The production capacity (P) usually
exceeds the demand (D). Some of the brake linings will therefore go into inventory that is gradually increased until production is idle and the inventory is being
reduced. When production needs to start again (release point), cost is incurred to
readjust machinery etc. These are _setup cost_.

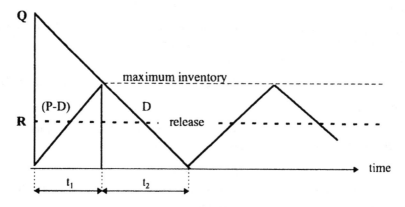

Figure 1-26
Inventory buildup.

The question now is
- How many units should be produced during a production run?
- When should a production run be started?

A typical production lot size model is shown in Figure 1-26.

Nomenclature for calculations:

Q = lot or batch size	t_2 = inventory depletion time
R = release point	C_U = production cost per unit
P = production rate per period (year)	T_L = lead time (to setup production)
D = demand rate per period	C_S = setup cost to produce a batch
t_1 = time period of inventory buildup	F_H = inventory holding cost factor

Optimum units in inventory:

$$Q_{opt} = [(2PDC_S)/(P - D)C_UF_H]^{1/2} \qquad \dots\dots\dots(18)$$

Holding cost = $\qquad (P - D)QC_UF_H/2P \qquad \dots\dots\dots(19)$

Annual setup cost = $\qquad (D/Q)/C_S \qquad \dots\dots\dots(20)$

The setup point is the lead time in days times the daily demand rate

$$R = (DT_L)/W \qquad \dots\dots\dots(21)$$

where W is the number of working days per period. The maximum inventory level is reached at

$$t_1 = (QW)/P \qquad \dots\dots\dots(22)$$

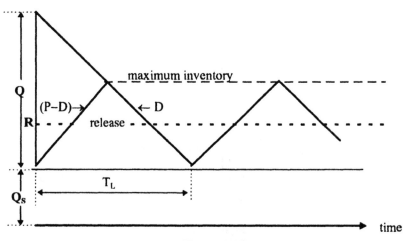

Figure 1-27
Inventory buildup -safety stock.

Total costs (excl. production cost) are obtained by adding equations (19) and (20)

$$TC = (P - D)QC_UF+/2P + (D/Q)/C_S \qquad \dots\dots\dots\dots\dots\dots(23)$$

With a safety stock included (see Figure 1-27):

If P is the production rate and C_S is the setup cost,

$$Q_{opt} = [(2PDC_S)/(P - D)C_UF_H]^{1/2} \qquad \dots\dots\dots\dots\dots(24)$$

and $\qquad\qquad TC = (P - D)QC_UF_H/2P + (D/Q)/C_S \qquad \dots\dots\dots\dots\dots(25)$

There are a great number of inventory models we could create. To set up those models requires differential and integral calculus. This is beyond the scope of this text.

Probabilistic Inventory Models

When uncertainties enter the decision making process such as assumed demand and lead times, the model includes the use of statistical probability distributions.

When both, demand rate and lead time are uncertain, techniques such as dynamic programming or computer simulation can be used. This will not be discussed here. When either the demand rate or lead times are uncertain, we have several methods that deal with those problems.

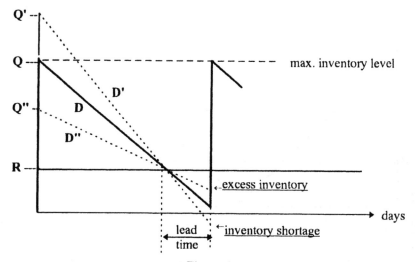

Figure 1-28
Inventory with uncertain demand.

There are three basic methods, the *normal probability distribution*, the *single period model*, and the *Poisson distribution*. Below is a short overview of those methods:

a) Normal Probability Distribution

Assuming the lead time of a cycle is known and constant but the demand is uncertain. The standard model will have inventory uncertainties introduced as shown in Figure 1-28 (broken lines). When demand exceeds supply, sales are lost because of an inventory shortage (D').

When supply exceeds demand (D"), an excess is created in inventory. For the purpose of dealing with uncertainties in demand we assume a normal probability distribution with its contours defined by the mean (μ) and standard deviation (σ). We assume a mean demand D_q based on judgment, experience or historical records. Our confidence in this estimate is then expressed in terms of a standard deviation S_q for a normal curve (Figure 1-29).

For example, we may believe that we need a stock of 2000 valves to install a mechanical system during 100 days. The accuracy of this prediction may vary by 10%. The valves are stocked in the warehouse at the site. The foreman picks them up once a week (5 days).

Compared to the basic model, D_q can now be substituted for D, and

$$Q_{opt} = \sqrt{\left(2C_O D_q\right)/H} \quad \text{items} \qquad \qquad \text{.......(26)}$$

If the average cost of a valve is $100.00, holding cost are 5% of purchase value and the paperwork etc. of ordering valves is $40.00, the optimum order quantity is

$$Q_{opt} = \sqrt{\left(2 \times 40 \times 20\right)/\left(0.05 \times 100\right)} = 320 \text{ valves and } (2000)/(320) = \mathbf{6.25}$$

6.25 orders will have to be placed on the average during the 100-day working period.

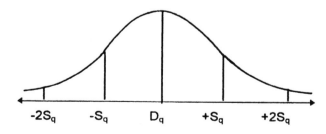

Figure 1-29
Uncertain inventory distribution.

With a 5-day lead time, the mean of the 5-day standard deviation is

$$\sigma = (D_q/W)(T_L) = (2000/100)(5) = \mathbf{100} \qquad\ldots\ldots\ldots\ldots(27)$$

The lead time standard deviation is the square root of the daily variance times lead time:

$$\sigma = [(S_Q{}^2/W)T_L]^{1/2} = [(2002/100)(5)]^{1/2} = \mathbf{44.72} \qquad\ldots\ldots\ldots\ldots(28)$$

rounded to 45.00, other standard deviations are

$$2\sigma = 90 \text{ valves and } 3\sigma = 135 \text{ valves etc.}$$

There is an equal chance (50%) that the demand will exceed R (a stockout will occur) or the inventory will accumulate R excess items. The question is, what is more desirable, a shortage of inventory or a surplus? This service level of 50% can be re-evaluated by referring to standard normal distribution tables.

If we are willing to accept a 90% service level (40% over the mean service level), the reorder point is average demand during lead time plus standard deviation

$$R = R(\text{mean}) + Z\sigma \qquad\ldots\ldots (29)$$

As discussed previously under 1.4.2 the value of Z can be found in tables. For 39.97% of the area under the curve, the standard deviation is 1.28 units (Figure 1-30). Therefore

$$R = 200 + 1.28(44.72) = \mathbf{257 \text{ valves}}$$

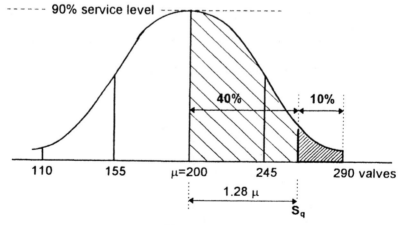

Figure 1-30
Inventory example - uncertain conditions.

When the inventory reaches a level of 257 valves, another 320 valves should be ordered. This so-called *safety stock* is

$$Q_S = R - \mu = 257 - 200 = \mathbf{57} \text{ valves} \qquad ....(30)$$

Those 57 valves are carried in inventory because of the uncertainty with the demand. The cost of carrying them must be added to the total inventory cost:

$$T_C = [(Q/2) + Q_S]H + (D/Q)C_O \qquad ....(31)$$

$$= (320 + 57)(5) + (2000/320)(40) = \mathbf{\$2135.00}$$

b) The Single Period Model

The construction environment usually has situations where single orders are placed at the beginning of the inventory period. At the end of the period, the inventory may still have surplus items in stock.

To calculate the optimum order quantity, incremental analysis is applied. It deals with the lost profit of understocking an item (U) or the cost of overstocking an item (O) It means that we are searching for a value Q_{opt} where the expected loss of overstocking is equal to the expected loss of understocking an additional item.

Assuming a mean quantity D_q as a probable demand and a standard deviation of S_q (Figure 1-31). The probability that the demand is smaller or equal to Q_{opt} is

$$P_D = U/(U + O) \qquad ....(32)$$

which is the area beyond the point Z. From a standard normal distribution table, Z can now be found (the ratio of the area under the normal curve is $0.5 - P_D$).

Assuming the cost of *understocking* valves is \$0.50 per valve and *overstocking* is \$2.00 per valve, then the probability for a demand of less or equal to the optimum quantity Q_{opt} is

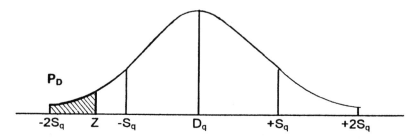

Figure 1-31
Single period inventory model.

$$P_D = (0.5)/(0.5 + 2) = 0.20 \text{ or } 20\%$$

The area under the curve is $0.5 - 0.2 = 0.3$ which equals a Z value of -0.84

$$S_q = [(2000^2)/100]^{1/2} = 200$$

and
$$Q_{opt} = D_q + ZS_q = 2000 - (0.84)(200) = 1832 \quad ....(33)$$

Please note, that the negative value for Z can be determined by investigating what incremental effect lower or higher demands have on the cost of overstocking or understocking. If the demand is likely to be *smaller or equal* to Q_{opt} and we increase the inventory by one valve, we have overstocked for an additional cost of $(0.5)(\$2.00) = \$1.00$. With the mean inventory of 2000 valves and an expected demand of more than 2000, we have understocked by $(0.5)(\$0.50) = \$0.25$ per valve.

If the expected demand is *less or equal to* $(Q_{opt} - 1)$ or to 1999 valves, the area under the probability distribution moves to the left and

$$Z = (1999 - 2000)/200 = -0.005$$

with a shift in area under the curve by 0.002 (from a standard normal distribution table) and a probability of overstocking Q_{opt} by $0.500 - 0.002 = 0.498$.

If we lower the inventory to 1999, we are understocking by $1 - 0.498 = 0.502$ with a demand larger than 1999.

Overstocking will then cost $(0.498)(\$2.00) = \$0.996$,

Understocking will cost $(0.502)(\$0.50) = \$0.251$.

A demand less or equal to Q_{opt} is therefore chosen for calculating Q_{opt}. At the level of 1832 valves, the expected loss of adding an additional valve is equal to the probable cost of *not* adding a valve (shortfall). If we assume a salvage value C_S, the overstocking cost (O) would be reduced by the set-up costs C_S.

c) *The Poisson Distribution*

If the standard distribution cannot be determined with fair confidence, we can use the Poisson distribution. With this distribution we only need to know the expected value. The cumulative probability that the demand is smaller or equal to Q_{opt} is

$$P_X = U/(U+O) = \frac{(D_q)^X e^{-D}}{X!} \quad ....(34)$$

To solve for x, Poisson distribution tables can be used. For those who do not have published tables handy, it is a simple procedure to use a spreadsheet. This will even help with related calculations. Using the average demand D for various values of X will generate P_X.

Table 1-13

Inventory - poisson distribution

Demand X	Probability P_X	Shortage X–22	Losses $(X–22)P_X$
22	0.0769	0	0
23	0.0669	1	0.0669
24	0.0557	2	0.1114
...	....	....	....
...	....	....	....
39	0.0001	17	0.0017
40	0.0001	18	0.0018
			0.9860

The average demand in our *example* was 20 valves per week. If the demand increases, and shortages must not exceed 5% of demand, then, starting with a demand of 22 valves

The expected loss is $\Sigma(X–22)P_X = 0.986$ and the losses in terms of weekly demand are $(0.986)/(20) = 0.0493$, which is very close to the 5% requirement (Table 1-13). We may not need more than 22 valves at the pick-up point.

Overview

Management Science has developed mathematical models that deal with the control of inventory. The goal is to *optimize* the cost of managing the physical assets that are temporarily put in storage. Inventory control is also applicable to the construction phase of capital projects and therefore a very important part of project cost management.

The integrated approach to cost management requires the timely availability of equipment and materials that must be delivered *just in time* to satisfy the installation schedule. Whereby the demand usually fluctuates at a certain level on an ongoing manufacturing production line, it builds up, peaks, and then diminishes at a construction site. In fact, some materials may not enter the inventory system at all. They are delivered exactly in time for installation on site. This needs a tight economic balance to minimize project cost.

Production line:

Construction inventory:

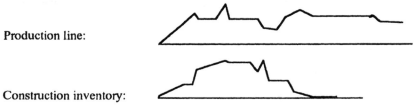

1.4.5 Optimization

A Project Manager who is confronted with a problem will recognize that there is always more than one way to a solution and there are often several solutions. A decision must be made by selecting the best, most favorable solution.

This is referred to as the *optimum* solution.

Optimization is dealt with in Jelen's book "Cost and Optimization Engineering" (5). Jelen distinguishes three categories of optimization:
- Physical
- Preferential
- Economic

①*Physical* optimization is strictly a logical process to reach the best possible solution, which can *mathematically be proven* to be correct. This can also be referred to as "Technical Optimization".

②*Preferential* optimization is subject to human emotional judgment and feelings. An example is the final choice for the best sounding melody among several songs or the best tasting wine in a competition.

③*Economic* optimization has to consider human values also, but they should preferably be expressed in quantitative terms.

Usually cost data are included, but the environment can be uncertain. Therefore, the decision a manager must make can involve more than a structural logical approach. This includes simulation and other probability considerations. Cost Optimization is applied in the disciplines of Cost Engineering and Project Cost Management.

We may have to overlap or combine any of the above categories for certain problems.

Technical Optimization

Assume we want to design a container that uses the least material for maximum volume. This is a pure technical optimization problem which is satisfied with a mathematical solution:

	Cube	Sphere
Surface Area	$6a^2$	$d^2\pi$
Volume	a^3	$d^3\pi/6$

For a unit volume of one liter, the surface area of the cube is 600 cm^2 and that of the sphere is 484 cm^2.

In this case there is no question that the technical decision is made in favor of the sphere. Other bodies are not considered for this exercise.

In management, however, the technical evaluation is usually only a part of the total picture. If we go back to the simplest example of the sphere, we find other considerations:

a)   The material used for the sphere does not have to be as thick as that of the cube by 10% say, because the configuration is stronger. This is a factor in favor of the sphere.

b)   The cube can stand on its own while the sphere needs additional support, which will add 20% to the cost. This is a factor in favor of the cube.

c)   The manufacturing process is considerably more complicated for the sphere which is estimated to add approx. 25% to the cost of production.

d)   Surveys indicate that a spherical design is more pleasing to the eye and sells better than a cube. This is in favor of the sphere.

e)   Shipping containers for the sphere need to be larger (d = 13cm) and need to be more protected from damage on transit than the cube (a = 10cm). This would add 15% to the cost of storage, handling, and shipping. This is in favor of the cube.

We can see now, that many factors have to be considered when making optimal decisions. The output of some factors are known with certainty while others contain uncertainties. Those can be dealt with by using probability calculations.

Cost Optimization

Cost Optimization is an integral part of Economic Evaluation. Below is a broad overview where optimization can best be applied (Table 1-14).

(Optimization and cost control problems are discussed throughout the text without specifically referring to it as such).

Table 1-14
Cost optimization

Categories	Optimize quality time and cost
Materials	Queuing, inventory, economic order quantity (EOQ)
Equipment	Lease vs. buy, availability, utilization, replacement
Labor	Productivity, methods, learning curve, creativity
Utilities	Conservation, availability, utilization
Supplies	Inventory, availability, EOQ,
Services	Economic, production, methods, utilization
Space	Layout, sequence/location, traffic, utilization

While optimization is *proactive,* such as selecting materials, planning facility layout, modifying processes, initiating method studies, defining standards, cost control is *reactive* to information received before action is taken.

The most difficult part of optimization is probably the formulation of the problem. Most optimization problems deal with *minimum* or *maximum* values over a specific range. The queuing model is a typical example where minimum costs are the optimal solution (see Figure 1-21).

When maximum values are considered to be the optimal solution, the shape of the curve is reversed. For example, the efficiency of a work force increases with experience (learning curve) but can decrease due to boredom or excessive overtime (Figure 1-32).

There are situations when the local optimum point on a curve is not the desired optimum. This is a situation that can easily be overlooked. Assume, the overtime requirement has suddenly been lifted. The rebound efficiency would have a strong effect on the learning curve (Figure 1-33).

The optimum has now shifted to point B. Point C is called the *deflection* on the curve. We could identify the extremes on the curve mathematically. This is called the *analytical method.*

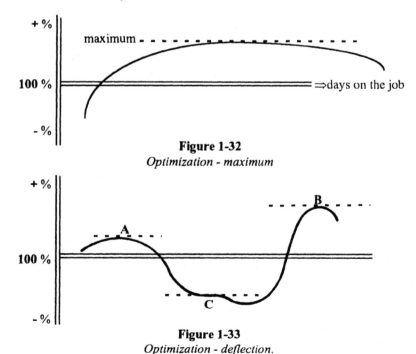

Figure 1-32
Optimization - maximum

Figure 1-33
Optimization - deflection.

If the curve has the function $Y = F(X)$, then $y' = \dfrac{dy}{dx} = \dfrac{df(x)}{dx} = 0$

Setting $Y = 0$, indicates a minimum when positive and a maximum when negative because as x increases on an interval, y increases if y' is positive and decreases if y' is negative.

For example, if $y = u^2 + 3$ and $u = 2x + 1$, then $\dfrac{dy}{du} = 2u$ and $\dfrac{du}{dx} = 2$

further, $\qquad y = \dfrac{dy}{du} \times \dfrac{du}{dx} = 8x + 4$; If $y = 0$, $x = -0.5$

The negative result indicates the *maximum* point on the curve. F(x) has a maximum value also if the second derivative f"(x) is negative and a minimum value if the second derivative f" is positive.

$$\dfrac{d^2 y}{dx^2} = + \text{ or } - \text{ (positive or negative)}$$

If the second derivative is zero, then the third derivative would tell us that we have a deflection point if + or − .

There are situations when a solution is not feasible even though the curve or the mathematical solution gives a correct answer (Figure 1-34).

A machine may perform best at high revolutions, say 100 Hz, but some attachments to the machine may not be able to take the stress, which in the graph (Figure 1-34) is 60 Hz.

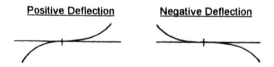

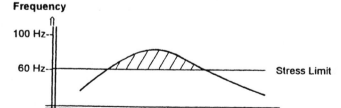

There are situations when a solution is not feasible even though the curve or the mathematical solution gives a correct answer, e.g.

Figure 1-34
Optimization - nonfeasible solution.

Q #1.42:
By analyzing cash flows for specific projects it was found, that income (inflow) during a typical year rises at

$y_1 = 2M^{0.5}$ and expenses (outflow) are $y_2 = 4M^{-0.5}$, where y = $/month.

The net cashflow is indicated as $2M^{0.5} + 4M^{-0.5}$ $/month

At what point in time is the net cashflow lowest?

A #1.42:
To find the point when the net cashflow curve changes direction, we simply set the derivative to zero and solve for M:

$$\frac{d\$}{dM} = 0.5 \times 2M^{-0.5} - 0.5 \times 4M^{-1.5} = \text{or } M^{-0.5} = 2M^{-1.5} \text{ or}$$

$$- 0.5 \log M = \log 2 - 1.5 \log M \; ; \; \therefore \; M = 2 \text{ months}.$$

In our example this would be the end of February.

We can solve the problem the hard way by calculating each month of the year (Table 1-15) and plotting the sum of the two curves (Figure 1-35).

Table 1-15
Optimization example

M	$2M^{0.5}$	$4M^{-0.5}$	$\Sigma\$/M$
1	2.00	4.00	6.00
2	2.83	2.83	5.66
3	3.46	2.31	5.77
4	4.00	2.00	6.00
5	4.47	1.79	6.26
6	4.90	1.63	6.53
7	5.29	1.51	6.80
8	5.66	1.41	7.07
9	6.00	1.33	7.33
10	6.32	1.27	7.59
11	6.63	1.21	7.84
12	6.93	1.15	8.08

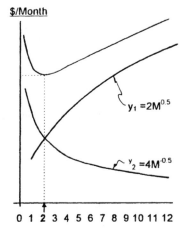

Figure 1-35
Optimization example.

CHAPTER 2
PROJECT INITIATION

The objectives of this chapter are:

1. To provide a total cost picture of a project from initiation to termination.
2. To discuss the elements of a decision-making process including the concept of risk management.

Most of us do not have the chance to work in all areas of project management. We usually specialize in cost estimating, scheduling, purchasing, constructing, designing, or other areas of project management.

This chapter shows how construction and operation fit into the various stages of a capital project. It gives an overview of how we decide to initiate a project and how we prepare to reach that decision.

There is also an example of a specific corporate decision-making process which relates the author's experience.

Introduction

For every problem under the sun there is a solution or there is none.
If there be one, try and find it; if there be none, never mind it.

Based on an old proverb

We will restrict our discussion to problems that can be solved. No project will be initiated without an investigation of its feasibility and profitability.

Because of various degrees of uncertainties there are many risks we have to take when we decide to initiate a project. Not all projects will be successful, some may even have to be canceled during the construction phase if it is more cost effective to cancel and write off the losses than to continue with an unsuccessful project.

The emphasis is therefore on a constant review and re-evaluation of the *total life cycle cost* of a project.

The risk aspect of the decision making process with a short introduction to decision analysis are also included in this chapter. Their application is widely used in industry. It should be mentioned, that corporations use different criteria when making decisions. What is shown in this chapter is only a guide, subject to modifications depending on the type of project and management style.

2.1. INCENTIVES TO START A PROJECT

What is a project?

- ♦ It has a beginning and an end.
- ♦ It has an objective.
- ♦ It is distinct from other projects.
- ♦ It uses multiple diverse resources.
- ♦ It requires a project leader and competent associates.
- ♦ It must be planned and its progress monitored.
- ♦ Its performance must be reviewed continuously.
- ♦ It is affected by internal and external forces.

What determines the need to initiate a project? Essentially, it is the desire to *maintain or increase the quality of life.*

We want to live well. We desire plenty of consumer goods with great variety and choices. An increasing population creates greater demands which in turn needs higher production.

This increases the rate of resource consumption and environmental pollution. Since many of those resources are limited, rationing of traditional resources and conversion to alternatives will eventually affect the way we create future projects. External (uncontrollable) influences on the capital expenditure process increase while the internal (controllable) portion decreases. This results in a more sophisticated application of project controls.

The latest trend is toward *"natural-resource"* accounting. Our economic indicators overstate the nation's economic growth by only measuring the market values of goods and services produced during a year. The more cars are sold and the more garbage is collected, the bigger, busier and more powerful the economy is. This is reflected in the Gross Domestic Product (GDP). There is no penalty deducted from GDP for clear-cutting a steep slope for example. Soil washes away and the gutted mountainside can never grow trees again. The records just show a logging boom. Often, denaturalization is not counted, therefore, it does not seem to count. In this case resource deterioration is a loss of domestic assets that is unrecorded and therefore *not* included in the economic reporting structure. Most of those costs are either ignored (pay-me-now-or-pay-me-later) or spread over the entire population of the country through taxation and deficit financing. In fact, governments *encourage* exploration for new resources as existing resources are used up by maintaining a depletion allowance.

Guthry (6) asked many years ago: "Is our monetary system capable to finance the huge investments necessary to sustain our present way of life?" Many doubts have been expressed that the quality of life may suffer unless we change our traditional economic concepts of waste and *increase our productivity* drastically.

The essentials of our capitalistic system is the maximization of profits. This *profit motive* provides the stimulus for continuous growth and prosperity in a materialistic sense. This creates competition among ourselves, which in turn imposes reasonable capital expenditures, low interest rates and an adequate return on investments. Beside the profit motive, we can now identify another incentive to start a project, i.e. *social needs.*

There is no expected monetary return on investment here. Public funds are obtained from pledged donations, bond issues, government grants and taxes from the local population. Social needs are satisfied by building schools, city halls, hospitals, parks, etc. Work is sometimes created in the public interest during periods of economic recession with no expected rate of return.

There are also projects started that do not meet social needs and have no profit motive. Those projects are executed with *unrestricted resources.*

Gold-plated swimming pools or pleasure boats for individuals are two examples.

To maximize *profit* for the benefit of private enterprise probably creates the strongest incentive to control cost. Owners' and managers' support for the installation of a proper cost control system is usually not a problem.

Controlling projects based on *social needs* is often more difficult. The "owner" is a bureaucracy. Politics often enter the picture. This creates additional external influences on a project, stifling a manager's control over his/her project.

2.1.1 Project Control Concepts

The total project control concept is concerned with cash flows during all phases of project development. There are three major areas of control (Figure 2-01):

> Marketing
> Construction
> Operations

Marketing cash flows include anticipated revenues from sales and cost of sales, promotion, market research etc.

The *construction* plan is concerned with establishing the capital budget and time schedules for building capital facilities. Control is based on the project *cost flow* during construction and the impact of acquisitions on the overall *cash flow* of the project.

The *operating* plan is also based on the scope used for the marketing plan. The operating plan includes estimates for raw materials, utilities operation and maintenance expenses. In addition, the cost of managing the facilities is also included.

Economic Control
ECONOMIC DECISIONS:
   GO/NO GO

Capital Control
FEASIBILITY STUDIES
CONSTRUCTION BUDGETS

Expenditure Control
PHYSICAL BUILDINGS
AND OPERATING BUDGETS

Before the decision is reached by the owner to go ahead with the project, the type of control is *economic*. *Capital* control is necessary during the physical building of the facilities. *Expenditure* control, sometimes called financial control is necessary during the construction phase *and* the operating phase of the project.

> **Project control systems are essentially a mechanism to capture data from the source of action, to process the data, analyze the information obtained from the data and continually forecast the outcome of the project.**

2.1.2 The Project Life Cycle

Corporate decision makers always deal with uncertainties when required to select a new capital project. The information on which decisions are based is usually subject to errors and bias. Those who do the initial estimates are facing two major problems:

Lack of Information

Precise information for new projects is never available. Records for similar existing facilities may help the estimators to make better forecasts of new facilities. Since estimates deal with predictions of the future, they will always be wrong. The question is by how much? Interest rates or escalation indexes can drastically affect the cost of a project, market conditions may become highly competitive, government policy may change, etc. It is therefore essential, that senior managers be informed by estimators what their level of confidence level is in the estimate.

Project Proposal Bias

Capital project data are not generated by unbiased observers with precise measurement instruments; rather, they are estimates made by lower managers with incentives to have their projects accepted. They may even shade their estimates to make their proposals appear better than competing ones. Those behavior factors should be taken into account when decisions are made. Although we cannot eliminate the bias and errors in estimates, we *can provide structures that help induce truthfulness.*

Clancy and Finn(7) based those structures on six stages during the total life of a project by recognizing that those stages will enhance the control over the project. *(Reprinted [in modified form] with permission of The Society of Management Accountants of Canada from an article which appeared in <u>Cost and Management</u>, by D.K. Clancy and D.W. Finn, July/August 1985 issue.)*

The Stages of a Capital Project

1) DISCOVERY STAGE

<u>Generation of Ideas:</u> Discovery is a human *creative* process. A team of method study experts can generate ideas that can determine which initial concepts are worth developing into a proposal. (Creativity and Method Study techniques will be discussed in Chapter 7). Opportunities are discovered sporadically.

2) REVELATION STAGE (Informal process)

Credibility Evaluation: There is still no rigid structure during this stage. Close associates with proven experience get together to develop schemes which they will use to lobby those who will actually make the funding decision. Proposals are informally revealed and supported by the majority of participants.

Specific individuals are connected with the proposed project. Economic evaluation including risks analysis is done (see Section 2.2.0).

3) SELECTION STAGE (Formal process)

Accept/Reject: Those who formalized the proposals will now have to convince the decision makers that the selected projects will be successful. That means; funds must be available, project must meet the criteria of the selectors, those who support the project proposal must have a good track record of previous successes. Senior management must comprehend the uncertainties involved, striving to reduce bias and personal influences.

Successful projects must be made to happen by the will, creativity and energy of project sponsors and the support of senior management. A rejected project is sent into oblivion. The cost incurred is written off as a loss. Senior management should strive to reduce the surprise of the rejection and provide a graceful exit to proposers.

4) INITIATION STAGE (Required resources)

Cost control and risk monitoring: Early success is important. This is a period of maximum risk and exposure to loss of cash, especially for projects with long time horizons. A periodic review of project status and cost performance must be done.

Cost control is very crucial at this stage. There are always inevitable problems, not foreseen in the proposal, that require total project cost management. The losses can be substantial if a project has to be canceled after resources had already been committed. Individual reputations are at stake. Formal identification with a rejected (or canceled) project will have a negative effect on the credibility of sponsors in getting future projects funded.

5) MAINTENANCE STAGE
(Profit and leadership maintenance, low risk, maximum gain)

Net Resource Generator: At this stage, the project should have passed the "risky" phase to become a generator of products. A part of resources of material, labor and management may have to be released or replaced to maintain a profitable operation. Cost monitoring is done periodically.

6) OBLIVION STAGE (Identification fading)

<u>Minimize losses</u>: Many projects reach oblivion after continued success over long periods of time. When a product is no longer in demand, the project designed to produce that product will fade. A decision must be made whether to modify or de-commission the plant. To be cost effective, the decision must be timely. Proper *cost control* methods should be used during the de-commissioning period.

Because the selection process influences the behavior of those preparing project proposals (estimating bias), proposers must be required to supply information about potential problems during initiation, maintenance, and even termination.

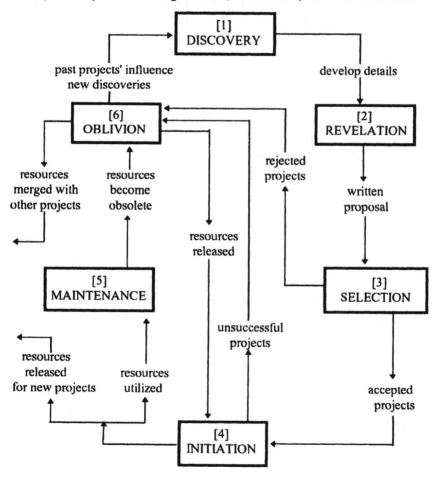

Figure 2-01
Project selection flow diagram.

In looking at the uncertainties surrounding a project, the decision maker (owner) must consider the entire life cycle of a project.

The project selection flow diagram (Figure 2-01) gives an overview of the life cycle stages as outlined by Clancy and Finn.

The decision to approve or reject capital expenditures is a major responsibility of management. Approval authorizes an immediate commitment and mobilization of resources such as money, people, materials, equipment. The capital expenditure decision can be very difficult. Many alternatives and the risks involved have to be considered. Project evaluation covers the total life cycle of a project with due consideration of the economic environment.

The other diagram (Figure 2-02) depicts the major phases during the life of a fictitious project.

The money spent to build the facilities is the *capital investment* in the project. This investment will have to be recovered during the *payout time*. At this time, the project has reached the *break-even point*. Any benefits beyond the break-even point are *profits*. Benefits include sales and capital recoveries such as land and salvage at the point of plant shutdown.

To make certain that the project is profitable, we have to estimate the cash flow and evaluate the return on the investment. It is the objective to recover outlays and expenses as soon as possible and thereby maximize the profitability curve (Figure 2-02).

The rough diagram (Figure 2-03) is not a scaled cash flow curve. It only highlights the basic stages during a project life cycle. The cumulative cash flow here is a *net* cash flow and money going *into* the project is negative and money received due to operating the plant is *positive*.

It would be ideal, if construction were on schedule and within budget, and the recovery period as short as possible to maximize the area above the profitability line.

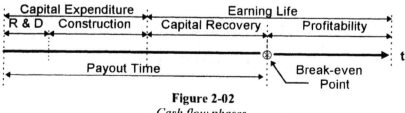

Figure 2-02
Cash flow phases.

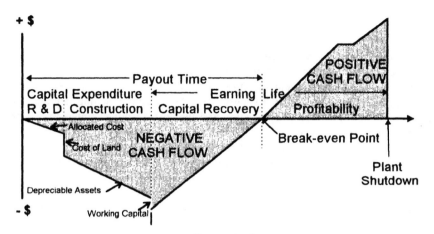

Figure 2-03
Project life cycle.

2.1.3 Project Evaluation

Our economic environment determines the viability of a project and the adequacy of the invested funds that are returned from project operations. The financial environment of a corporation determines the capability to provide the funds for capital projects. All economic evaluations are based on the following premise:

A viable project is one where the total receipts add up to more than the original total cost

Otherwise there is no net productivity.

To satisfy ourselves that there is a net positive productivity we need to measure it quantitatively in order to compare diverse projects (apples and oranges) such as 50-year old hydro electric plants and 2-year old computers. The solution is to look for a measure that is a ratio of the yield per annum.

According to the economist Samuelson:

"A capital or investment project's net productivity is that annual percentage yield which you could earn by tying up your money in it."
(Quote from "Economics", Paul A. Samuelson, McGraw-Hill, 1951, page 573.)

If you buy apples for $80.00 and spend $20.00 in total to process it, and sell the juice for $110.00 a year later, the interest yield is 10% per annum.

There are basically three major areas to evaluate in a project:

Capital Investment
Operations
Profitability

In general, the *capital investment* includes the value of fixed assets, leased assets, working capital. Benefits accrue from sales and other revenues. Costs are incurred for procurement of equipment and materials, consumable supplies, labor, supervision, utilities and overheads. Expenses include marketing, selling, administrative and technical services, depreciation and the cost of money (interest). Taxes are paid on the net income.

Corporations have their own guidelines on an acceptable rate of return and on the formulae they use to derive at the profit per dollar invested. This is then used as a basis for the GO/NO-GO decision for a future project.

Plant operations are activities that convert raw materials into sellable products. Relating the investment dollars to the net annual sales income we obtain

$$\text{Investment Ratio} = \frac{\text{Total Capital Investment}}{\text{Net Annual Sales}}$$

The investment ratio indicates the payment time necessary to reach the break-even point. We must also compare the cost of raw materials to the production cost per unit. This is the

$$\text{Conversion Factor} = \frac{\text{Product \$ / unit}}{\text{Raw Material \$ / unit}}$$

Depending on the type of industry the above ratios can vary considerably.

Profitability measures the financial effect on the capital expenditure decision (total income compared to total outlay) in terms of a "rate of return". Traditionally, company guidelines give a "minimum expected annual rate of return" on the investment. There are several methods to calculate profitability. The most common are:

Payout Time

To determine the break-even point, it is customary to calculate a cumulative cash flow. Simply speaking, where the curve crosses the base line is the *break-even point*. So called "payout time" calculations are relatively simple and may or may not include interest. The disadvantage of this method is, that the period beyond the break-even point (the profit period) is not emphasized.

If payout time does *not include interest*, it is assumed that operating costs are recovered immediately, thereby not affecting the flow of cash.

Payout time *with interest* adds an investment charge for the previous year to the cash flow. This includes interest on the working capital, but here again, no delay in the payment of capital.

To smooth the cash flow, *rolling averages* can be used (see also Chapter 5, cost reporting). Jelen (5) calls this the "equivalent maximum investment period".

Return on Original Investment (ROI)

This is simply the average annual profit during the earning life of a project divided by the original investment, including working capital.
If

I = Investment Capital
B = Benefits (annual net profit)
E = Expenses and various other costs
T = Taxes
C = Cash Flow

then

$$ROI = \frac{C}{I} \text{ usually expressed in percent,}$$

where $$C = B - (E + T)$$

ROI calculations should be used for short duration projects and only for preliminary evaluations.

Return on Average Investment (RAI)

will give entirely different results than ROI calculations. Here, we are taking the average *outstanding* investment into consideration; i.e. depreciation. This of course will affect the cash flow because the amount of depreciation is deducted from the original investment. Therefore,

$$RAI = \frac{ROI}{\text{Average Outstanding Investment}}$$

Because the results are different, we must not mix the two calculations when making comparisons.

For larger projects, the time value of money should be taken into account. In this case, we may look at two common methods

Net Present Value (NPV) and Discounted Cash Flow (DCF)

NPV is used when the discount rate is determined by corporate monetary policies (fixed rate of return)

$$NPV \text{ Index} = \frac{\text{Present Value of Net Cash Benefits *}}{\text{Present Value of Outlays}}$$

* = after income tax

DCF is used to find the rate of return by setting the present value of all cash flows to zero. A trial and error calculation is required:

$$\text{Zero} = \sum_{n=1}^{n=n} \frac{\text{Cash Flow for Year n}}{(1+r)^n}$$

Assuming we have a company investing \$ 100 000 in a facility to establish a new production line. The estimate over a 5-year period indicates sales of \$ 50 000 in the first year and a 15% increase every year thereafter. The original investment is depreciated on a straight line basis. Income taxes are 34% and the cost of capital is 15%. Variable expenses are 30% of sales (Table 2-01). What is the cash flow in k\$? What is the rate of return?

$$\text{Return on Investment} = \frac{(337 - 46.2 - 101)/5}{100} = 18\,\%$$

Taking the time value of money and depreciation into consideration

$$\text{Return on Investment} = 123.2/100 = 23\,\%$$

Table 2-01
Cash flow #1

Years:	0	1	2	3	4	5	Total
Sales		50.0	57.5	66.1	76.0	87.5	337.1
Variable expense		−15.0	−17.3	−19.8	−22.8	−26.2	−101.1
Fixed expense		−20.0	−20.0	−20.0	−20.0	−20.0	−100.0
Total expense		−35.0	−37.3	−39.8	−42.8	−46.2	−201.1
Taxable income		15.0	20.3	26.3	33.2	41.2	136.0
Tax @ 34%		−5.1	−6.9	− 8.9	−11.3	−14.0	−46.2
After tax income		9.9	13.4	17.3	21.9	27.2	89.7
Depreciation		20.0	20.0	20.0	20.0	20.0	100.0
Investment	−100	-	-	-	-	-	−100.0
Cash flow	−100	29.9	33.4	37.3	41.9	47.2	89.7
Present worth	−100	26.0	25.2	24.6	24.0	23.5	23.2

Discounted Cash Flow

$$\frac{(100)}{(1+r)^0} + \frac{29.9}{(1+r)^1} + \frac{33.4}{(1+r)^2} + \frac{37.3}{(1+r)^3} + \frac{41.9}{(1+r)^4} + \frac{47.2}{(1+r)^5} = 0$$

If $r = 0.25$, $-100+23.9+21.4+19.1+17.2+15.5 = -3.0$
If $r = 0.23$, $-100+24.3+22.1+20.1+18.3+16.8 = +1.5$
By proportion, rate of return $= 23 + (2 \times 1.5)/4.5 = 24\%$

2.2 DECISION MAKING

Better decisions are made in the long run, when a decision maker attempts to explicitly and objectively assess the various impacts of alternatives on the future well being of the organizational entity for which a decision is required.

Even though experience is a powerful factor in many cases, a "rule-of-thumb" decision is considered almost random in nature and not as likely to lead to the type of decision that is possible through an objective decision analysis.

2.2.1 Considering Risk

Good management needs to consider risks encountered while executing a project. If management is overly conservative, many potentially acceptable projects will be turned down. If the judgments are overly optimistic, corporate profitability will suffer.

Considering risk has become an accepted part of cost control.

We should recognize that it is extremely *risky* to budget projects, especially large ones, on the basis of conceptual estimates made with a low percentage of project definition and design completion. Risk analysis techniques should be used to evaluate undefined items, scope growth potential, process and design status, schedules, regulation changes, procurement, productivity, start up and management skill. Contingency factors should relate to the risk evaluation in order to get a more accurate estimate of final cost.

Defining Risk

The Oxford Dictionary defines risk as "the possibility of meeting or suffering harm or loss (by) exposure to this." However, we would not voluntarily take any risk unless there is the possibility of a *reward*. Therefor, the consequences of risk-taking must be evaluated before action is taken..

AACE International (2) in "Standard 10S-90" defines risk as

the degree of dispersion or variability around the expected or "best" value which is estimated to exist for the economic variable in question, e.g. a quantitative measure of the upper and lower limits which are considered reasonable for the factor being estimated. *(A practical application of the above is called "range estimating" - to be discussed later in Section 3.4.)*

PMI(3) [Exposure Draft 8-94] have devoted a separate section to this subject. They divide so-called Project Risk Management into four major functions:

Risk Identification; Risk Quantification; Response Development; Risk Control.

Almost all activities on a project can be *identified* as having different degrees of risk involved. Quantifying those risks or uncertainties is a major step toward the response development, which is sometimes called *Risk Mitigation*. Mitigation is a defense action against possible risks. It moderates the severity of risk. This may include:

♦ Avoidance (hire non-union labor to avoid strikes)
♦ Reduction (security guard vs. fence around property)
♦ Transfer also called *deflection* (contract clause includes more responsibilities by others)
♦ Sell-off (have "weasel-out" clause in liability insurance)

> **We cannot "manage" risks, but we can use management methods to identify, analyze, evaluate or assess risks, and to respond to uncertainties.**

Risk Analysis is the application of probability calculations to *identify* the *magnitude* of uncertainty.

Risk is always with us in our world of uncertainties. When we take our car out for a drive, we are running the risk of having an accident. How many of us are conscious of this fact and use the car only when absolutely necessary?

Assume that a city with the population of 2 million people have 91 fatal accidents per year. This means statistically that every fourth day there is a fatal accident. We have now the risk *identified*. An individual driving in this city has to expect to be killed at a rate of one in eight million for every day the car is used. This is risk assessment. Even though the chances of a fatal accident are real, we do not stop driving for two main reasons:

1) the chance of being killed is remote (exposure)

2) to drive is an activity considered advantageous or even necessary in our society (reward).

We can reduce this "average" risk factor by

1) keeping our car in top mechanical condition

2) avoiding to drive when impaired

3) driving defensively.

This results in a *mitigated risk* for the driver. When we consider the risk of one life in eight million as being acceptable, we express an *attitude* toward that risk. This attitude very much depends on our *perception* of danger.

The great popular fear of exposure to radiation has probably been a major factor for our crippled nuclear industry. The horrors of Hiroshima may have had an influence on our perception of radiation dangers. If the terrible bomb had not been developed first, would we still be afraid of nuclear generating plants? Workers in an American nuclear station are exposed to $\frac{1}{20}$ the radiation dose as people outside the station are to natural sky and earth radiation (0.33 REM).

How about nuclear accidents? The consequences of an accident vary with the design of the station. The CANDU reactor for example uses heavy water (deuterium) for both, moderator and cooler. Even though more expensive, it is considerably safer than fast reactors which use enriched fissile fuel and graphite moderators. And yet, in the eyes of the public, *all* nuclear stations pose the same danger. We therefore need to be aware of the difference between perception and *actual statistical data* of risk involved.

💣 *There once was a fellow who was afraid of flying because of the possibility that there may be a terrorist's bomb on board. When he was told that the chances of having a bomb on board are one in fifty million, he was still not satisfied. However, the chance of having two separate terrorists plant each a bomb on board of the same plane are so slim, that it may never happen. This satisfied him and his famous last words were: "I will take my own bomb on the plane. This, together with the terrorist's bomb will reduce the risk to almost zero".*

It appears, that the future will give us a "greener" technology. Greater environmental consciousness calls for more innovative solutions to project management. It will become more difficult to run successful projects because of additional restrictions that are uncontrollable to the project manager. Interest groups will lobby politicians to intervene with technological progress and innovations, introducing legislation that will probably result in bigger legal problems and higher risks factors.

More skills than ever will be needed by project management during the competitive 1990s and beyond. We may be moving from a large technical content towards a more legal content. Management needs to convince others that the project is good for society in all respects. Communication and negotiation skills will become very important.

2.2.2 Decision Analysis

When a project manager is confronted with choices, he or she must make decisions based on the best probable outcome. Very often there are *uncertainties* and thereby risks involved when making decisions.

In practice, decisions are made by instinct, by precedence, experience, or decision analysis. The latter focuses on a flow of *good* decisions, resulting in the best outcome.

Not all new theories developed by mathematicians are equally successful in practice. There is the "chaos theory" one reads about and also the trendy "fuzzy logic" and the "neuron networks." Unless we fully understand those, we must be careful using them. Specific problems require the proper management science tools for their solution.

UNCERTAIN

Furthermore, not all decisions lend themselves to the algorithm of the decision theory. There are many situations where problems can not be solved mathematically or when it may not be worth the effort. In situations where the outcome of an action is automatic or can easily be programmed, it is not necessary to apply decision analysis. Basically, there are two types of decision criteria:

Decisions under certainty
Probabilistic decisions

Decisions under Certainty

There are potential courses of action we can follow. If it is a choice of either/or, those courses of action are called *alternatives*. For each alternative in turn may exist one or more *states of nature* over which the decision maker has no control. The outcome of this combination is the payoff.

If the probabilities of the various payoffs cannot be determined, the decision criterion depends on the *optimistic* or *pessimistic* attitude of the decision maker. An optimist uses the maximax criterion, whereby the alternative with the highest possible payoff is selected. If the decision maker selects the alternative that has the highest among the worst outcomes (decreasing state of nature), it is called the *maximin criterion*. There is also a minimax regret decision, whereby the difference between the optimum payoff for a given state of nature and the actual payoff, the so called opportunity loss, is minimized. That means, the <u>max</u>imum opportunity loss (regret) for each state of nature for all alternatives is found first. The lowest (<u>min</u>imum) of those maximum losses is then chosen as the best alternative.

Example: Let there be three investment schemes: Invest $100; twice that amount; ten times that amount. There is an even chance for a return of *five times* or *1.5 times* the investment or *lose five times* the original amount.

Alternative	States	of	Nature
	+ 5 ×	+ 1.5 ×	- 5 ×
1000	5000	1500	- 5000
200	1000	300	- 1000
100	500	150	- 500

Maximax decision: The $ 1000 alternative would be chosen, because it results in the highest payoff ($5000).

Maximin decision: The highest among the worst outcomes is a loss of $500. Therefore, the $100 alternative would be chosen. This is a conservative decision.

Minimax regret: For each "state of nature" column subtract the payoff from the highest payoff value for each alternative:

Alternative	States	of	Nature
	+ 5 ×	+ 1.5 ×	- 5 ×
1000	0	0	4500
200	**4000**	1200	500
100	4500	1350	0

The $200 investment would be chosen because it has the lowest maximum opportunity loss ($4000).

Probabilistic Decisions

When we are able to assess probabilities with each outcome, we can apply a more structured decision making process. We are applying the law of averages to obtain the *expectation* of an outcome to occur or not to occur. The expected value of X is

$$E(X) = \Sigma XP(X)$$

where P(X) is the probability for X to occur. This is true only in the long run, similar to flipping a coin.

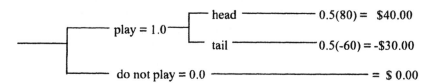

The outcome is 0.50 (fifty/fifty) for heads or tails. If we win $ 80.00 for flipping the head, and we lose $ 60.00 for flipping the tail, and nothing if we decide *not* to flip, then over the long run we would average

$$E(X) = 80(0.5)+(-60)(0.5) = \$ 10.00$$

This result can also be represented in form of a tree.

We cannot always base our decision on the statistics of past performance. We may have to use "subjective" probabilities that are based on personal opinion or many years of experience and our state of mind.

A manager purchasing major equipment for a construction project is well advised to consult the manager of operations to decide what equipment to buy, because the maintenance and repair cost play an important role in the decision making process. But how do we know what those cost will be several years from now? The reputation of the supply company and the experience of the buyer will have a bearing on the probability assessment.

Example: The construction manager in conjunction with the operations manager propose that a generator from company A has a 5% chance that it will not need any repairs for the next ten years. This compares with a 10% chance for supply company B. For various repair cost, the probabilities are considered as shown in table 2-02.

Based on this calculation, company B has an edge over company A with all other considerations being equal. We must not forget that the manager operating the plant has no control over frequency and cost of repairs (states of nature). The subjective assumptions made may be the best at the time but may proof to be wrong later on.

The best decisions based on subjective probabilities may not necessarily result in the best outcome.

In fact, there is a _cost_ associated with uncertainties. Below is an example that will show the cost of uncertainty and also common expressions used for this type of problem.

Table 2-02
Probabilistic decision example

Repair Cost (x)	Company A Probable. P(x)	Company A Value xP(x)	Company B Probable. P(x)	Company B Value xP(x)
$0	0.05	0	0.10	0
$2000	0.20	400	0.30	600
$4000	0.65	2600	0.50	2000
$6000	0.10	600	0.10	600
Expected	cost	**$3600**		**$3200**

Another Example: A facility is to be built to produce pumps. The question is how large a facility should be built? Market research showed a minimum demand of 3000 pumps per year and a maximum demand of 6000 pumps. The subjective probabilities at kilo intervals are assessed as follows:

Demand	Probability
3000	0.10
4000	0.30
5000	0.40
6000	<u>0.20</u>
	1.00

Fixed cost per year is 200 k$ with a variable cost of $50 per pump. The selling price is $150 per pump. The payoff would therefore be

$150×(number of pumps sold)-200 k$ + $50×(pumps produced)

If we now tabulate the payoffs, keeping in mind that the demand cannot be controlled (states of nature). If the production is lower than the demand, the number of pumps sold will equal the number of pumps produced. For example if the demand is 5000 pumps and there were only 3000 produced, the payoff is

$150×(3000) - 200 000 - 50×(3000) = 100 k$

This is summarized in the Payoff Table 2-03 below. We will now introduce *probabilities*. Each result is multiplied by its assigned probability and the total payoff summarized for each production (see table 2-04):

It appears that we should build a facility with the capacity of producing 5000 pumps per year. The total of 225 k$ is the expected value under uncertainty, sometimes called expected monetary value or EMV.

This value was obtained by our assignment of high probabilities to the <u>sales</u> of 5000 pumps. But we know, that we have no control over future sales (demands), i.e. if only 3000 pumps are sold, the profit is zero.

Table 2-03
Payoff #1

(in k$)	States of Nature			
Production	3000	4000	5000	6000
3000	100	100	100	200
4000	50	200	200	200
5000	0	150	300	300
6000	-50	100	250	400

Table 2-04
Payoff #2

States of Nature

Production	3000(0.1)	4000(0.3)	5000(0.4)	6000(0.2)	Total
3000	10	30	40	20	100
4000	5	60	80	40	185
5000	0	45	120	60	**225**
6000	-5	30	100	80	205

That means our decision to produce 5000 pumps may not necessarily result in the best outcome. To reiterate,

there is a cost of being uncertain!

Our calculation indicates what profit we might expect, based on our subjective assignment of probabilities. Our profit expectation is 225 k$ if we produce 5000 pumps. There is no way that we can tell that the demand will be as stated. Assuming we are able to come up with the best decision every time, the *expected value under certainty* will be

$$(0.1)(100)+(0.3)(200)+(0.4)(300)+(0.2)(400) = 270 \text{ k\$}$$

This is the result if our predictions came true every time. The difference between the two values 270 - 225 = 45 k$ is *the cost of being uncertain* (or the maximum amount we may be willing to pay for "perfect information")

Anything we do under uncertain conditions such as projecting into the future may have an outcome that is not the best result. Using decision analysis however will require us to think in a more structured way and enter information into an equation that we may otherwise not even have considered. A somewhat different format is used when we apply probabilities to cash flows. The example under 2.1.3 can be used to solve a basic engineering economics problem. Those problems usually involve short duration cash flows with range estimates expressed as most likely, low, and high.

We said in chapter 1 that the expected value of a frequency distribution is the *mean* of the distribution. An approximation of the mean for a range estimate is generally considered to be

$$\text{Mean of Estimate} = \frac{(\text{low}) + (2 \times \text{most likely}) + (\text{high})}{4}$$

Repeating the problem under 2.1.3: A company considers investing $ 100 000 in a facility to establish a new production line (table 2-05). The estimate over a 5-year period indicates sales of $ 50 000 in the first year and a 15% increase every year thereafter. The original investment is depreciated on a straight line basis.

Table 2-05
Cash Flow #2

Years	0	1	2	3	4	5	Total
Sales		50.0	57.5	66.1	76.0	87.5	337.1
Present Worth		43.5	43.5	43.5	43.5	43.5	217.4
	Discounted Income after Tax **143.5**						
Variable Expenses		-15.0	-17.3	-19.8	-22.8	-26.2	-101.1
Present Worth		-13.0	-13.0	-13.0	-13.0	-13.0	-65.2
	Discounted Expenses, Tax deducted...........						**-43.0**
Net Cash Flow	-100	26.0	25.2	24.6	24.0	23.5	**23.2**

Income taxes are 34% and the cost of capital is 15%. Variable expenses are 30% of sales. What is the cash flow in k$? What is the rate of return?

Investment = [Sales] - [Expenses] - [Net Cash Flow] = k$ **77.2**

Net return on investment = 123.2/100 = **23%**

This is an acceptable return on investment? How accurate are those figures? Assuming that the above values are based on a preliminary estimate with uncertain information having the following ranges:

Sales   + 25% ; - 35%
Expenses   + 30% ; - 20%
Investment   + 35% ; - 5%

We can now calculate the high and low values and the mean:

	most likely	high	low	mean
Sales	143.5	179.3	93.3	140.0
Expenses	- 43.0	-56.0	-34.4	-44.1
Investment	-77.2	-104.2	-73.3	-83.0

A commonly used confidence interval used in range estimates is 80%, i.e.

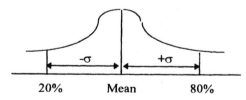

20%   Mean   80%

What we now need is the standard deviation of the distribution. If we consider a normal distribution for ease of calculation and an 80% confidence level (see Chapter 1),

$$\sigma = 1.282$$

Applying this to our values, we subtract the mean from the high estimate and divide by 1.282. *(Texts differ substantially. Jelen suggests* σ =(high-low)/8. *AACE Skills and Knowledge uses* σ = (high-low)/2.65).

We now add the standard deviation to our table:

	Most likely	High	Low	Mean	S. D
Sales	143.5	179.3	93.3	140.0	30.78
Expenses	-43.0	-56.0	-34.4	-44.1	9.23
Investment	-77.2	-104.2	-73.3	-83.0	16.56
		Totals		**12.8**	**36.15**

The total standard deviation is calculated by taking the square root of the sum of variations:

$$\text{Total S.D.} = \sqrt{30.82 + 9.22 + 16.62} = 36.1$$

Subtracting the standard deviation from the mean will give us the net present value at 20% probability, adding it to the mean results in the net present value at 80% probability:

$$\text{Low value} = 12.8 - 36.1 = 23.3$$
$$\text{High value} = 12.8 + 36.1 = 48.9$$

plotted on the graph below (figure 2-04):

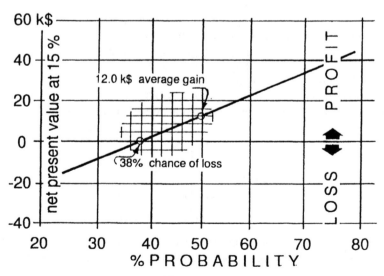

Figure 2-04
Probability -profit/loss example.

The point estimate gave us quite a favorable return on our investment. By introducing uncertainties, we find, that there is a 38% chance that we may incur losses and an even chance we may gain a profit of $ 12 000 net present value.

2.2.3 Decision Trees

Assuming we are on a boat in an unknown river system deep in the African jungle and going upstream to find a village situated at one of the tributaries, where we can get needed supplies. But we are lost and do not know, which tributary to take. The first decision we have to make when we come to the first fork in the river system is: Do we turn left or right? We do not know; therefore the chance of being correct is 50%. The same is true for the next decision point. But when we come to the third fork we notice an empty bottle floating down the right tributary. This *could* mean there is a strong possibility, say 90%, the village is on that tributary.

What are the chances to find that village?
Rules:

The sum of parallel branches add up to 1.0.
The probabilities along each path are multiplied with each other.

Therefore, the answer is $0.5 \times 0.5 \times 0.9 = 22.5\%$
Here is another example:
(Based on the fabled game of Truel.)
Ancient European tribes had a cruel method to solve serious disputes: They would give each contestant two stones, place them 20 paces apart and let them throw these stones at each other, one at a time. Whoever gets hit, loses the dispute.

There was once a handsome young suitor with the name *Attila* who came from a distant village. He fell in love with the daughter of the headman, who would rather give his daughter to the rich old *Brutus*, or to *Celtus*, the best but very cruel warrior of the tribe.

"O'K" said the headman "let's solve the problem with our stone method. Here ,are the rules: Attila starts, then Bruce, and then Celtus. When it is your turn, you can throw at either of the other, but you do not have to throw, you can pass. However, if you pass, you have forfeited one stone. Only one of you can win, and he will get my daughter."

The contest did not seem fair. Celtus was known to hit a target 16 times out of 16 throws in previous war games. Brutus was pretty good too, he hit the target 12 times out of 16 throws. But Attila, being still quite young and not as experienced in war games, has a record of only hitting 10 out of 16 throws.

The tribe has now gathered around the clearing to watch the contest. The girl looks sadly at Attila who has been given the first stone. He now has three choices: Throw the stone at Brutus, throw it at Celtus, or not throw the stone at all.

Attila hesitates, looks at the girl and decides *not* to throw the stone! He is now a sitting duck.

Brutus immediately picks up his first stone. He is mad at Attila who attracted the girls favorable attention. Attila shall pay for this!

But hold it! - Brutus' face turns white. He realizes, if he hits Attila, Celtus has the next turn and will surely not miss him. He must try to knock out Celtus. He aims at Celtus and throws the stone with all his might!...

Next day, the wedding took place. Attila was the happiest man on earth. What made him decide to forfeit the first throw?

He applied the decision tree technique!

The chances of hitting
(probabilities) were:
    Attila   = 10/16
    Brutus = 12/16
    Celtus = 16/16

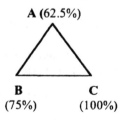

Before we draw a decision tree (Figure 2-05), we will assign labels to the decisive actions of the contestants. Attila *passes* the first throw (A_1 - p); Brutus *misses* or *hits* Celtus with his first throw (B_1 - C_m) or (B_1 - C_h). If he hits, Attila has the next turn. He either misses or hits Brutus (A_2 - B_m) or (A_2 - B_h). If he misses, Brutus has a good chance to win. If he hits, he will get the girl.

What are his chances that this will happen?

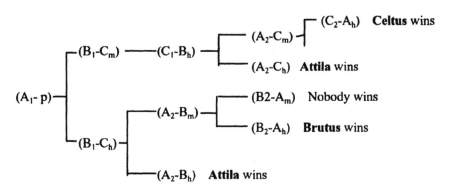

In terms of probabilities:

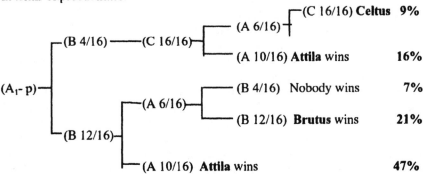

Figure 2-05
Decision tree game.

This indicates that Attila has a 12 to 7 chance (47/28) of winning if Brutus hits Celtus. Even if Brutus misses, his chances are still better than Celtus' (16/9).

We should not be fooled by the simple examples above. In reality, the process of decision analysis by means of the decision tree can become quite complex.

The AACE (Cost Engineers' Notebook - Economic analysis) (2) identifies specific features:

Decision Nodes are represented by square boxes
It could be construed as the point of a "fork" where
a decision must be made.

Actions indicate activities that are the outcome of the
decisions. They are shown as branches emanating
from the decision node.

Event Nodes sometimes called event forks or chance
nodes are represented by small circles.

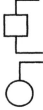

These are points where possible outcomes of the decisions are evaluated.

Events (outcomes) are the branches leaving these event forks. They are beyond the control of the decision maker.

The decision tree diagram has the advantage for us to be able to
- identify outcomes for each alternative in the decision process.
- estimate probabilities associated with each alternative
- assign cash flows in form of profit or loss.
- keep the sequence of decisions (actions) or outcomes (events) in proper order.

The tree can grow at the following sequence:
- Indicate the initial decision.
- Assign the probabilities.
- Assign the cash flows.
- Calculate the outcomes.

The grown tree can now be analyzed:
- Do a backward pass (fold back the tree).
- Do a sensitivity analysis if necessary.

For example, let's assume a manufacturing company considers to build an assembly plant. Several preliminary design schemes have been prepared and their estimates submitted to the CEO. The finance committee reported that an investment over 50 M$ is *not* an option. Following that, design schemes for under 50 M$ were reviewed in more detail. The highest of those was estimated to be most likely 45 M$, optimistically 10% less, pessimistically 20% more. This was for a fully automated plant of a sophisticated design capable of producing one Million units per year. A less sophisticated automated plant producing .8 Mill. units/year was estimated to cost 40 M$, - 8%, +14%. Two partially automated plants were also considered with estimates of 30 M$, -5%, +10% and 24M$, -5%, +10%, turning out 0.6 Million units/year and 0.4 Million units/year respectively.

Marketing studies show, that the highest demand for the product will not exceed 0.8 Million units/year. Low demand will not be less than 0.4 Million units/year; both having the same chance of occurring.

The initial decision is the consideration to build:

Do not build

Build

If the company does not build there will be no further branches on that side of the tree. Other branches will be needed on the "build" side of the tree. That means we estimate the capital cost of the project and the restrictions on raising that capital.

If the cost is higher than the money that can be raised, only the favorable estimate (< 50M$) is considered. This< 50 M$ branch will now continue to grow (Figure 2-06).

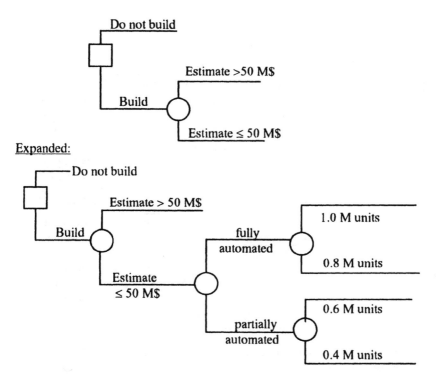

Figure 2-06
Decision tree - assembly plant - #1.

Here, we had to make another decision based on the design of the automated plant. It is either fully automated or partially automated. Each of those branches will have events indicating the annual production capacities of the various design schemes. This is followed by the market distribution of the units (Figure 2-07).

The most favorable path can now be calculated. The branches having range estimates will need the mean and expected values which is at an 80% confidence level (Table 2-06)

Table 2-06
Expected values - assembly plant

units(M)	optimistic	pessimistic	mean	S.D.	EV20%	EV 80%
1.0	40.5	54.0	46.1	6.2	39.9	52.3
0.8	36.8	45.6	40.6	3.9	36.7	44.5
0.6	28.5	33.0	30.4	2.0	28.4	32.4
0.4	22.8	26.4	24.3	2.1	22.2	26.4

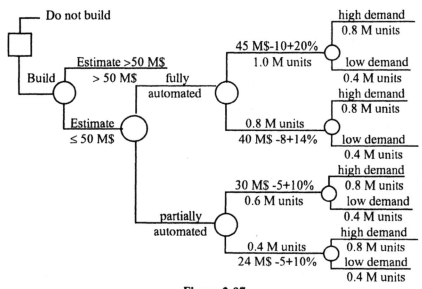

Figure 2-07
Decision tree - assembly plant - #2.

Since there is a 20% chance (one in five) that the estimate may fall outside the range, the probabilities are

low estimate (optimistic)	= 0.2	$Z = 1.282$
high estimate (pessimistic)	= 0.2	S.D. = (high - mean)/Z
expected value (mean)	= 0.6	

Let the unit production be *proportional* to the number of units produced. The net unit cost for 1.0 Mill. units is $10 per unit. If the selling price is $20, the income after tax etc. is $10 per unit per year. For ease of calculation we assume that escalation and discounting over a ten-year plant life is included in the figures below (Table 2-07): We can now calculate the net income (M$) for each demand, high and low, by applying the mean of the distribution (Table 2-08):

Table 2-07
Net Income - Assembly Plant

Plant Capacity	Cost/unit	Net Income
1.0 Mill. units	10.00	100.00
0.8 Mill. units	11.00	95.00
0.6 Mill. units	13.00	85.00
0.4 Mill. units	15.00	75.00

Table 2-08

Mean Net Income - Assembly Plant (M$)

Production	0.8 Mill.units	0.4 Mill.units	Mean
1.0 Mill. units	80	40	60
0.8 Mill. units	76	38	57
0.6 Mill. units	68	34	51
0.4 Mill. units	60	30	45

Fully automated plant:

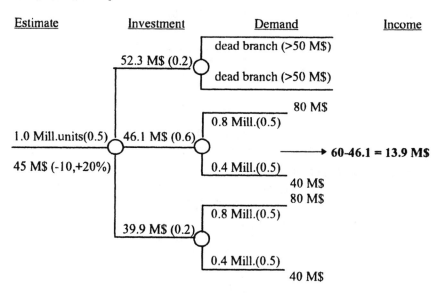

Figure 2-08

Decision tree - assembly plant - #3.

Entering the investment, the net income and the outcome (profit), this is the final tree (Figures 2-08 to 2-11). Please note, that we dropped the demands that exceeded the supply. The calculation of the outcome is done by "folding back" the tree, e.g.

For 1.0 Mill. units the mean income is $80(0.5) + 40(0.5) = 60$ M$.
The investment is $52.3(0.2) + 46.1(0.6) + 39.9(0.2) = 46$ M$.
Folding back, income minus investment is $60 - 46 = $ **14 M$.**

The same calculations apply to the other schemes.

Fully automated plant:

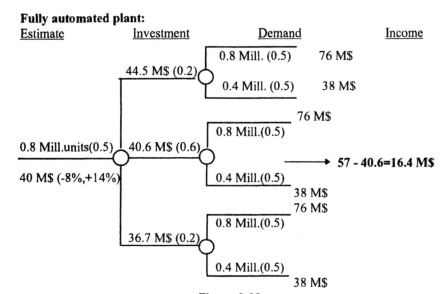

Figure 2-09
Decision tree - assembly plant - #4.

Partially automated plant:

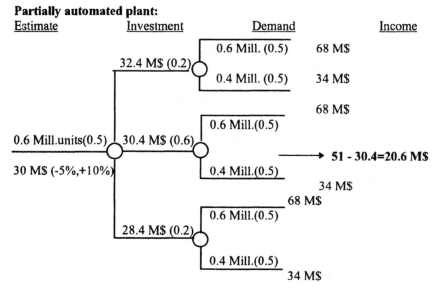

Figure 2-10
Decision tree - assembly plant - #5.

Partially automated plant:

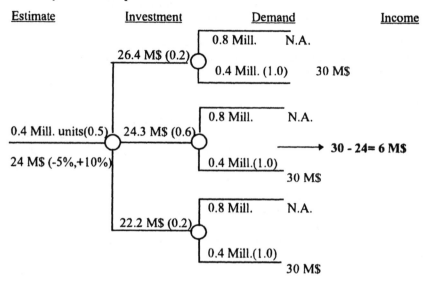

Figure 2-11
Decision tree - assembly plant - #6.

The partially automated plant that produces 0.6 Mill. units per year is the most profitable plant under consideration.

To summarize, a decision tree is a graphic display of the decision-making process. It helps us to organize our thoughts and to visualize the path to follow to meet an objective. Generally, the development of a decision tree can be much more involving than shown in the simple example above. The assignment of probabilities of outcome is also a function of risk preferences. Few decision-makers are risk neutral. Some are "risk-seekers," others are "risk-avoiders."

2.2.4 Corporate Decision Making Processes

To ensure financial effectiveness and sound business decisions, a large corporation must have well defined Management Control Policies and Procedures. This will help their employees to perform decision analysis, interpret this analysis and make recommendations to the level of organization that makes the final decision.

Decision making in smaller companies may not be as structured or formalized. Less effort may be expended, depending on the impact the decision has on the company's future. It should be obvious though,

**the more management science capabilities exist in a com-
pany, the more effective it is in a competitive environment.**

The decision making process outlined below (31) includes the planning phase, a problem analysis, followed by an evaluation of various alternatives, ranking and selecting the best option for final approval.

Planning

Evaluation Effort: Greater evaluation efforts are justified for *large scale expenditures*. This, of course is a relative term. Significant expenditures could be 1 M$ for a large corporation or $ 10 000 for a smaller company. Even a program costing $ 100 000 for a large corporation could have a serious impact on customers' relation for example, therefore justifying a full evaluation effort.

Cost Effectiveness: A full analysis can be expensive. The benefits resulting must exceed the cost of the decision analysis by a comfortable margin.

High Degree of Uncertainty: The greater the risk involved, the more evaluation effort will be required. A smaller amount of evaluation effort may be appropriate if

+ documentation is available for previous decisions made under similar circumstances
+ one alternative is obviously superior to others
+ only one feasible option exists. This includes overriding regulatory requirements
+ the time available is a priority restriction. Under some emergency circumstances, an immediate decision may become necessary

Responsibilities: Organizational responsibilities and signing authorities are usually well established in most companies. An example of responsibilities for decision making, where each participant has a defined role in the process can be outlined as shown in Table 2-09 below:

Decision Analysis

Before evaluations take place, the problem must be analyzed, taking into account the technical aspects, the results required (objectives), any restrictions and the consequences of alternatives.

Technical Analysis: For large or complex engineering type studies a technical analysis is necessary early in a decision process. By identifying feasible alternatives, a technical analysis establishes a basis from which cash flow estimates can be derived and assessed.

Table 2-09

Decision Making Responsibilities

General Responsibility$\Rightarrow$	specify parameters to regulate the execution of a task.
Direct Responsibility $\Rightarrow$	direct the execution of a task.
Specific Responsibility$\Rightarrow$	execute the task.
Must be Consulted $\Rightarrow$	must be contacted for advice and information prior to decision.
May be Consulted $\Rightarrow$	may be asked for advice and information.
Must be Notified $\Rightarrow$	must be notified after completion of task.
Must Approve $\Rightarrow$	must approve or veto decision.

The technical analysis should preferably be confined to the study of physical input and output relationships expressed in non-monetary terms. *Detailed documentation* is a requirement.

Required Result: It is the physically measurable result that is necessary and must be produced by all alternatives under consideration. The majority of decisions a corporation may be faced with originate as a direct result of its operation which in turn is based on its Goals and Objectives. In questioning the need for a specific result the analyst should not be influenced on past requirements, but on the necessity to achieve an essential end product. For example, a production expansion decision will provide the basis for design decisions, which provide the basis for supply and construction decisions.

The following questions should also be asked:

- What is the impact on the corporation if the result is *not* achieved?
- What is the appropriate timing?
- What are the consequences of *delaying* achievement?
- Should the scope be enlarged or narrowed?
- Are the specifications too limiting to allow for innovative alternatives?
- Are the requirements clear or ambiguous (leading to a wasted effort)?

To further clarify the purpose of an activity in relation to corporate goals and specific objectives of the organization, the required results are categorized. Depending on the type of organization, categories could be

- level of service (standards, research, distribution)

- laws and regulations (environmental, safety, quality)
- cost control (productivity, efficiency, optimization)
- net positive cash flow (profitability, new ventures, by-product sales, recycling)
- other results (corporate relations, image, external support)

Restrictions: Restrictions limit the freedom of decision makers to *trade off* competing corporate objectives when developing alternatives. Restrictions state, in effect, that in trading off a certain objective such as greater safety in the interest of another, say lower cost, the decision maker can go so far, but not further. If an alternative does not meet all identified restrictions then it is eliminated from further consideration.

Alternatives: Alternatives are different ways of achieving a required result. Every effort should be made to identify and develop all worthwhile alternatives that accomplish the required result. The emphasis should be on *developing* alternatives, not evaluating them to avoid the possibility that the best alternative might be dismissed prematurely. When developing alternatives, the consequences of implementing them must also be considered to include future conditions and opportunities.

Evaluations

After having established a set of feasible technical alternatives, we can now compare them on the bases of financial, economy-wide and quality evaluation.

Financial Evaluation: Three assessments are involved:
- economic impact
- corporate financial impact
- risk assessment.

The financially preferable alternative would be obtained by minimizing
- long term cost (maximize long term profit)
- uncertainty exposure (risk)
- adverse financial impacts.

Furthermore, decisions must be
- made from the point of view of the corporation as a whole, local objectives take second place
- based on those items they can influence
- realistic, representing what is considered most likely to occur. Inaccuracies of assumptions and estimates must be recognized and assessed (ranges)
- based on differences of alternatives

Before assessments can be made, it should be determined, what amount of effort is appropriate to the evaluation. This is similar to the criteria described under

Planning above. These are the *evaluation parameters*, which include the identification of

- ◆ amount of effort
- ◆ required results
- ◆ any restrictions (constraints)
- ◆ all feasible and realistic alternatives.

Furthermore, all important *Assumptions* made by the analyst and the type of expenditure should be identified. This *Expenditure Classification* is needed to understand the general nature and scope of the decision problem, especially when interpreting the results.

Expenditure decisions could be classified either as cost minimization (with revenues included or excluded) or profit maximization (the difference between revenues received and costs incurred). In addition, the *Evaluation Period* over which all significant cash flows are identified must be included before assessments can be made. This is sometimes called the "study period". This period usually terminates at a calendar date when

- ◆ the major assets involved reach the end of their economic lives and technological obsolescence is an important consideration
- ◆ cash flow differences between expenditure alternatives are so small, they will become insignificant at their present value
- ◆ contractual arrangements prescribe a time interval (rent or lease)
- ◆ corporate policy determines a specific duration. Work processes and programs are often limited to a five-year life.

It often happens that one or more of the assets used in alternatives under consideration still have some value left. Such value is called "terminal" value or "salvage" value. This should be included (usually at the end) in the cash flow if an active market exists.

In case of replacement cost, the terminal value is usually considered to be

$$\text{Terminal Value} = \text{Replacement cost} \times \frac{\text{service life remaining}}{\text{total service life}}$$

We now come to the stage where *economic assessments* can begin. The economic assessment analyzes the effect on the cash flows of the corporation as a whole, not on individual organizational units. The following are the basic steps in performing an economic assessment:

1. Estimate total attributable (also called incremental, relevant and avoidable) cash flows and the timing of such cash flows. Common cash flows with same amount and timing should be included.

2.   Discount total cash flows by an acceptable discount rate to arrive at the net present value. Rates are usually set by the corporation's financial department.
3.   Rank the alternatives.

The estimating effort includes the investigation, analysis, consultation, testing and research that are needed for the preparation of cost estimates. The purpose of this effort is to reduce the risk that an inferior alternative will be chosen. The cost of the effort itself must be recognized and related to its potential of reducing that risk. In other words, the cost of increasing the estimating effort must be less than the benefits of mitigating the risk.

Only estimates of future cash flows should be used, thereby limiting decisions based on variables the decision itself can influence. It is very important, that only *most likely* estimates be prepared. Target estimates that are set low or high to achieve a purpose, project cost estimates that include non-incremental items (depreciation, interest, overheads) or optimistic and conservative estimates should not be used without adjustments.

Finally, it should be mentioned again, that all attributable cash flows must be converted to their financial equivalence in the "present", i.e. discounted or compounded to a standard point in time.

Companies usually have standard form sheets or computerized programs to calculate the net present value of alternative cash flows. The calculation itself is very simple.

Assessment of Risk and Uncertainty:
Some uncertainties can be quantified while others can not. It has been recognized more and more that net present value calculations based on most likely values only is, in many cases, not adequate.

An assessment which employs probabilistic data provides the most likely NPV, the expected value of the NPV and the range of the NPV, thus improving both the quality and scope of the information for decision making. Depending on the size of the expenditure and the complexity of the evaluation, we may categorize risk assessment techniques under four headings:

Qualitative Analysis involves the application of judgment and reflects the views of the evaluator. It is applicable to minor expenditure decisions.

Sensitivity Analysis identifies variables with the most significant impact to vary, one at a time, over selected ranges. Changes in the result, particularly the minimum and maximum values are summarized for consideration by the decision maker. Sensitivity analysis is applicable to expenditure decisions of median magnitude.

Project sensitivities are used to identify the important variables affecting a project appraisal. In its simplest approach, "what if ..." questions are being asked when looking at project components with large uncertainties. How do we determine "large uncertainties" ? This is a matter of consensus among experts. Their estimates are based on *subjective probabilities*.

When we discussed linear programming under 1.4.1 we assumed one solution to a specific problem (maximum or minimum). In practice there is usually uncertainty about the value of objective-function coefficients. There is also uncertainty about the constraints.

How much leeway in setting up the objective-function coefficient or the right hand side of a constraint can a manager allow before the previously established "optimal" solution will change an acceptable amount? The calculations involve an *incremental change* approach. The effect of changes in *each* coefficient on the outcome must be investigated. That means it will be determined how sensitive the "optimal corner point" is to those changes.

Decision Tree Analysis describes an alternative within a tree structure, whereby events are linked in a *sequential* manner with discrete probabilities assigned to each outcome event. It is useful for describing the inter-relationships of the major decision variables. The application of a decision tree is a logical, mathematical one. It does not take into account the *lateral* thinking process (see Chapter 7 on creative thinking). This can be a disadvantage in some situations.

Monte Carlo Simulation usually requires the construction of a computer model. The model would have a number of random variables with pre-determined probability distributions from which a random sample is generated. This in turn is then statistically interpreted. This Monte Carlo simulation is a major evaluation effort and should only be used for very large expenditure decisions.

We are now in the position to assess the corporate financial impact. As defined previously, a company's financial evaluations have a corporate perspective, using future cash flows based on external transactions.

What impact will this have on *Line Organization Budgets* which are based on cost flows (accounting cost)?

Program Budgets are based on estimates of *internal* allocation and organized along departmental responsibility lines. Attributable cash flows cut across these lines and have a *corporate* perspective. It is therefore not unusual to find that a least cost alternative from the Corporation's point of view has an adverse effect on specific line organization budgets. This may result in priority ranking and cancellation of some programs.

A corporate's financial preference is normally to minimize cost and maximize profit in decisions, while paying due regard to risk and corporate impact.

Economy-wide Evaluation

Certain corporations in certain circumstances will base their evaluation also on the impact to the community. In Canada for example, corporations controlled directly by the Government (Crown Corporations) have to consider the potential impact of their decisions on the Provincial or Federal economy and social community.

Qualitative Evaluation

The qualitative evaluation is intended to include those aspects of an expenditure decision which cannot be quantified in dollar terms. This may include (but not necessarily always and completely) safety, environmental impact, quality, training, labor relations, customer and community attitudes.

Rather than allocating evaluation efforts to dollar estimates in such cases, evaluation efforts might be better utilized by analyzing the situation in non-monetary terms. For example, it may be less expensive to release expert employees during a slow business cycle and later do rehiring. However, quality evaluation may indicate the need to maintain long-term human resource capabilities in specialty areas.

Selection

Selecting the preferred alternative involves trading-off decision criteria when different rank orderings arise amongst evaluations.

Trade-offs

No specific guidance can be given on how to make trade-off judgments. Good management experience and intimate knowledge of corporate goals is the key in making the best decisions. A helpful approach is to indicate by "weight" what is considered the relative importance of each decision criterion.

The following factors may be weighted:
♦ The current external environment (attitudes of others toward company, product acceptance)
♦ The political situation (political "correctness")
♦ The corporate strategy (priorities, direction)
♦ The uncertainties inherent in the decision situation (not quantifiable decisions).

Assuming an alternative requires large capital investment vs. one which requires more labor input. Given the uncertainty associated with future events, a high level of resource commitment implies a reduction of flexibility in the future, resulting in less options later. This can only be evaluated by judgment and utilized in the final selection process.

Overall Ranking

There are three common methods to "weigh" the decision criteria:
- assign a percentage value
- order by level of importance (1,2,3...etc.)
- use a scoring scale, say from 1 to 10.

Such approach does not in itself make the decision. It only provides the framework for applying judgment. If the ranking established in each of the evaluations is multiplied by the weights, this overall ranking measures the overall performance of the alternative.

Cost Benefit Analysis

Decision analysis usually looks for known desired results, that is *benefits*. When those benefits are considered in conjunction with *resources* (costs), a trade-off decision is needed. A changing environment and limited resources require that constraints, uncertainties and conflicting objectives be addressed with a view for compromise. Those involved in the programming and budget process of a corporation will need to formulate the strategy and the work program (Figures 2-12 and 2-13)

Documentation and Approval

Documentation is an integral part of an efficient management system. Documentation identifies planning, directing and controlling activities in the management process. It gives guidance to those who have to analyze, review, recommend and approve the outcome of financial evaluations.

Below is a sample document which gives an overview of decision analysis requirements. This is meant to be used as a guide to the evaluator (Figure 2-14).

Table 2-10
Trade-off Implications

CONSTRAINTS	UNCERTAINTIES	CONFLICTING INTERESTS
Technical specifications	Technological changes	Minimizing short-term cash needs vs. long-term costs
Borrowing limitation	Market fluctuations	
Limitation in product price	Variability in sales forecasts	Cleaner environment vs. higher cost of production
Remain competitive in the long term	Changes in raw material	Office automation vs. job preservation

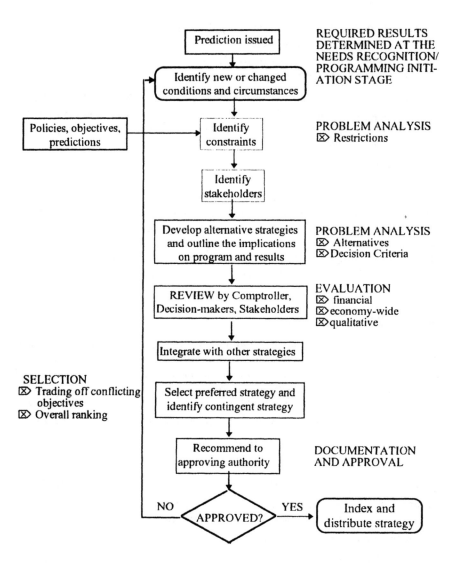

Figure 2-12
Cost benefit strategy.

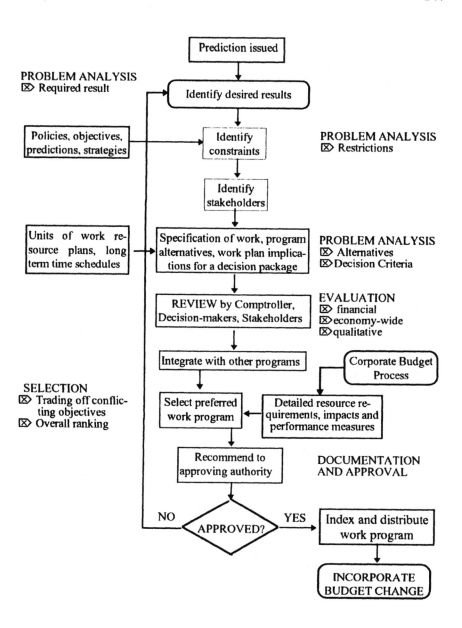

Figure 2-13
Cost benefit work program.

Project Title Project No.

Background/Purpose

- What is the purpose of the report
  - make recommendation
  - discussion/information
  - request approval

- Background
  - what is the required result
  - list restrictions
  - why is the project being carried out now
  - what alternatives were rejected and why
  - what are the key assumptions and rationale
  - project type (cost minimization, etc)

Results of the Financial Evaluation

- Economic Assessment
- Risk/Uncertainty Assessment
- Corporate Impact Assessment
- Identify financially preferable alternative
- Give rationale if all or part of FE not performed

Results of Economy-Wide Evaluations

- Give rationale if all or part of EWE not performed

Results of Qualitative Evaluation

Recommendation

- Decision criteria (financial, qualitative, economy-wide)
- Rationale

Monitoring

- What variable should be monitored and who should do the monitoring

Figure 2-14

Decision making documentation.

CHAPTER 3
THE COST PLAN

The objectives of this chapter are:

1. to provide an overview of the planning required for the design and construction of a capital project.

2. to discuss the concept of an integrated project organization and work breakdown structure.

3. to introduce the basics of scheduling and capital cost estimating.

The planning phase establishes a base that performance can be measured against.

Introduction

If you don't know where you are going, you may end up somewhere else !

Unknown

To avoid serious delays and cost overruns, an adoption of modern management systems to plan and build projects is important. This is especially true during the planning stages, where decisions have a large impact on the outcome.

Companies can minimize delays and cost overruns by installing a more precise and detailed scheduling system. Better scope identification during the design stage will result in better estimates. For legal considerations and good will, contracting practices are getting more complicated. A thoughtful and meticulous preparation of contract documents, matching the owner's objective and resources with those of the contractor, is an important part of the planning process.

In addition, the cost plan must make allowances for quality, safety, constructability, work organization and methods.

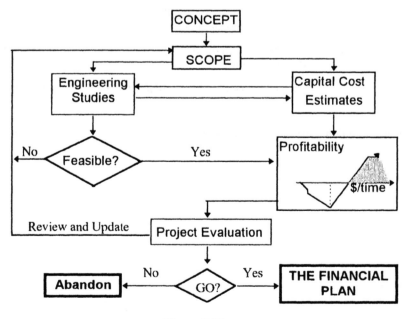

Figure 3-01
Financial planning.

3.1. CAPITAL PROJECT OVERVIEW

All projects start with a concept. This concept is further developed and evaluated as to feasibility and economy. Engineering studies establish the feasibility of a project. Economic evaluations provide the profitability criteria.

This results in an evaluation model showing the difference between benefits (revenue) and the sum of cost (variable) and expenses (fixed). The sensitivity between the break-even point and the capital investment in regard to scope (capacity) is then tested for several alternatives. The resulting optimum balance produces a financial plan (Figure 3-01).

At this stage, the project is approved and all scope and cost data flows into the project plan, which now becomes a reality (Figure 3-02).

3.1.1 The Planning Cycle

A carelessly planned project will take three times longer to complete than expected. A carefully planned project will take only twice as long. International Systems Inc

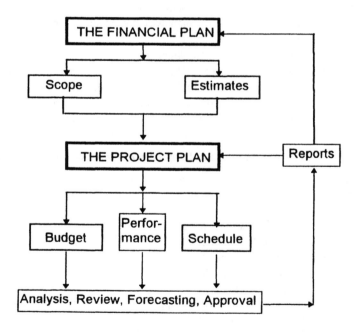

Figure 3-02
The financial plan.

Financial planning is done at the corporate level. This usually involves long range planning for several projects. The planning for *all* projects seems to be performed by using the *same* standardized techniques and procedures.

Depending on their size and duration, the accuracy of the cost estimate can vary substantially. Information for the project plan is primarily obtained from the owner, based on existing data bases, from the design engineers, from suppliers' records and by subcontractors. A project manager may inherit a plan that is just a projection of the corporation's current operations and is leaning toward the financial "bottom line."

In practice, no project is alike.

A project manager must get involved early in the process. Construction work has a high degree of unpredictability often underestimated by the strategic planner. Obviously, the greater the uncertainty the more difficult it is to realize the plan as defined. Uncertainties can be reduced by starting the planning process early and using an active, dynamic approach to information gathering.

A very important consideration is the use of an appropriate *project management software*. Unfortunately it is common to find several different systems in the same organization, including partially implemented systems, or

- ◆ the implementation plan was rushed to deliver results unreasonably early
- ◆ expected results were undefined
- ◆ users and management became frustrated and abandoned the plan
- ◆ the planner tried to implement every function of the system at the same time, causing inconsistent results and unstable conditions
- ◆ the plan tried to bring the entire organization on-line simultaneously with a wide-area network and client-server database.

There is a difference between *planning* and *the project plan*.

Planning is an ongoing, iterative activity. At a certain stage, the planning activity is "frozen" into a plan containing these major ingredients:

- ◆ The Project Objective
- ◆ The Statement of Work
- ◆ Project Requirements
- ◆ Work Breakdown Structure
- ◆ Milestone Schedule
- ◆ Capital Budget

The initial plan is documented and approved by the responsible authority (owner). Some programs need approval by both, the customer and the contractor. After approval of the selected scheme a more detailed design takes place. Major equipment is purchased and contracts are awarded.

Detailed design and active construction continue while design changes affect procurement and construction work. At completion of installation, field tests are made and the final start-up of the facilities takes place.

If we had to categorize the *phases* of a construction project, it would look like this:

- Concept
- Definition
- Acquisition
- Construction
- Start-Up

The above breakdown may have different terminology or it may be broader or more restrictive, depending on what literature one reads or who is in charge of the construction project. Furthermore, the size and type of the project will determine how much overlap or even fusion there is among any of those phases.

The graph below (Figure 3-03) represents construction phases for a large project of about three years duration. If the project is an apartment complex, it is likely that several similar ones have been built before. The design, drawings, duration and procurement information are basically available *before* construction starts. The concept will already have been established before the plan was approved. All information input is relatively accurate (Figure 3-04):

Let us now assume we need to build a nuclear power station with a ten to fifteen years design and construction duration. There may be scope changes due to technological advancements or different government regulations on safety and on the environment while building the facilities. This means there is a very high degree of uncertainty in the original plan.

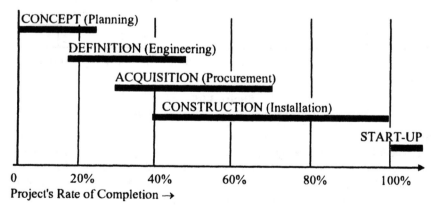

Figure 3-03
Construction phase - average duration.

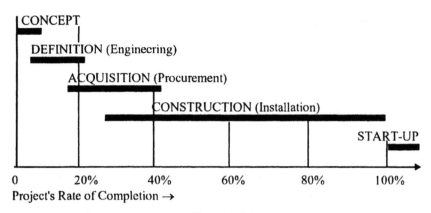

Figure 3-04
Construction phase - short duration.

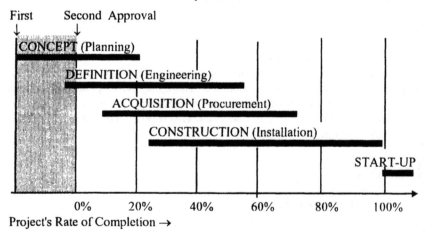

Figure 3-05
Construction phase - long duration.

Any prototype project such as developing a new defense initiative or fusion development falls into this category. The concept may change as planning becomes more detailed. Many design schemes may find better alternatives at a later phase than the initial project plan indicated. The design and procurement stage will overlap considerably with the installation of the capital plant (Figure 3-05).

There could be more than one approval for work charged to a complex capital project of such a very long duration. In this case a work order is given for preliminary engineering, site investigation and specific environmental studies.

The project manager would intimately be involved in the first and second approval planning process. Funds would be approved in stages. The first approval would only cover preliminary engineering in preparation for the second approval stage. If the project is not viable, a second approval will not be given. Expenditures will be written off as expenses to current corporate operation.

The Role of the Project Manager

The Project Manager would be a mature individual with strong communicative and interpersonal skills and leadership qualifications. In some way the term "Project Manager" is not really describing the total individual because complex organizations staffed with motivated professionals need *leaders* first and managers second. We talk about labor leaders, political leaders, scout leaders, not labor managers, political managers or scout managers. Professionals seek involvement and participation at their work place.

Their managers must *lead* the way to reach a goal. This requires *vision* related to individual roles. That vision must not be "owned" by the manager alone but instilled into all members of the project team. Vision is "what"; planning is "how, who and when". The project manager must have the talent to structure an organization that is related to both, the task and the people. The larger the organization the more important it is for the leader to play a symbolic role by exemplifying openness, trustworthiness and consistency.

Project managers are also chosen for their *knowledge and experience* in engineering, procurement and construction. In addition, they must be able to manage aspects relating to finance, cost engineering, environmental and regulatory requirements, labor problems, client and public relations.

Because the manager has large responsibilities for the coordination and integration of multiple activities performed by functional and outside organizations over which he/she has no authority, his/her task is very difficult.

The Project Objective

This is the project scope, outlining required outputs, resources and timing. The statement of objectives for the initial project plan should be very short, usually not covering more than one or two pages. In addition to configuration, budget and timing, objectives statements may also be made as to good customer relations, environmental goals, security and labor relations.

The Project Scope

To achieve the project objective, we need to define the project scope which puts limits and boundaries (inclusions and exclusions) on the work. PMI (3) in the PMBOK Exposure Draft (Aug. 1994) defines *scope* as

⇨ *"The sum of the products and services to be provided as a project."*

This includes the start and finish dates of the project and the estimated final cost. It includes constraints for standards, quality, configuration, organizational relationships, responsibilities, points of internal and external interaction, etc., all structured in a level-by-level fashion of the work breakdown structure. This is the scope baseline which includes a statement of work.

As the project evolves, design reviews and engineering changes will take place, affecting the configuration of the capital facility. Scope statements will now become more detailed. This requires good *scope* management.

The Statement of Work

The work that has to be performed to meet the objective is described in broad terms only during the initial planning phase (investigative engineering). This will set the design basis for the plant and related construction facilities including socio-economic and environmental impacts. It will include applications for construction and operating permits.

During the early planning stage, the statement of work for a thermal plant *for example*, deals mainly with basic studies, site layout, outside contacts, site preparation, and plant configuration (civil, mechanical, electrical, instrumentation), procurement, contract administration etc.

The Project Team

All technical phases of engineering must be properly administered and coordinated. The project manager must build a project team that plans the work, contributes in organizing the resources, meets budget and schedule commitments and ties together the efforts of contractors, designers and construction forces.

Example of a Project Plan

(This example is based on the planning and preliminary engineering for a thermal-electric project (31). Actual data and names have been changed and greatly condensed to fit the intent of this text.)

The Planning Group of the corporation has indicated a need to add 3000 Megawatts to the electric power system. Optimization studies, which included comparable cost of sites, generation facilities, operators' colony, system facilities as well as any differences in annual cost of operation due to system losses, coal costs, water temperature, dredging and any other factors, concluded that a six-unit plant with 500 kW switching be built on Lake "Somewhere" for an estimated cost of xxx M$. Construction to be started two years from now in June 19xx and completed six years later.

Below is a statement of the civil engineering portion of a project manager's work.

Basic Studies:

All major events over the project life cycle including key management decisions will be sequenced on a consolidated schedule, showing approximately 100 activities. The critical path method will be used for planning and scheduling. The Project Engineer will call meetings with experts to develop and monitor programs of *hydrological and geological investigations* which are needed for

1. the design and operation of plant
2. monitoring the effect of plant emissions on the environment
3. measuring the effect of vacuum priming on dissolved oxygen content in the discharged water or rainwater runoff from the coal pile.

Hydrological investigations include
- ♦ Surface- and subsurface currents
- ♦ Ice observations
- ♦ Aquatic weed growth
- ♦ Water temperature
- ♦ Water analysis (sampling)
- ♦ Wind effects on currents and weed dispersion
- ♦ Water levels and lake bottom contours
- ♦ Marine biology studies
- ♦ Turbidity measurements
- ♦ Lateral drift and others

For *soil investigation*, the project engineer, in conjunction with civil, geological and research functional groups will outline a program and negotiate a contract with a consulting firm. The investigation should include
- ♦ Contour lines
- ♦ Properties of bedrock (including water seepage)
- ♦ pH value, sulfates, etc. in groundwater
- ♦ Bearing value of bedrock
- ♦ Shear strength data on overburden
- ♦ Extent of rock stripping
- ♦ Geological profile
- ♦ Site Layout:

The initial site layout includes the creation of a task force under the chairmanship of the project manager with representatives of functional design engineers and the thermal operations group to plan the location of major features relative to each other. This covers *access by road, rail and ship.*

To cope with heavy loads, existing roads may need to be improved or new ones built and care must be taken in arranging the junction between the access road(s) and public highways, railway crossings, bridges etc. The shortest direct way of delivery has to be found but also must the timing of excavation, underground installations and other major construction activities in specific areas be observed and correlated.

Rail siding into the station loading bay and transformer area is an advantage for the handling of heavy equipment. The transmission right-of-way may be used to locate the rail line. The possibility of coal delivery by rail should be considered in case of long interruptions in ship deliveries. For waterborne coal supply, the layout must provide for dock facilities based on plant fueling requirements. Close liaison with shipping companies is essential to establish type of dock, its size, orientation, depth of dredging etc.

Plant Location

A detailed study must be made to determine the most economical plant location and related facilities. Major factors are:

Foundations: Investigate the impact of soil conditions and geographical faults on the position for the station.

Station Level: The cost of operating the condenser cooling water pumps will restrict the level of the condenser, thereby determining the basement level of the station. This in turn will dictate the grading of the site. Design each section of the system to obtain the least head loss without spending more for construction cost than the saving in head warrants.

Coal Handling: When coal is received by ship, deliveries take place only during navigation periods. That means large storage and handling facilities which provide a secure amount of storage.

Ash Disposal: The burned coal is reduced to about 10% ash by weight. The disposal area at the site must be sized to contain furnace ash and fly ash for a extended period of time.

Settling Lagoon: For an isolated plant, provision must be made to clean up domestic effluents.

Switch Yard: Provide for the most economical place for the switch gear, relay buildings and microwave tower.

Construction Facilities: The layout should include buildings and service facilities adequate for construction crews, contractors and on-site office staff. Investigate, if camp facilities should be provided and if the inclusion of a ready-mix concrete plant is more cost effective than purchased deliveries.

Other Features: Work is required to include Administration Building, permanent yard drainage, chimney, discharge ducting, oil storage, combustion turbines, precipitators, ash disposal piping, transformer locations with transmission egress, maintenance buildings, parking area, minor structures and the area on the lake side of the powerhouse, such as forebay, tempering and discharge channels and recirculation conduits.

All the above is incorporated in a general site plan for the purpose of showing the approximate location of the main site features. This plan is constantly updated and used as a background for discussions during Site Layout Task Group meetings. It forms a base for more detailed design drawings.

Similar to the civil engineering portion of a project manager's work will be the mechanical and electrical engineering portion of the work resulting in basic flow diagrams and equipment layouts.

A manual is then produced with the initial statement of work to be performed to meet the project objective.

This manual is the first issue of a series of more detailed and comprehensive design manuals.

Project Requirements

In addition to the general statement of work which lists the features of the electric generating station, a manual of project requirements is issued by the project group. The requirements have been set mainly by task force meetings, by correspondence and sine qua non by corporate and functional departments. They are formal documents numbered by a subject index based on a work breakdown structure, issued to all contributors to the project work (for examples see Tables 3-01 and 3-02). Project Requirements are descriptions of basic decisions reached and design parameters to be observed. The object is to foster communication and induce comments on decisions reached.

Table 3-01
Project requirement - example 1

Index: 01370	Subject: Technical Specifications	
Requirement: Elevations used in Engineering Specifications and Drawings pertaining to the Generating Station are to be based on the official Geodetic Survey datum.		
References: Corporate Policy	memo of 19..-04-23 by _____	
Description: Issued for engineering and construction purpose.		
Revision: #0	Date: 19..-05-15	Signed: ....

Table 3-02
Project requirement - example 2

Index: 23200	Subject: Water Intake
Requirement: A 6.5 meter diameter concrete lined tunnel 550 meter long will be constructed to carry cooling and service water for the first four units plus tempering water for all units. There will be a provision for interconnecting forebays.	
Rev.#1: The future tunnel shall be designed to ensure that the head loss in the intake system for all units will not exceed 0.30 meters.	
References: Minutes of task force meeting 19..-10-31	
Description: Revised to include max. head loss.	
Revision: #1	Date:	Signed: ....

These requirements are not necessarily static; they may be revised when condi-tions change. Constant additions to the manual take place. This gives interested and contributing groups the broad concept and guiding principles on which to base their work.

The project requirements are obtained by screening several alternatives of the work to be performed. In some cases historical or exploration data will be used. For new schemes, method studies may have to be made to optimize project and design requirements. For example, some design alternatives for the cooling water intake are

  ♦ underground tunnel (secure installation)
  ♦ open cut into the rock (high maintenance)
  ♦ conduit on ground surface (risky)

After evaluating various alternatives and design schemes, the following decision was reached at a project coordination meeting:

Build a water passages model for $60 000, scale 1:36. Show the intake conduit, forebay, pump well, mixing water and discharge channels. Use this model to determine a satisfactory mixing scheme, investigate hydrau-lic phenomena in the forebay and conduit under various load conditions.

Below is an excerpt of a typical coordination meeting (names and other irrelevant data have been changed or omitted)

ABC PROJECT DEPARTMENT   April 18, 19..
FILING MEMORANDUM   File: XY\ABC-23000

ABC GENERATING STATION
Cooling Water Scheme
Minutes of Meeting - April 14, 19..

The principle of a by-pass arrangement for limiting temperature rise of discharge water
was discussed.
Those present were
        Messrs. ....   Two representatives from Operations Div.
                              Six representatives from Design Departments
                              Three representatives from the Project Team
        Thermal pollution requirements for the generating site dictate a maximum circulat-
ing water temperature rise through the condenser of 10°C at nominal full load.
Two alternatives were studied by the design departments:
Alternative #1:
A 12 000 square meter twin shell axial condenser with center outlet boxes; circulating
water flow = 1 100 000 Liter per minute with a temperature rise of 8°C at nominal full
load.
Alternative #2:
(a) 15 000 square meter single shell axial condenser, circulating water flow =662 000 Liter per minute with a temperature rise of 14°C at nominal full load.
(b) A 473 000 Liter per minute mixing flow per unit through open shell between forebay
and discharge channel to mix with the flow of the single shell axial condenser effluent of
(a), resulting in a circulating water temperature rise of 8°C between inlet to the forebay
and discharge to the lake.
According to a cost breakdown study by Design (attached to the minutes) the mixing
water scheme is *more economical*. It also mixes air into the oxygen reduced condenser
discharge water.
Decision reached:
Alternative #2 was adopted. Condenser specification can be prepared on that basis.
Studies shall proceed on the tempering water system.
Action to be taken:
Intake study by Civil Design shall compare 8°C and 10°C temperature rise, as it affects
the size of the intake. Mechanical Design group to proceed with system design on the
same basis. Model studies to be performed by the Hydraulic Department to determine the
most effective mixing scheme.

With the use of model results and computer simulation, the installation of an underground tunnel was considered the best alternative and became a project requirement.

The Project Team

TEAMWORK

Project coordination meetings play a proactive, dynamic role in collecting pertinent information. They bring functional managers, designers and outside experts together to trade and verify information, thereby contributing to the planning process.

The best intentions of the project team can be frustrated if the functions and the purpose within a corporation are not clearly defined and fully supported by line management. The project mission must not be compromised when dealing with other project stakeholders. It is important to identify where joint work though open personal contact will give maximum payoff for the project.

Some may argue that too much time, effort and money was spent at the preliminary stage. We will learn later when we discuss control systems that the "cost prevention" influence is highest at the concept and early design stages. Therefore, increased effort during the earlier stages of the life cycle will improve the chance of project success.

By "increased effort" is meant the gathering and communicating of information needed for the preliminary plan, *not* necessarily increasing design details. Furthermore, the project team is not just *gathering* information, but is an *active* group that collects, exchanges and verifies information, preparing a process-oriented plan that includes not only time, cost and configuration, but also methods, logistics and site operations.

Before we put the first spate into the ground or start detailed designs on which specifications for major purchases are based, uncertainties are very high for a long duration or prototype project. Statistically based decision-making models do not seem to work too well on those projects. They are best used where the future can reasonably be projected from past data.

Application for Approval

The work order taken out to do preliminary engineering work had the purpose to gather enough information for the corporation to decide on the final release of funds. The preliminary engineering is sometimes called "final feasibility study" or "second phase project evaluation".

Not all projects will survive this phase. The corporation may decide to interrupt the effort and the expense on unfavorable projects by terminating them early. This is the advantage of the systematic approach of project cost management.

In our example, there was a budget with a stated deadline to do the preliminary engineering. The objective was to produce a time and cost estimate with a calculated probable accuracy (range estimates). This was done on time with the publication of a formal report requesting the release of funds for project execution (final design, acquisition and construction).

The decision to approve this project plan and release the funds is also based on alternatives such as "make or buy", i.e. import power and delay construction etc.

The planning process continues at an accelerated pace after approval.

3.1.2 Types of Projects

AACE International defines a project as

"an endeavor with a specific objective to be met within the prescribed time and dollar limitations and which has been assigned for definition or execution."

PMI defines a project as

"any undertaking with a defined starting point and defined objectives by which completion is identified. In practice most projects depend on finite or limited resources by which the objectives are to be accomplished."

There is a huge range of projects in regard to purpose, size and execution. When you build your own garage, you are the owner, looking after financing, material purchasing, supervising the construction or even building the facility yourself. You may have a contractor come in to put the roof on. But this is a typical small project, where the planning and the design are completed before construction starts. On large projects, especially the prototype, design and construction proceed in parallel. Obviously, the risks involved vary with the type and duration of projects. Generally, there are three major types of projects:

1. Heavy construction
2. Building construction
3. Miscellaneous projects

Heavy construction projects require heavy and/or highly specialized equipment and usually have large uncertainties. If a pipe has to be laid on the ocean bottom at a depth never attempted before, the owner will probably not be able to obtain a fixed price from a contractor to do the work. The owner will have to assume most of the risk. The following are examples of heavy construction projects:

+ Dam erection, canal work
+ Hydroelectric development,
+ Fossil and nuclear generating plants,
+ Highway and railway work, bridges,
+ Airports and runways,
+ Tunnels and shafts, dredging,
+ Ports, wharves, jetties, breakwaters,
+ Pipelines and related pumping stations,
+ Missile installations, other military work,

All of the above projects contain uncertainties of various dimensions. They are generally contracted on a cost-plus or some form of unit-price basis.

Building construction projects are usually planned and to a large extent designed and specified before construction starts. The owner will usually ask for a fixed price to be quoted by the contractor. Some larger specialized buildings are done on a cost-plus or lease-back contract. They include:

+ High-rise offices, apartments
+ Hotels, theaters, schools
+ Shopping centers
+ Residential buildings
+ Hospitals
+ Post offices etc.

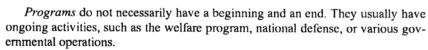

Miscellaneous projects are non-construction projects that fit the definition of a typical project such as the development of a new product (R & D project), the installation of an inventory control system (Information services project), or the production of a new airplane or automobile (manufacturing project).

Those projects are supported by various types of industries and institutions, some of which are

+ Governments
+ Pharmaceutical Companies
+ Financial Institutions
+ Defense Industry
+ Educational Organizations
+ Hospitals
+ Telecommunication Industry etc.

Programs do not necessarily have a beginning and an end. They usually have ongoing activities, such as the welfare program, national defense, or various governmental operations.

Productivity and cost control is also applicable to ongoing work of a defined scope which often includes a series of different programs. The NASA space program is an example of such work.

3.1.3 The Project Organization

Human Resources must be organized in an appropriate way to perform the tasks in a timely manner and to meet a unique objective.

When analyzing the above statement, we may reach the following conclusion:

Unique objective: Every project is unique and distinct from other projects. There is a very large range of projects, each with a different purpose, size and definition. Because of this uniqueness, project organizations will also have to be formed and maintained in a unique way. A residential building project will probably be designed and controlled by the Head Office functional groups. There may or may not be a project coordinator monitoring and reporting to a line manager. On larger, more complex projects a project manager is appointed with delegated authority. This type of project organization will control its own resources. Therefore, an organization formed to execute a project can range from a truly functional form to a separate and distinct project team.

Perform the tasks timely: The tasks that need to be done at the various stages during the life cycle of a project are not routine. The project organization is "time dependent"; it has a life cycle of its own. The organization grows, peeks, and closes down when the tasks are accomplished.

Organized in an appropriate way: There is no such thing as a standard project organization that can be rigidly applied. To be most effective, it must be organized to match the project manager's leadership style. Who is in a better position to assemble the team than the manager himself? There certainly will be conflict with line management when assigning staff to the project team. However, the functional departments are well advised to transfer capable individuals to current projects.

Much has been written about project organizations and the relationship with their parent organization, for example, *Project Management, Harold Kerzner, Van Nostrand Reinhold, 1989, Third Edition. Project Management, Jack Meredith & Samuel Mantel, John Wiley & Sons, 1989, Second Edition, etc.*

This text will only give an overview of the *type* of organizations, not the cultural, social-economic and psychological issues of organizations dealt with in the texts mentioned above.

Corporate enterprises have strategic business objectives and goals for commercial operations that determine the structure of their organization. This includes repetitive functions such as sales, billings, inventories and other financial requirements.

Corporate management is concerned with information and decisions on the overall planning and direction of a total enterprise. This is a typical owner or *functional* organization.

If the owner's organization is in business to build production plants, the organization may look like shown in Figure 3-06.

If the owner does *not* have the expertise to design and build a facility, a building task group or project control group can be formed. This group will hire an architect/engineer to do designing, estimating and scheduling, and will award the construction contract. It will approve designs, progress payments, change orders and contractors' claims (Figure 3-07).

It can be seen that the traditional organization is a hierarchy type structure. It shows levels of authority and an upward reporting flow. This is a convenient way to depict the *skeleton* of a specific organization.

It does not show non-formal communication and personnel interactions that are necessary for the proper functioning of a group of people. It is, however, a very effective *planning entity*.

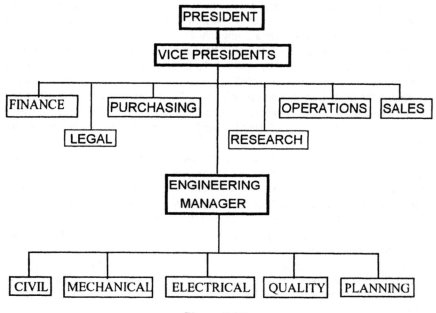

Figure 3-06
Owner's organization.

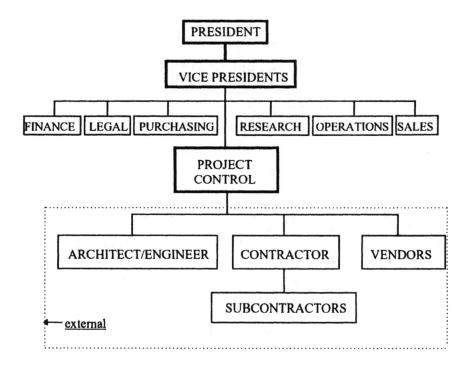

Figure 3-07
Project control group.

The Architect/Engineer organization is highly specialized in the engineering disciplines such as Civil, Structural, Geotechnical, Hydraulic, Mechanical, Electrical, Instrumentation & Control, in order to do the overall project design. In addition, specialists such as environmentalists, economists, financial analysts, legal advisors, may also be part of the organization. The series ISO 9000 standards will give the A/E organization an incentive to add a quality assurance function.

A large contractor with many projects will have typical head office functions under a General Manager, serving all projects. Such functions are bidding, marketing, equipment maintenance, shop operations, estimating, engineering and technical supervision, and business management.

A Construction Manager is appointed for a specific project, heading a site organization such as shown in Figure 3-08.

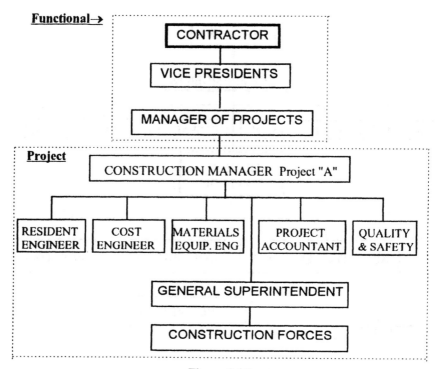

Figure 3-08
Contractor organization.

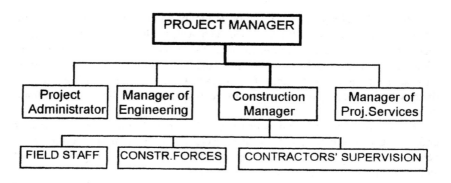

Figure 3-09
Task force organization.

Table 3-03

Project task force breakdown

Project Administrator:	Manager of Engineering:
Records Keeper	Plant Design Engineer
Typing Pool	Electrical Engineer
Contract Administrator	Instrumentation Engineer
Procurement Manager	Structural Engineer
	Civil Engineer
	Specification Engineer
	Chief Draftsman
Construction Manager:	Manager of Project Services:
Security & Personnel Officer	Scheduling Engineer
General Superintendent	Estimating Engineer
Resident Engineer	Cost Control Engineer
Planning and Control Engineer	Spec. & Materials Engineer
Project Accountant	
Construction. Materials Engineer	
Safety Superintendent	

If the owner is in business to design, build ("make" instead of "buy"), and operate construction facilities such as electricity generating stations or process plants, a *pure* project task force, headed by a project manager, may look like shown in Figure 3-09. This organization can be further sub-divided (Table 3-03).

A project starts with a few people, then expands, needs more and more people until it reaches a peak and then releases staff and ends again with very few people. This results in a typical "bell curve" type of manpower loading.

Obviously, this is not the most efficient way to operate.

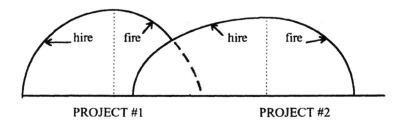

PROJECT #1   PROJECT #2

Figure 3-10

"Hire/Fire" organization.

A more effective organization is based on a *matrix* concept. The corporate divisional organization is somewhat duplicated by the project organization. The project draws on corporate resources to staff its organization with technical and service personnel. There is a special Project Team relationship with the parent organization (Figure 3-11).

When many projects are being build, the design & construction forces will move to other projects that are in earlier stages of completion. The advantage is that all project personnel are dedicated to each individual project and yet are still employees of the corporate structure.

Upon completion of their assignment, employees return to the corporate structure. This preserves continuity, adds knowledge and experience to individuals and contributes to a relatively steady employment situation (Figure 3-12).

Certain support functions are maintained by the corporate office. This depends on the size and type of project. Procurement for major equipment, personnel hiring and training, inspection services are a few of those functions.

A matrix organization faces great challenges. It is one of the most complex organizational forms. The functional departments provide staff for the projects' requirements. They also control design and costing standards and in addition may do major purchases in behalf of projects. Merit reviews and promotions are the responsibility of functional managers.

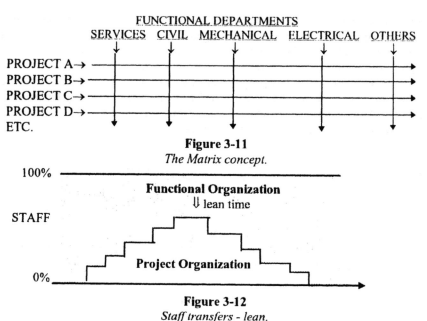

Figure 3-11
The Matrix concept.

Figure 3-12
Staff transfers - lean.

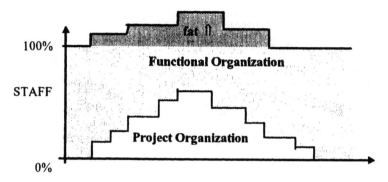

Figure 3-13
Staff transfers - fat.

It is extremely important that authority and responsibility of the managers for the two organizations be well defined and understood. There is a danger, that the "release" of staff to the project organization reduces the functional organization to such an extent that functional managers make up for their "losses" by creating an excessive amount of positions, thereby driving up the overhead (Figure 3-13).

If the owner is in business to design, build ("make" instead of "buy"), and operate construction facilities such as electricity generating stations or process plants, a *pure* project task force, headed by a project manager, may look like shown in Figure 3-09. This organization can be further sub-divided (Table 3-03).

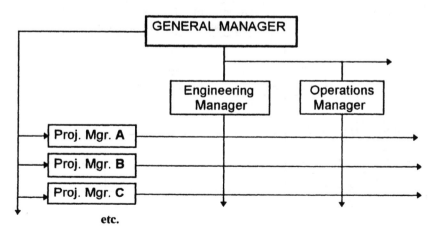

Figure 3-14
Pure matrix organization.

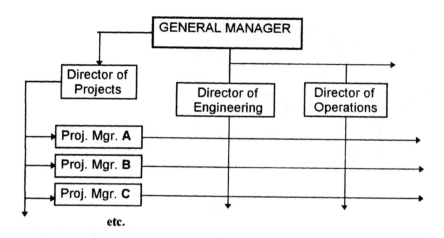

Figure 3-15
Director of projects organization.

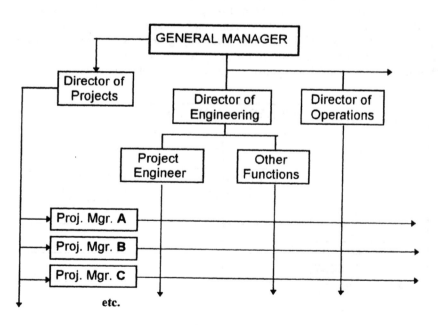

Figure 3-16
Project engineering organization.

A matrix organization may be considered to have three basic forms:

1) The pure matrix structure, whereby an engineering functional manager as well as the project manager report directly to the General Manager of the corporation. This is applicable where there are only a few small projects (Figure 3-14).

2) When a corporation has many large projects, the organization may expand to include a Director of Projects. This will relieve the General Manager of the work overload of having to deal directly with each project (Figure 3-15).

3) By placing project engineering into a project office, the project engineer is looking after all technical aspects, whereas the project manager looks after time and cost management. Communication will have to be very strong between the project engineer and the project manager (Figure 3-16).

In summary we can say that there is a wide variety of organizational forms, structured to meet the objective of the project, the needs of the team members, and the goals of the company.

A project organization is therefore not rigidly structured but a "living" entity responding sensitively to imposed changes and new predictions. This is especially true for long duration projects where the environment initially conceived is very often different from the environment in which the project is completed.

Q #3.01: The Research Department of a corporation is being scaled down in size. A brilliant development engineer at a high salary level is being rewarded for her services by appointing her as project engineer for a construction project. Is this selection justified?

A #3.01: Probably not. Just because she is available and a technical specialist who progressed up through the ranks is not justification for a transfer into a management position which needs leadership, talent to motivate others and the skills to control time, cost, performance and knowledge of contractual requirements.

Corporations may use a "dual-ladder" system. This allows line specialists to be transferred without loss in pay to a "consultant" position, assisting the project manager with specific assignments. It is essentially a familiarization and training position. After exposure to her new environment, the incumbent may prove to be a talented candidate for a future management position.

3.2 THE WORK BREAKDOWN STRUCTURE

W. Garner in his book "Uncertainty and Structure as a Psychological Concept" wrote:

"....the search for structure is inherent in behavior, people in any situation will search for meaningful relations within the variables of a situation, and if no

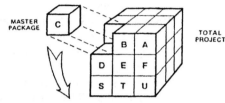

such relations can be perceived, considerable discomfort occurs..."

This search for structure has given us a simple tool called a

Work Breakdown Structure (WBS).

The WBS will help us plan, organize, and control any project, large or small. The WBS is simply a systematic approach to subdividing an effort into its components and/or end products. This approach allows the project to be broken down into *manageable portions*. A complex project is now fragmented into its component parts. Inversely, the parts are summarized into the project's objective.

The more complex a project is in scope and the larger it is, the greater the demand for a well-structured data collection and control system. For example, without the use of a Work Breakdown Structure, the first network planning techniques which were invented in the mid-sixties, this could not have been done (22).

3.2.1 The Work Package Concept

To organize the scope of the work better, several systems, physical areas, activities, organization units, contracts, etc. may be grouped together to form manageable portions of the project work. It is a hierarchical breakdown, starting from the top down.

With the project subdivided into manageable tasks, everybody will know their responsibility, what results are expected, and how much each task would cost. We recognize three simple rules in designing a WBS:

Rule #1:

> **All work packages or tasks at a given level shall be
> comparable in terms of completion time and cost.**

Figure 3-17 is a simple example with only a two-level breakdown.

The completion time for all motor parts at level two is the completion time for the motor at level one. The related costs must add up in a similar manner.

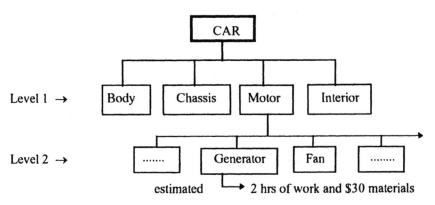

Figure 3-17
Work packages - rule #1.

Overhead cost may be a separate package at level one otherwise it must be distributed over each package.

Rule #2:

Work packages must have a definable output and a specific product that must be generated for the task to be complete.

These are the deliverables when building a car. *No overlap is allowed!* The belts that drive the fan and the generator can only be included in either package, not in both. Also the motor mount is either in the motor package **or** in the chassis package (Figure 3-18).

Rule #3:

Every work package must have a definable beginning and end. Those who are responsible for work packages must have the work properly scheduled.

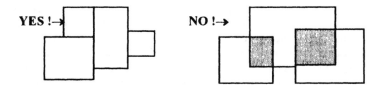

Figure 3-18
Work packages - rule #2.

A huge project like a nuclear power station (14) may be broken down into 20 major chunks of work, such as

- Reactor-Boiler,
- Turbine Generator,
- Electric Power Systems,
- Instrumentation and Control, etc.

Each of those big chunks in turn may have about 16 sub-chunks such as Spent-fuel-bay Cooling System and Boiler-feed System and Motor Control Equipment etc. Again, each of the sub-chunks may have a further breakdown if too big to be managed by one control engineer.

At specific nuclear power projects each of those chunks were called *work packages*. This is the generic term for the hierarchy of the work breakdown structure. The work packages used an "alphameric" code. The big chunks were identified alphabetically as the first character in a field of three, e.g.

- Master Package C   = Reactor Building and Fuelling Duct
- Forecast Package C13 = Reactor Building, Civil Work

How we divide the total effort and how we name the work packages is a matter of preference. This specific example is used only to demonstrate how work packages can be designed and numbered.

While planning a large project we need to determine which work will be performed by contract and which will be done "in-house". A contract will usually be a separate package which, in case of a lump sum contract, needs no further breakdown.

The components of a WBS need not to be of similar size, but each must be a unit that can be measured and controlled by the organizational hierarchy. Those work packages and the organization structure are interrelated. A *responsibility code* is added to each forecast package. This validates management coordination and control in meeting the objectives (schedule, cost flow).

Each package contains specific cost elements that are identified by a "Subject Classification Index" (SCI). In this example, it is a list of 5-digit numbers.

The WBS is analogous to a pyramid, triangular in shape. During the early planning stages (concept) the larger packages are defined first. They are based on project requirements. As the project is specified in more detail, adjustments to existing package descriptions and the creation of lower level packages become necessary and must be done promptly. When all contracts have been awarded and the remaining work has been identified, the final breakdown is achieved.

Planning adjusts packages from top to bottom. During the execution of a project, costs are collected in each package through a matching cost accounting system. Those costs are summarized from the bottom up (Figure 3-19). Several lower level work packages summarize into the next higher level, culminating in total project cost.

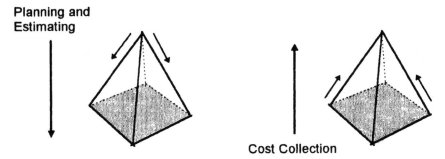

Planning and Estimating

Cost Collection

Figure 3-19
The WBS pyramid.

One work package is distinct from another. It is essential that a full description of *inclusions and exclusions* be provided to avoid overlaps (see rule #2 above).

Each work package description should contain

- ◆ A work definition
- ◆ A definitive schedule
- ◆ An estimated cost (budget)
- ◆ A cost element identity (coding)
- ◆ Single organizational identity (for accountability)
- ◆ Means of determining work progress
- ◆ Cost account traceability

Responsibility Assignment

The package concept can only be successful if a hierarchy of responsibilities are assigned to each of the packages. Each package is managed by one person in the organization. This individual then becomes a mini-project-manager and assumes responsibility for the design, procurement, estimates, schedules, reporting of progress and variance explanation. Of course, on smaller projects there may only be a few packages. The breakdown may not be by system, but by discipline or activity. Medium size projects may have a combination of those.

The reason for having one person in charge of each package is, that the cost estimate, the schedule, procurement, etc. can be integrated and an audit trail established through a hierarchy of reports. The responsibility coding for the various packages is attached to the package number.

The Cost Streams

Certain categories of costs flow through various packages and should be identified in such a way that they can be summarized separately.

This creates a cost matrix, whereby labor cost, engineering cost, material cost etc. contained in each work package are called cost streams (Figure 3-20).

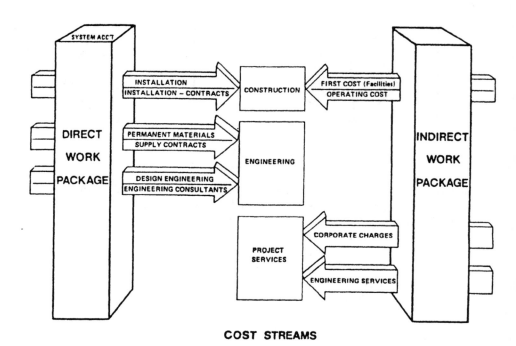

COST STREAMS

Figure 3-20
Cost streams.

3.2.2 The Code of Accounts

The *cost code structure* is directly related to the funding, estimating, accounting and reporting procedures established for each project. The *code of accounts* should be developed for each project within the cost structure, recognizing the uniqueness of a project. It is used to distribute budgets, collect workhours, accrue costs and establish forecasts for each element in home office and field office. The code of accounts should be standardized as much as possible.

Codes can be represented in numerals, in alpha form or a combination of both. The amount of digits depends on the size of the job and the level of control needed. With the event of better computer memory, a more detailed breakdown should not pose a problem if handled properly.

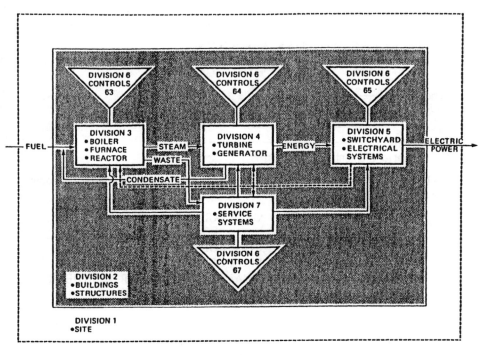

ELEMENTS OF A GENERATING STATION

Figure 3-21
Work breakdown elements.

Figure 3-21 is a typical code structure that depicts seven master packages under which all direct work is estimated and collected. We can see that the high level coding follows a certain logic. It shows the INPUT (fuel), the PROCESSES that are housed in buildings and structures (shaded area), and the OUTPUT (electric power), all within the site area and off-site (access). In addition to the above, division 8 deals with construction, division 9 with engineering and support services. Division 0 (zero) was chosen for corporate overheads and administration.

Depending on the size of the job, codes may have as many as 20 or even more digits. The lowest sub-division, which is defined in scope and content, will have a number of accounting codes. Those codes will be used to *collect* cost. They are sometimes called

System Accounts.

Those accounts are *mutually exclusive* to a package. As an *example*, lets have a look at the five digit SCI code of accounts. This system was used for the construction of nuclear stations in Canada and Korea.

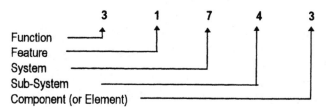

Each five digit number now becomes a potential cost account. Each cost account has a headline, for *example:*

  30000 REACTOR, BOILER & AUXILIARIES
     3**1**000 REACTOR
          31**7**00 REACTIVITY CONTROL UNITS
              317**4**0 IN-CORE FLUX DETECTORS
                   3174**3** VERTICAL FLUX DETECTOR ASSEMBLIES

another *example:*

  20000 BUILDINGS AND STRUCTURES
     2**1**000 REACTOR BUILDING AND FUELLING DUCT
     21**1**00 REACTOR BUILDING
          211**2**0 SPECIAL CONTAINMENT PROVISIONS
          2112**5** METALLIC LINERS

It should be noted, that the first example shows a clear system breakdown as designed, the second example is a breakdown by area on the project plan. Other breakdowns may address tasks instead of systems or areas, or a combination of those.

Figure 3-22 depicts an example how a system classification index can be used for the identification of documents. The System Classification Index number 31700 is the WBS code from the previous page, i.e. REACTIVITY CONTROL UNITS. NK30 identifies a particular nuclear generating station. The document is the first (serial #1) size A1 (ISO standard) revised drawing.

The assignment of numbers to the index is performed by specific disciplines; although, no set of numbers, or even a particular number, is for the exclusive use of one organizational group. Certain engineering departments are the custodians and advisors on the subject matter or system content.

It is because of their expertise in a particular field that the disciplines involved formulate a logical breakdown of the categories and organize the knowledge of the subject. This includes a content description of inclusions and exclusions to avoid duplications.

With a comprehensive coding, we have now identified all the elements of a specific project. Estimates will be based on those elements which in turn will be the approved budget. Whatever work is done on those elements, the incurred cost will be identified by the cost accountant and collected in the proper slot. This structure will form the nucleus for assigning responsibility, for estimating, cost collection and control. When there are several similar projects underway, a *master index* manual should be maintained from which each project draws the codes it needs to publish its own *project index* manual.

Any codes missing in the master index, but needed by the project, would result in the updating of *both* manuals (Figure 3-23).

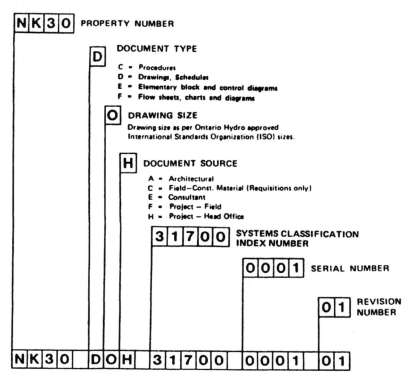

Figure 3-22
Document identification.

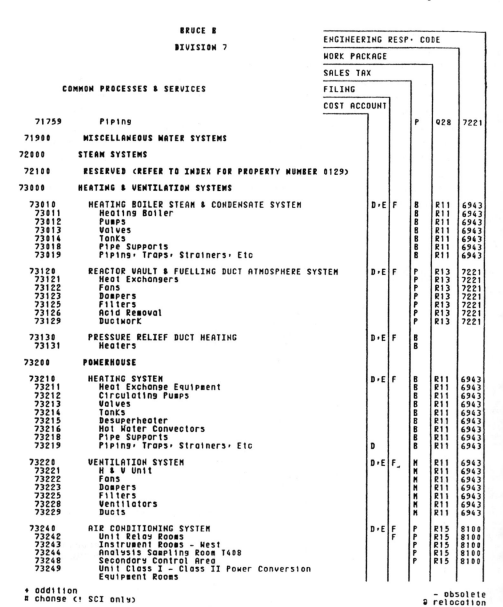

Figure 3-23
System subject index.

A proper numbering system will permit the costs of like or related items to be collected together, and analyzed in relation to budgets so that trends in any part can be examined. Expenditures are then sorted and summarized into related groups and categories.

The above examples have a five-digit breakdown. This was developed and used by Ontario Hydro in Canada. The index is supplemented by various codes:

Cost Account:
It indicates under which number the costs are collected and summarized. For example, the powerhouse heating system uses 73210 as a cost account. The cost include valves, tanks, piping, etc. The code "D" denotes d̲irect costs such as equipment and material. "E" denotes e̲ngineering cost such as design and procurement. Engineering time is charged to this account number.

Filing:
This indicates under which index number documents are labeled and filed.

Sales Tax:
It identifies the taxing authorities, i.e.

♦   F = Federal taxes apply
♦   P = Provincial (State) taxes apply
♦   M = Municipal taxes are levied
♦   E = Tax exempt

Work Package:
Heating Boiler Steam and Condensate Return System and the Powerhouse H&V System are one WBS package called R11.

Responsibility Code:
This identifies the individual or the group responsible and accountable for a specific work package.

Q #3.02: What is the definition of a WBS and what is its purpose?

A #3.02: The WBS is a product-, service- or task-oriented "tree" of project components which sub-divides an effort into its hierarchical parts and defines the scope of the project. Each descending level of the WBS represents an increasingly detailed definition of its element.

Its coding structure is conducive to cost collection and control because specific authority and responsibility can be assigned. Furthermore, it lends itself to a standard data base that is useful to future projects.

3.3 SCHEDULING

Scheduling is *time planning* for a project. It estimates the time frame in which various tasks need to be performed in order to meet a pre-determined completion date. It is sometimes called *time estimating*.

3.3.1 Project Duration

The completion date for a project is usually determined by the process of project evaluation in the concept stages of the life cycle. With all known facts at that point in time including historical data from other similar projects, an *optimum* project schedule is established.

What is an optimum duration?

If we *speed up* a schedule (reduce the duration) from an assumed "normal" position, premiums must be paid on overtime, shift differentials, early delivery charges and possibly penalties for lower productivity. There may also be extras to contracts. All those are additional *direct* costs.

If we *stretch* the schedule (increase the duration), we pay more for interests, overheads, site services, because they have to be paid even when less direct costs are incurred. Those are additional *indirect* costs. The cost for *shortening* a project duration is called

(a) speedup cost

The cost for *stretching* the schedule is called

(b) delay cost

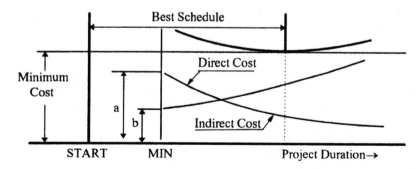

Figure 3-24
Optimum schedule duration.

Plotting estimated direct and indirect costs separately on a $/time graph will give us a minimum cost point on the resultant curve. Its projection is the optimum schedule duration (Figure 3-24).

If the owner should decide whether it is more important to have the facility produce ahead of time by spending extra capital (speedup cost), or if a speeding up of the work may be the proper course of action.

3.3.2 The Bar Chart

The easiest method of scheduling is the use of a calendar and a list of activities. In some cases, especially for small projects, it is not necessary to use sophisticated scheduling methods. Bar charts are easy to read. They are quite sufficient to plan smaller projects. They can also be used for high level or "milestone" reporting for larger projects. See Table 3-04 for an example.

Many cottages of similar design may have been built by the same contractor. There is no need for a sophisticated scheduling system. A "Gantt Chart" will give the builder the planning tool needed (Figure 3-25). There are 55 working days from March 15 to May 31.

Table 3-04
Building a cottage

	Activity	From	To	Duration
A	Clear the site	Mar 15	Mar 19	5 days
B	Excavate	Mar 22	Mar 23	2 days
C	Foundation	Mar 24	Mar 29	4 days
D	Septic tank	Mar 25	Mar 29	3 days
E	Backfill	Mar 30	Mar 31	2 days
F	Framing	Apr 01	Apr 15	10 days
G	Roof	Apr 16	Apr 20	3 days
H	Dock	Apr 08	Apr 16	6 days
I	Windows, Doors	Apr 20	Apr 26	5 days
J	Cladding	Apr.26	Apr 28	3 days
K	Plumbing	Apr 26	May 05	8 days
L	Wiring	Apr 28	May 05	6 days
M	Insulation, Drywall	May 06	May 14	7 days
N	Interior Finishing	May 10	May 28	15 days
O	Landscaping	May 20	May 26	5 days

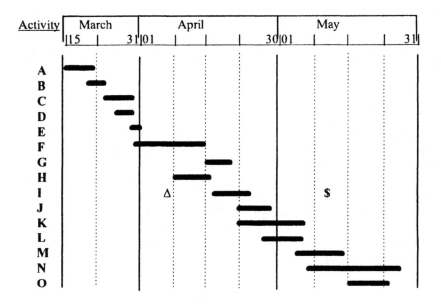

Figure 3-25
Bar chart.

A broken line is sometimes used to interconnect activities. This introduces a certain "logic" to the schedule. In our example, the foundation work cannot start until the excavation is finished:

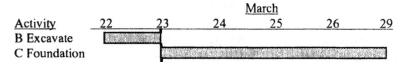

In addition to activity durations, "point-in-time" events such as delivery dates and scheduled payments can also be added. The symbol for deliveries is usually a capital delta (Δ) and payments are denoted by the dollar sign ($).

The planner will have to consider trade classifications, particularly when there is an overlap of activities. The framing and the dock building both need carpenters. If those two activities are scheduled at the same time, more carpenters will be needed.

The time scale for a bar chart may be based on consecutive working days or on the calendar. Durations may or may not include a time allowance.

For example, if the starting date for activity K-Plumbing is April 26 and the finish date is May 5 with a duration of 8 working days, it may be possible to do the work in six days. The bar chart will show this time allowance as

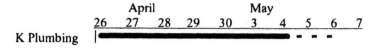

So far, we discussed the bar chart for *planning* a job. It is considerably more difficult to *control* the project by just using the bar chart. If we are falling behind within one activity, how would that affect the end date?

The calendar/activity bar chart is not sophisticated enough to analyze connections among activities that are critical in reaching the overall objective of the project, i.e. finishing on time and within budget.

3.3.3 Process Scheduling

Manufacturing projects often combine individual processes within a production cycle. The assembly line is a typical example. There are repetitive activities down the line to fit the various components. When a plastic toy is assembled, each component is scheduled to arrive at a specific time. The time required to assemble all parts determine the scheduled *cycle time* for the product. The activities on the assembly line are usually *sequential*. Representing this on a bar chart may look like this:

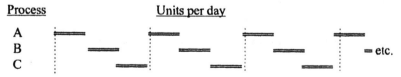

The above three processes are repeated until x units are produced every day. If process B is delayed, it will affect the outcome of the whole line. In other words, the operation is controlled by the weakest link. The target production rate is achieved by regulating the rate of travel of the assembly line. Assuming a manual operation, the process time can be adjusted by increasing or decreasing the number of operators serving the line.

We thus have three major components for the production model:

♦ production rates
♦ cycle times
♦ balanced resources

What has this to do with construction? Some construction projects have repetitive, sequential activities such as laying a pipeline, slip or climb type structures, highways or shore protection.

Consider laying an underground pipe line. There is a certain rhythm for each length of pipe; clearing, excavating, sand bedding, pipe laying, welding, wrapping, backfilling and seeding. Those activities would be sequential (sub-cycles) on a bar chart. Each will have a duration (production time). We would also look at the deployment of resources (men and machines) that need to be balanced.

If we plan the job so that the "start-to-start" and "finish- to-finish" events for each activity is the operation sub-cycle time which we assume to be two days, we may have a schedule that looks as shown in Figure 3-26.

Activity or Task	days	work force
Clearing	2	2
Excavating	2	4
Sand Bedding	2	3
Pipe laying	2	6
Welding	2	3
Wrapping	2	2
Backfilling	2	3
Seeding	2	1
Total	16	24

Work on pipe #6 can start as soon as the clearing crew is finished for the specific section of pipe #5, and work on pipe #7 can start as soon as the clearing crew becomes available. In order to minimize idle time,

the balancing of resources is of utmost importance.

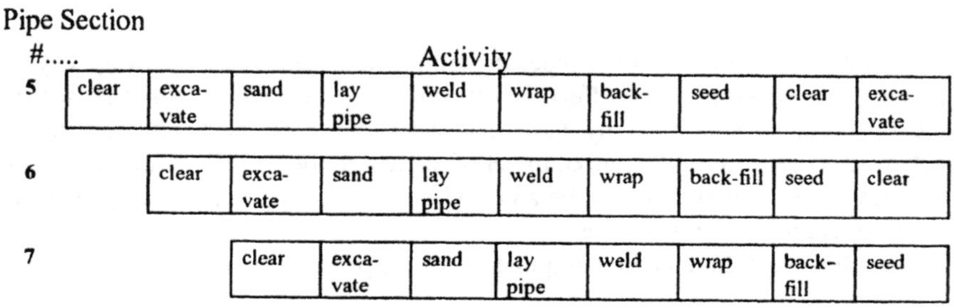

Pipe Section
#.....   Activity

5	clear	exca-vate	sand	lay pipe	weld	wrap	back-fill	seed	clear	exca-vate

6		clear	exca-vate	sand	lay pipe	weld	wrap	back-fill	seed	clear

7			clear	exca-vate	sand	lay pipe	weld	wrap	back-fill	seed

Figure 3-26
Operation sub-cycle time.

Days →

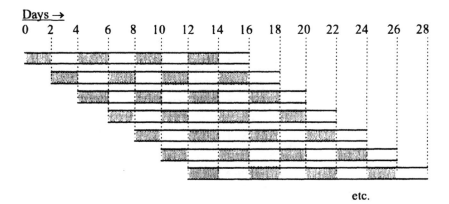

etc.

Figure 3-27
Two-day sub-cycle.

With a two day sub-cycle, five pipe sections are completed in 24 days. Obviously, with the start time of pipe #6 scheduled to be two days after pipe #5 has started, the finish time will also be scheduled to be two days later (Figure 3-27).

Two days is therefore the operation sub-cycle time.

Each pipe section is scheduled to take 16 working days. The work flow for this repetitive construction will have a production goal of one pipe section every two days.

Assuming the terrain changes from a wooded area to an open landscape. The clearing operation may then take less than one day. Easier access will also impact on other operations. In this case we may want to re-balance the activities to three days by lumping together clearing and excavating.

This will affect the other activities too because we need to maintain the resource balance. It is possible that we may end up with less activity categories but more resources and a shorter cycle time.

For example:

Activity or Task	days	work force
Clearing and excavating	3	6
Bedding and pipe laying	3	12
Welding and wrapping	3	6
Backfilling and seeding	3	5
Total	12	29

With a three day sub-cycle, five pipe sections are completed in 24 days (Figure 3-28). The production has not changed. We will later learn that efficiency increases to a certain extent for repetitive work (learning curve).

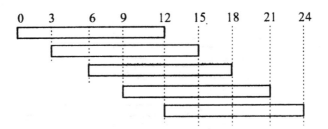

Figure 3-28
Three day sub-cycle.

We can also allow for a limited idle time or we can put the first activity *outside* the production cycle, i.e. the clearing would move ahead faster than the laying of the pipe.

There are many possibilities to plan a smooth production. The costing aspects are foremost when planning this kind of job.

3.3.4 Network Planning

Two major systems use networks in planning and scheduling (Figure 3-29).

♦   The Critical Path Method (CPM)
♦   The Program Evaluation & Review Technique (PERT)

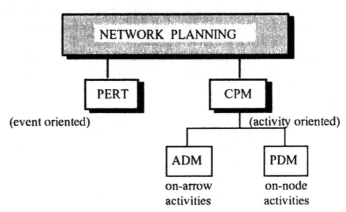

Figure 3-29
Network planning methods.

CPM requires a single, "normal" time input, whereby PERT is more sophisticated as it requires a probability input (optimistic, most probable, pessimistic). The expected (mean) time is then statistically calculated. PERT is best applied for projects where activities are well defined but with great uncertainties in time requirements (development work).

The critical path network in form of an arrow diagram (ADM arrow diagram) is easiest to understand by the practitioner. Flow paths can readily be identified and followed in relation to project activities.

Those not directly involved in the construction of flow diagrams seem to prefer the precedence diagram method (PDM). It has an apparent simpler "look" to it. PDM is best applied in repetitive work or where many concurrent activities take place. Whether the arrow diagram, which was developed first, or the PDM precedence diagram is used depends mainly on individual preference rather than effectiveness or efficiency of any one system.

Below is a short description of the

Arrow Diagram Method

The ADM network begins with an *event* called the *origin*, which may represent the start of the project (forward method) or the completion of the project (backward method). Events are usually depicted by a circle.

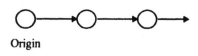

Origin

In some cases a square is drawn for origins or special major events. Lines, representing *activities* are drawn, starting from the origin. These lines terminate with an arrow → and a circle representing the next event. Those circles are also called *nodes* on the arrow diagram. The event at the start of an activity is a *predecessor* event, and that at the conclusion of an activity is a *successor* event.

Time flows from a predecessor event to successor event(s) as indicated by the arrow. In the forward method, time flows are from left to right.

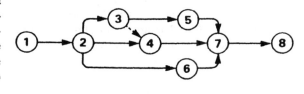

As each activity is added to the network, its *relationship* to other activities is determined. Those relationships are either *sequential* or *concurrent*. Performing activities in sequence requires that the start of the successor activity depends upon *completion* of the predecessor activity. Activities performed concurrently must be *independent* of one an-other.

No!

Independent activities may have a common predecessor event or a common successor event, *but not both.* This prohibition makes an arrow diagram easier to analyze.

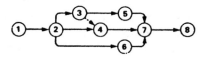

When there are parallel activities which have logical links between events, it is depicted by a *dummy*. A dummy does not represent work; it cannot have resources assigned to it (even though in some cases it may have a duration) A dummy is shown with a dotted line and serves only as a connector between events. Its sole purpose is to indicate logic.

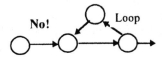

Loops are not allowed. Loops are caused by tying activities back up-stream. Loops that are not detected during the layout of a network can cause a computer to "lock up".

Hammocks are not part of the actual network. They are spanning several more detailed activities. They are used to simplify a network for those with no need for details. It may be said that they give a summary of elapsed time between events.

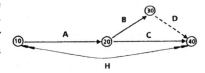

To plan a project is to construct a model of future activities.

We start with the initial event and add the activities we had previously listed within the WBS. Seldom do we have to start from scratch. The information that needs to be assembled to draw a network diagram forces us to "think" through the project in detail. We must be able to "visualize" interdependencies of activities, called the *logic* of the diagram. A "must precede" exercise shows the logic of diagram construction (Figure 3-30 and Table 3-04).

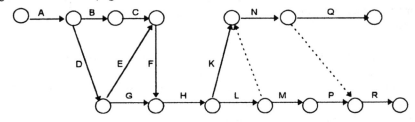

Figure 3-30
"Must precede" exercise.

Table 3-04
"Must precede" exercise

A	must precede	B & D		K	must precede	N
B	must precede	C		L	must precede	M & N
C & E	must precede	F		M	must precede	P
D	must precede	E & G		N	must precede	Q & R
F & G	must precede	H		P	must precede	R
H	must precede	K & L				

There are usually records available from similar jobs that tell us what the crew size and trade mix are and how long it may take to install specific equipment.

There are also commercial catalogues available such as the MEANS "Construction Costs" (Figure 3-31).

152 | Plumbing Fixtures

		152 400	Pumps	CREW	DAILY OUTPUT	MAN-HOURS	UNIT	MAT.	LABOR	EQUIP.	TOTAL	TOTAL INCL O&P	
465	0660	8.455 L/s, 745.7 W, 51mm discharge (2")		Q-1	2.80	5.714	Ea.	765	135		900	1,050	465
	0680	76mm discharge (3")			2.70	5.926		800	140		940	1,100	
	0700	264.95 L PE tank, 0.757 L/s, 372.85 W, 51mm discharge (2")			2.60	6.154		610	145		755	890	
	0710	76mm discharge (3")			2.40	6.667		665	160		825	970	
	0730	5.49 L/s, 521.99 W, 51mm discharge (2")			2.50	6.400		785	150		935	1,100	
	0740	76mm discharge (3")			2.30	6.957		830	165		995	1,175	
	0760	8.455 L/s, 745.7 W, 51mm discharge (2")			2.20	7.273		865	175		1,040	1,225	
	0770	76mm discharge (3")			2	8		905	190		1,095	1,275	
480	0010	PUMPS, SUBMERSIBLE Sump											480
	7000	Sump pump, 3 048mm head, automatic											
	7100	Bronze, 1.388 L/s, 186.425 W, 32mm discharge		1 Plum	6	1.333	Ea.	230	35		265	305	
	7140	4.291 L/s, 372.85 W, 32mm or 38mm discharge			5	1.600		375	42		417	475	
	7160	5.931 L/s, 372.85 W, 32mm or 38mm discharge			5	1.600		515	42		557	630	
	7180	6.626 L/s, 372.85 W, 51mm or 76mm discharge			4	2		490	53		543	620	
	7500	Cast iron, 1.451 L/s, 186.425 W, 32mm discharge			6	1.333		109	35		144	175	
	7540	2.209 L/s, 248.567 W, 32mm discharge			6	1.333		123	35		158	190	
	7560	4.291 L/s, 372.85 W, 32mm or 38mm discharge			5	1.600		252	42		294	340	
490	0010	PUMPS, WELL Water system, with pressure control											490
	0020												
	1000	Deep well, multi-stage jet, 158.97 L tank											
	1040	33 528mm lift, 18.144 kg dischg., 0.316 L/s, 559.275 W		1 Plum	.80	10	Ea.	445	265		710	890	
	2000	Shallow well, reciprocating, 94.625 L tank											
	2040	0.316 L/s, 248.567 W		1 Plum	2	4	Ea.	390	105		495	590	
	3000	Shallow well, single stage jet, 158.97 L tank,											
	3040	4 572mm lift, 18.144 kg discharge, 1.01 L/s, 559.275 W		1 Plum	2	4	Ea.	380	105		485	580	

Figure 3-31
Crew size and trade mix.

(The "Means Building Construction Cost Data" is published by the R.S.Means Company, Inc., 100 Construction Plaza, P.O. Box 800, Kingston, MA 02364. It is mainly used for estimating purposes).

The first line tells us that it needs a plumber and apprentice (Q-1) or two workers approx. six hours to install a pump of the size indicated (8.5 Liters per second and 0.75 kW electric power).

Restraints

Any condition which affects the start of an activity and determines the order in which each activity is placed in the diagram is called a *technological restraint*. A restraint can be *physical* (formwork *must* precede the pouring of concrete), it can be a *resource constraint* (labor, equipment and material must be available), or it can be a *policy constraint* (order of work dictated by higher authority). Physical restraints are prerequisites, they cannot be changed. Resource assignments are more flexible. If the initial plan does not fit the intended time period, some policy constraints could be removed and/or resources modified.

Coding

There is a convention in coding logic diagrams. This is especially useful when networks are computerized. Only simple diagrams can be calculated without a computer's help.

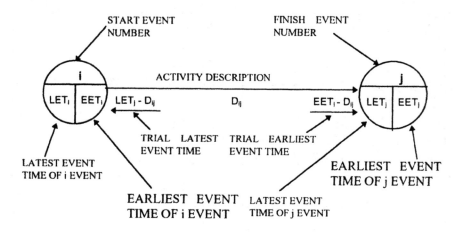

Figure 3-32
Logic diagram coding.

Each node is divided into three sections (see Figure 3-31). The predecessor node receives a START EVENT number **i** and the successor node receives a *finish event* number **j**.

(Why .i. and .j. ? Pioneers of the early scheduling age may still remember the time when logic diagrams were produced in the drawing office, in pencil, because of the many changes in dependencies and logic. The .i. was easily changed into a .j. by adding a short line below the i. Changing the .j. into an i, the lower part of the j was simply rubbed out.)

Activities have a *description* and a time *duration* code **Dij**. There are two more codes applied to the arrow diagram:
> Earliest Event Time (EET) and Latest Event Time (LET).

Earliest Event Time

The start event of a network cannot take place earlier than a specified time. The condition for earliest time can apply to any activity in the network. After time estimates are made for each activity of the network, a *path* of events is traced out by starting with the origin event *forward*, then proceeding to its successor event, then to another successor event etc., until the terminal event is reached.

The *longest path* from the origin to the chosen event is the earliest event time (EET) for the event. This applies to both, the earliest start time and the earliest finish time. The AACE terminology, based on a vast majority of surveyed membership, is using the following terminology:

Early Start (ES)   The earliest time an activity can be started.
Early Finish (EF) The earliest time an activity can be finished.

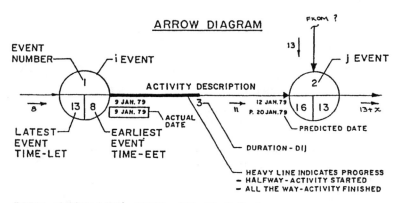

TOTAL FLOAT = LET j − EET i − Dij = 16 - 8 - 3 = 5

Figure 3-33
Arrow diagram coding.

Latest Event Time

The latest time at which an activity can be started or finished without causing a delay in the terminal event is the Latest Event Time (LET) for the origin event. This is determined by tracing a *backward* pass from the terminal event to the event in question. The *longest pass* of all backward passes determines the latest start time and the latest finish time. The accepted terminology is:

Late Start (LS) The latest time an activity can start without disrupting the schedule.

Late Finish (LF) The latest time an activity is allowed to be finished without disrupting the schedule.

Figure 3-32 is an example of the markings of an arrow diagram. For simplification, only one activity is shown with the predecessor and successor nodes numbered.

For a better understanding of the terminology above, networks can be drawn on a time-scale basis (i.e. days) showing (ES), (EF), (LS) and (LF) at a glance (Figure 3-33). This is not practical for computerization because it needs to be redrawn every time a revision is made. A standard computerized network revises *numbers* only when changes in duration are made.

Let us now construct a simple arrow diagram (Figure 3-35), using the precedence Table 3-05 as input. But first we should be aware of references to activities and events. This includes the flow direction when tracing different paths.

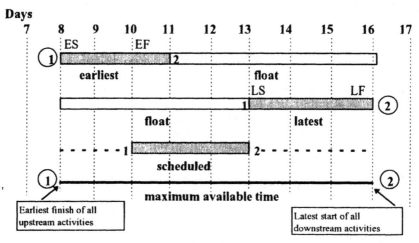

Figure 3-34
Time scaled events.

Table 3-05

Arrow diagram construction.

Event #	Activity	Preceding	Duration	EETi
5 - 6	F = Framing	-	10 days	13
6 - 8	G = Roof	F	3 days	
6 - 10	I = Windows	F	5 days	
10 - 12	J = Cladding	I	3 days	
8 - 12	K = Plumbing	G	8 days	
8 - 11	L = Wiring	G	6 days	
12 - 13	M = Drywall	J & K & L	7 days	

When this is done, we enter the *trial* early event times for each forward path. This can be indicated by an arrow → or a square box (Figure 3-36).

Path F – I – J – M adds up to 25 days
Path F – G – K – M adds up to 28 days
Path F – G – L – M adds up to 26 days

This means, the earliest time G or I can start is after the earliest finish of all upstream activities plus activity F. Therefore, event 6 will show an EET of 13 + 10 = 23 days. Similarly events 8 and 10 will have an EET of 13 + 13 = 26 and 13 + 15 = 28 respectively. Event 11 has an EET of 13 + 19 = 32.

The EET for event 12 has three path durations, 18, 21, 19.

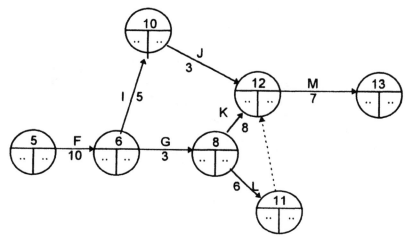

Figure 3-35
ADM example - 1.

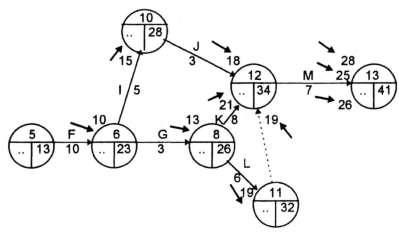

Figure 3-36
ADM example - 2.

The longest of those is path F - G - K with an EET = 13 + 21 or EET = 34, which will be entered into event 12. The final EET is 13 + 19 = 41.

Because event 13 is shown as the final event, the latest event time LET is also 41 days. We can now do the backward trial paths (Figure 3-37):

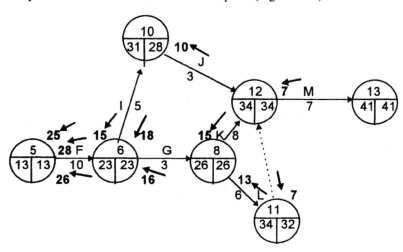

Figure 3-37
ADM example - 3.

Path M-J - I -F	reads 41–7	= 34–3	= 31–5	= 26–10		= 16
Path M-K-G-F	reads 41–7	= 34–8	= 26–3	= 23–10		= 13
Path M- L-G-F	reads 41–7	= 34–0	= 34–6	= 28–3	= 25–10	= 15

Again, events 8 & 6 & 5 have more than one backward path total. The highest number applies, i.e.

$LET_8 = 41 - 15 = 26$; $LET_6 = 41 - 18 = 23$; $LET_5 = 41 - 28 = 13$

To calculate the *total float* for each activity we use the formula

$$TF_{ij} = LET_j - EET_i - D_{ij}$$

For activity J, for example, the total float is $34 - 28 - 3 = 3$ days.

The Critical Path

The *critical* path is the *longest* network path. Its length determines the minimum time required for completion of the entire project. *Critical events* are those events that lie on the critical path. If a project has to be expedited, the accomplishment of at least one of the critical events must be expedited. Any delay in a critical event delays the completion of the project. The following steps are used to find the critical path:

1. Draw the network according to the rules.
2. Make a forward pass of the whole network, calculating the earliest event time (EET) for each event with zero for the start event (ES). Set the latest event time (LET) equal to the earliest event time (EET) for the terminal event (LF).
3. Make a backward pass starting with the terminal event's latest event time (LET), subtracting activity durations along all passes. The path that needs to start the earliest is the longest path and the calculated start time is the latest event time (LET) of the start event.
4. Highlight this critical path.

It should be noted, that there must be *at least one* critical path in any network. All activities along the critical path have zero float. This path goes through all nodes having LET = EET.

The arrow diagram shown above (Figure 3-37) gives a step by step procedure on the calculation required to determine a critical path, which is marked \\. Forward and backward passes are marked with arrows near pertinent nodes at the end or the beginning of an activity:

This time scale diagram is shown to better visualize how durations can "float" between two events. This is the essence of planning.

> **Planning defines *what* is to be done and *how long* it takes. As soon as we decide *when* we do it, or to put *dates* on activities, it becomes a *schedule*.**

Basic Computations

A calculation chart compliments the arrow diagram. Those calculations are initially done by hand, but there is a programmable algorithm that allows the planner to try out various schemes in order to obtain the optimum schedule duration and to allow for imposed constraints. First of all, we need to number the events such that the predecessor event i is always a *lower* number than the successor event j (i < j). This also holds true for multiple successor events. Therefore, for the terminal event **n**

$$EET_j = \max_i \Sigma(EET_i + D_{ij}) \qquad E_1 \leq j \leq E_n$$

This is the forward path which we manually calculated in our previous example, i.e.

$$E_0 = 5 \; ; \; E_n = 13$$

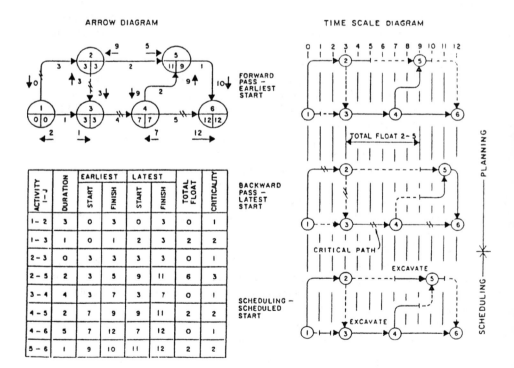

ACTIVITY I–J	DURATION	EARLIEST		LATEST		TOTAL FLOAT	CRITICALITY
		START	FINISH	START	FINISH		
1– 2	3	0	3	0	3	0	1
1– 3	1	0	1	2	3	2	2
2– 3	0	3	3	3	3	0	1
2– 5	2	3	5	9	11	6	3
3– 4	4	3	7	3	7	0	1
4– 5	2	7	9	9	11	2	2
4– 6	5	7	12	7	12	0	1
5– 6	1	9	10	11	12	2	2

Figure 3-38
ADM critical path.

The algorithm can be presented as follows:

$\Rightarrow$ $\underline{i = 5}$ For i = 5 and j = 6, EET6 = EET5 + D5,6 = 13 + 10 = **23**

$\Rightarrow$ $\underline{i = 6}$ i = 6 has two successor events, j = 8 and j = 10. The program selects the lower event first. EET8 = EET6 + D6,8 = 23 + 3 = 26, then the next higher number EET10 = EET6 + D6,10 = 23 + 5 = **28**

$\Rightarrow$ $\underline{i = 8}$ i = 8 has also two successor events, j=11 and j=12. The program selects the lower event number first EET11 = EET8 + D8, 11 = 26 + 6 = 32, then the next higher event number EET12 = EET8 + D8, 12 = 26 + 8 = **34**

$\Rightarrow$ $\underline{i = 10}$ i = 10 has only one successor event, j = 12 EET12 = EET10 + D10, 12 = 28 + 3 = **31**. But EET12 was calculated previously to be 34. The program selects the larger of the two: $\Rightarrow$ EET12 = **34**

$\Rightarrow$ $\underline{i = 11}$ i = 11 has only one successor event, j = 12. EET12 = EET11 + D11,12 = 32 + 0 = 32 (the dummy has no duration). EET12 was calculated previously to be 34. The program selects the larger of the two. $\Rightarrow$ EET12 = **34**

$\Rightarrow$ $\underline{i = 12}$ i = 12 has the final successor event j = 13. EET13 = EET12 + D12,13 = 34 + 7 = **41**

A similar algorithm is applied for the backward path, i.e.:

$$LET_i = \min_j (LET_j - D_{ij}) \; ; \;\; E_0 \leq i \leq E_n - 1$$

This differs from the forward path such that the smaller of the LET_i is selected from multi-predecessor events. Starting off by using the "zero-slack" convention, that is

$$EET13 = LET13 = \textbf{41}$$

and starting with $(E_n - 1)$

$\underline{i = 12}$	LET12	=	LET13 – D12,13 =	41 – 7 = 34
$\underline{i = 11}$	LET11	=	LET12 – D11,12 =	34 – 0 = 34
$\underline{i = 10}$	LET10	=	LET12 – D10,12 =	34 – 3 = 31
$\underline{i = 8}$	LET8	=	LET11 – D8,11 =	34 – 6 = 28
$\underline{i = 8}$	LET8	=	LET12 – D8,12 =	34 – 8 = 26

LET8 was previously calculated to be 28 days. The *minimum* time applies for the backward path, LET8 = **26**

$\underline{i = 6}$	LET6	=	LET 8 – D6,8 =	26 - 3 = **23**
				(minimum LET6)
$\underline{i = 6}$	LET6	=	LET10 – D6,10 =	31 – 5 = 26
$\underline{i = 5}$	LET5	=	LET 6 – D5,6 =	23 – 10 = 13

Having estimated the earliest and latest event times (sometimes called occurrence times), we have to identify if our daily activities are start or finish times.

So far our "time" intervals have been in days. It does not matter, what time units we use, as long as we are consistent throughout the network. We must state, if our calculations are based on " end-of-times " or "start-of-times".

If EET5 = 13 is based on start-of-day, then LET13 must also be based on start-of-day 41, i.e.

| 12 | 13 | 14 | .... | 39 | 40 | 41 |

If the network calculations are base on end-of-day, EET5 and LET13 are shown thus

| 13 | 14 | 15 | .... | 40 | 41 | 42 |

It is a matter of preference which system we use during the planning stage, as long as we are consistent and do not change in mid-stream. The system to use is probably covered by company procedures. Purchased software may also dictate which system to use.

Schedule Constraints

Assuming the network we have just completed (events 5 to 13) is *one* package of the work breakdown structure. There may be ten or twenty or more packages comprising the total project.

We now have to consider how they relate to each other and how key events (milestones) affect this combination of networks. In other words, we are putting constraints on the total project network.

Not all packages are sequential. Some packages have more activities than others and may be parallel to activities in other packages.

Our example may have the following packages:

Package		Activity	Duration
P-1	Substructure	A to E	13 days
P-2	Superstructure	F to M	28 days
P-3	Dock	H	6 days
P-4	Interior Finishing	N	15 days
P-5	Landscaping	O	5 days

At first let us have a look at package P-1. The package P-1 can be depicted in the logic diagram shown in Figure 3-39.

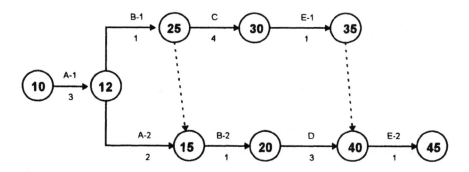

Figure 3-39
Example - package P-1.

	Activity	Duration
A-1	Clear access and building area	3 days
A-2	Clear dock area	2 days
B-1	Building area excavation	1 day
B-2	Septic tank and bed excavation	1 day
C	Form and pour foundation	4 days
D	Install septic tank	3 days
E-1	Backfill building	1 day
E-2	Backfill septic tank	1 day

The logic requires us to split some activities. This creates eight events for package P-1. But package P-2 starts with the event number 5. It is good practice to number events initially in blocks of 10 or more (similar to what every programmer does). All event numbers for future packages will now be multiplied by 10. This will allow more detailed planning later on.

The first event of package P-2 (now event 50) has an EET and LET of 13 days. Since the total duration of package P-1 is only 11 days (earliest finish) we must assume that event 50 is an event of "special significance." It could be a time constraint such as the earliest delivery of framing lumber (resource constraint) or a fixed milestone date (policy constraint).

To combine the packages into a total project network we must choose an event that interfaces the packages naturally with each other. This is called the *interface event*.

> **The Interface Event is an event of special significance in a network restraining or restrained by the activities in another network.**

If we "hammock" packages P-1 and P-2, the event 35 becomes a natural interface event (Figure 3-40).

All other packages can now be interfaced in a similar manner

Schedule Dates

Milestones are expressed in calendar dates, for *example*:
   deliver the turbine on 1996-11-15 (ISO standard date format) or
   complete the roof on 1994-10-10 etc.

Duration must therefore be converted into calendar dates.

Calendar dates are in the computer memory where allowance is made for non-working weekends and statutory holidays. Here, the start-of-day is the initial date and the end-of-day is the completion date.

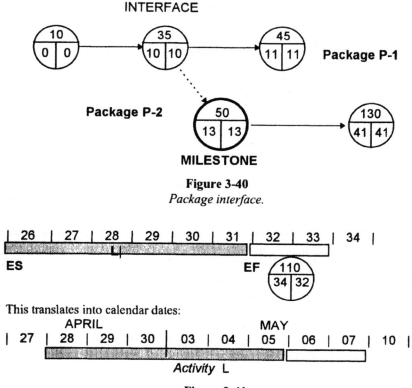

Figure 3-40
Package interface.

Figure 3-41
Package interface - scheduled.

If we are using activity L of package P-2 with a *duration* of six days as an example, we can depict it as shown in Figure 3-41.

The above early start ES = 26 and early finish EF = 32 for activity L will have a scheduled start date of April 28 and a finish date of May 5. This schedule allows for a two-day weekend (May 1 and May 2). Since we have float, the schedule dates can be shifted within any six days inside the period from April 28 to May 7. Those dates are now entered into an activity schedule with the following headings:

Events	System	WBS	D(i-j)	ES	EF	LS	LF	TF	Crit.
8-11	Wiring	P-2	6	Apr.28	May 06	Apr.30	May 07	2	3

... etc.

This activity schedule will list all events and can also be summarized by work package and by milestones.

Uncertain Completion Times (PERT)

The completion dates for some types of projects are very uncertain. This includes prototype and research projects. High risk projects such as dam building and tunneling also fall into this category.

A network scheduling and control model, dealing with those uncertainties, was devised in 1958 by the US Navy Special Projects Office. It is called PERT for Program Evaluation and Review Technique. It was designed to reflect some of the uncertainties encountered when estimating activity duration (please review 1.3.2 Probability Distribution). The creators of PERT developed a probability distribution with a three-point input.

The time estimate includes an optimistic (o), a pessimistic (p), and a most likely (m) assessment of an activity duration.

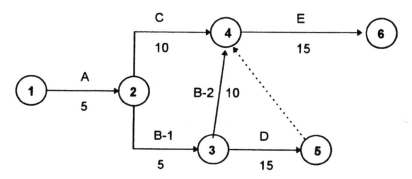

Figure 3-42.
PERT foundation example - 1.

Those three-point estimates can be represented by a beta (ß) probability distribution. We can now convert uncertainties into an *expected* activity time (E_t) and a *variance* of completion time (σ^2), which is a measure of dispersion (refer to Chapter 1):

$$E_t = (o + 4m + p)/6; \quad \sigma^2 = [(p - o)/6]^2$$

Example: A site has to be prepared before the excavation takes place and the concrete foundation can be poured. The logic diagram, showing the most likely duration may look somewhat like shown on Figure 3-42. The earliest completion time is 40 days.

To calculate the expected activity time and the variance, we need to list the 3-point estimates (Table 3-06).

The arrow diagram is then revised to show the *expected* time for each activity: The earliest expected completion time is now 39.5 days, and above is a list of the earliest and latest times (Table 3-07).

The critical path is A, B-1, D, E. Any delay along this path will delay the completion time. We know, that we cannot be certain about the 39.5-day completion time.

Because of this uncertainty, we will now have to apply the PERT probability distribution.

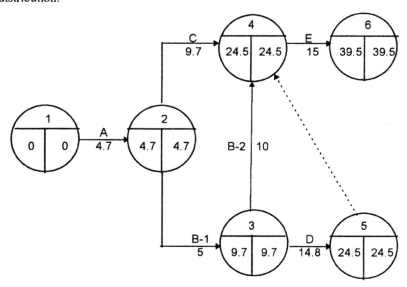

Figure 3-43
PERT foundation example - 2.

Table 3-06

PERT - foundation example 1

D(i-j)	Activity	Duration in days			E(t)	Variance
		m	o	p		
1 - 2	A Prepare site	5	2	6	4.7	44%
2 - 3	B-1 $^1/_3$ Excavation	5	4	6	5	11%
2 - 4	C Build fence	10	7	11	9.7	44%
3 - 4	B-2 $^2/_3$ Excavation	10	8	12	10	44%
3 - 5	D Formwork	15	12	17	14.8	69%
4 - 6	E Pour concrete	15	15	15	15	0%

Table 3-07

PERT foundation - example 2

Activity	ES	LS	EF	LF	Critical	Float
A	0	0	4.7	4.7	yes	0
B-1	4.7	4.7	9.7	9.7	yes	0
B-2	9.7	14.5	19.7	24.5	no	4.8
C	4.7	14.8	14.4	24.5	no	10.1
D	9.7	9.7	24.5	24.5	yes	0
E	24.5	24.5	39.5	39.5	yes	0

Assuming that the *probability* for taking more time than expected is unique for each activity, in other words we are treating each activity independently, then the *variance* of the *overall* completion time along the critical path **C** is the sum of individual activities **i** along the critical path:

$$\text{Variance } (\sigma_C)^2 = \sum (\sigma_i)^2$$

or in our example
$$\text{Variance} = 0.44 + 0.11 + 0.44 + 0.44 + 0.69 = 2.12$$
and the standard deviation is

$$\sigma = \sqrt{2.12} = 1.46$$

We discussed in Section 1.3.2 that the distribution formed from independent samples of a sequence of numbers will approximate a normal distribution. Our project completion time distribution will be normal with a mean $\mu = 39.5$ and a standard deviation $\sigma = 1.46$.

If we accept the risk that the project may exceed "x" days 10% of the time (80% confidence) we must determine the number of standard deviations x is from μ which is

$$Z = \frac{(x - \mu)}{\sigma}$$

Below is an excerpt from a standard normal distribution table:

Z	0.00	0.01	......	0.07	**0.08**	0.09
1.1	.3643	.3665	......	.3790	.3810	.3830
1.2	.3849	.3869	......	.3980	**.3997**	.4015
1.3	.4032	.4049	......	.4147	.4162	.4177

....

It shows us that Z = 1.28 for an 80% confidence level. Therefore,

$$1.28 = \frac{x - 39.5}{1.46} ; \quad \therefore x = 41.4 \text{ days}$$

With all individual activities taken into account, there is a 10% chance that the project will exceed 41.4 days (Figure 3-43).

Of course we could apply different confidence levels, say 90% which would result in

$$1.645 = \frac{x - 39.5}{1.46} ; \quad \therefore x = 41.9 \text{ days}$$

This statistical feature of estimating durations has limited applications. Even though it may not be widely used, the PERT system has the practical advantage that it forces individuals to produce a more realistic estimate. Those who produce single point estimates have a tendency to "bury some fat" with their time durations.

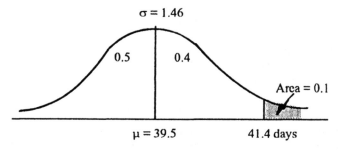

Figure 3-44
PERT foundation example - 3.

Q #3.03: Sketch a network diagram in which the predecessor event has the number 15, the concurrent successors have numbers 20 and 21. Activity A between 15 and 20 has a duration of 6 days and B has 10 days.

A) add activity C with a duration of 8 days in sequence with B. Start of C is dependent on the completion of A. The terminal event number is 22.

B) What is the earliest time at event 22?

A #3.03: a) diagram:

b) 18 days.

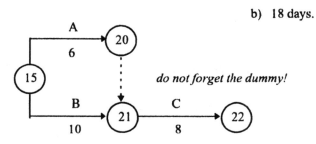

do not forget the dummy!

Q #3.04: a) What is the weakness of the bar chart (Gantt) over the logic diagram?
b) What has to be added to a logic diagram to become a schedule?

A #3.04: a) There are no interdependencies for various activities.
b) Dates.

Q #3.05: Automobiles arrive at a service station for gasoline. Services provided by the station include cleaning the windshield, checking the tires, battery, oil and radiator fluid. Sufficient personnel are available to perform all services simultaneously. The windshield cannot be cleaned while the hood is raised. Customers are charged only for gasoline and oil.

Activity	Duration	Description
A	1	Raise hood
B	2	Check tires
C	5	Add gasoline
D	2	Check radiator
E	1	Check battery
F	2	Check oil
G	3	Compute bill
H	1	Lower hood
I	2	Clean windshield
J	1	Collect payment

a) What are the immediate predecessors?
b) Label the nodes and draw the arrow diagram.
c) What is the shortest service time?

A #3.05: a)

Activity	Duration	Nodes	Predecessors
A	1	1 - 2	--
B	2	1 - 7	--
C	5	1 - 9	--
D	2	2 - 3	A
E	1	2 - 5	A
F	2	2 - 4	A
G	3	9 - 7	F, C
H	1	5 - 6	D, E, F
I	2	6 - 7	H
J	1	7 - 8	B, G, I

b) There are many ways to produce a flow diagram. It may look something like this:

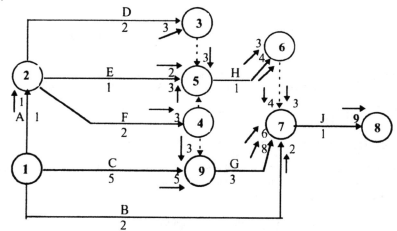

c) For the customer to be served in the shortest possible time, the critical path has a duration of 9 minutes.

Q #3.06: The diagram below has the following activities and durations:

Activity	Description	Duration
A	Paint walls	3 days
B	Paint doors	2 days
C	Install hangers	4 days
D	Install pipe	5 days
E	Install tank	5 days
F	Paint the tank	6 days

a) Which activities are most critical to finish the project on time?
b) Show A, B, E, F on a time scale network with both, EET and LET.
c) Since activities A, B and F are performed by the same trades, how would you schedule your work?

A #3.06:

a) Activities E and F are on the critical path.

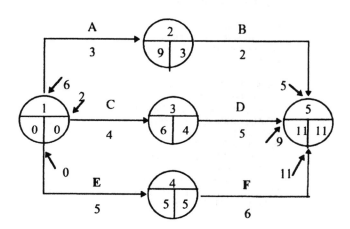

b) Time scale network:

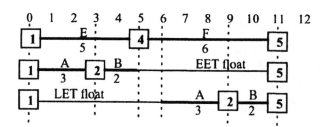

c) Work schedule:

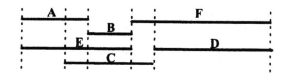

Q #3.07: You are considering the construction of your house. After some consultation, the following steps and times have been estimated to complete the project. Estimate the logic and produce the arrow diagram.

Job	Description	Days
A	EXCAVATE, POUR FOOTINGS	4
B	POUR CONCRETE FOUNDATIONS	2
C	ERECT FRAME AND ROOF	4
D	LAY BRICKWORK	6
E	INSTALL DRAINS	1
F	POUR BASEMENT FLOOR	2
G	INSTALL ROUGH PLUMBING	3
H	INSTALL ROUGH WIRING	2
I	INSTALL AIR-CONDITIONING	4
J	FASTEN PLASTER BOARD AND STRIPS	10
K	LAY FINISHED FLOORING	3
L	INSTALL KITCHEN EQUIPMENT	1
M	INSTALL FINISHED PLUMBING	2
N	FINISH CARPENTRY	3
O	FINISH ROOFING AND FLASHING	2
P	FASTEN GUTTERS AND DOWNSPOUTS	1
Q	LAY STORM DRAINS	1
R	SAND AND VARNISH FLOORS	2
S	PAINT INDOORS AND OUTSIDE	3
T	FINISH ELECTRICAL WORK	1
U	FINISH GRADING	2
V	POUR WALKS AND LANDSCAPE	5

A #3.07:

Job Name	Description	Prede-cessors	Succes-sors	Time (days)
A	EXCAVATE, POUR FOOTINGS	--	B	4
B	POUR CONCRETE FOUNDATIONS	A	E,C,Q	2
C	ERECT FRAME AND ROOF	B	I,H,D	4
D	LAY BRICKWORK	C	O	6
E	INSTALL DRAINS	B	G,F	1
F	POUR BASEMENT FLOOR	E	I	2
G	INSTALL ROUGH PLUMBING	E	J	3
H	INSTALL ROUGH WIRING	C	J	2
I	INSTALL AIR-CONDITIONING	C,F	J	4
J	FASTEN PLASTER BOARD AND STRIPS	G,H,I	K	10
K	LAY FINISHED FLOORING	J	L,M,N	3
L	INSTALL KITCHEN EQUIPMENT	K	S	1
M	INSTALL FINISHED PLUMBING	K	S	2
N	FINISH CARPENTRY	K	R	3
O	FINISH ROOFING AND FLASHING	D	P	2
P	FASTEN GUTTERS AND DOWNSPOUTS	O	U	1
Q	LAY STORM DRAINS	B	U	1
R	SAND AND VARNISH FLOORS	N,S	--	2
S	PAINT INDOORS AND OUTSIDE	L,M	T,R	3
T	FINISH ELECTRICAL WORK	S	--	1
U	FINISH GRADING	P,Q	V	2
V	POUR WALKS AND LANDSCAPE	U	--	5

Arrow diagram:

This arrow diagram is one of many possible layouts. An experienced scheduler may be able to improve on the one shown here. The simpler a diagram looks, the more efficient is the layout.

To trace the critical path, we need to mark the following nodes

1 - 2 - 3 - 4 - 7 - 8 - 9 - 10 - 11 - 12 - 14 - 16 - 18.

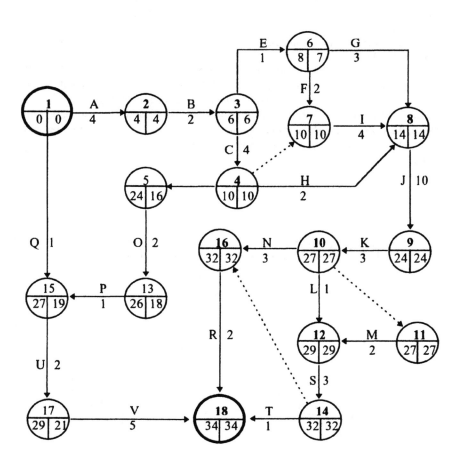

3.3.5 Precedence Diagramming (PDM)

It was not long after the development of the arrow diagram method (1957/58) that an attempt was made in 1964 to review the activity-on-arrow concept. The activity relationships that are shown on an arrow diagram are finish/start relationships. In practice, however, many activities overlap and need to be sub-divided. In our previous example, the formwork for a foundation can only start after five days of excavation has been done (Figure 3-42). We therefore need to either break down excavation into two activities B-1 and B-2 or introduce a *constraint dummy* (Figure 3-45).

Please note, that this constraint dummy has a 5-day duration (which is not the average dummy).

This means that one has to wait five days before activity D can start. Some believe that PDM makes planning more "work item friendly". It is an activity-on-node system. Work items are shown in square boxes (Figure 3-46).

In addition, there are four other types of dependencies, the start-to-start factor (SS_{ij}), the finish-to-finish factor (FF_{ij}), the finish-to-start factor (FS_{ij}) and the start-to-finish factor (SF_{ij}). Those dependencies are shown on the next page below Figure 3-46.

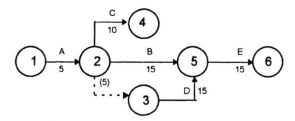

Figure 3-45
ADM constraint dummy.

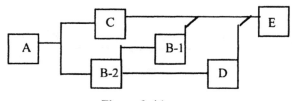

Figure 3-46
PDM demonstrated.

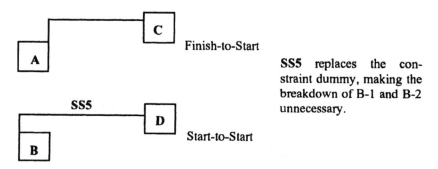

Finish-to-Start

SS5 replaces the constraint dummy, making the breakdown of B-1 and B-2 unnecessary.

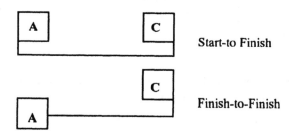

Start-to-Start

Five time units from start of activity B must elapse before activity D can start. The duration of this constraint is called the *lag factor*.

Start-to Finish

Finish-to-Finish

Coding the Node

Even though there is no apparent standard, the coding of the work item is often done as shown in Figure 3-47.

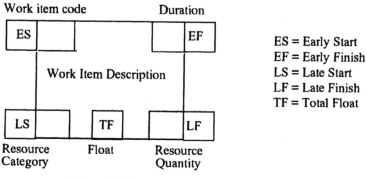

ES = Early Start
EF = Early Finish
LS = Late Start
LF = Late Finish
TF = Total Float

Figure 3-47
Coding the node.

How do we show precedence?

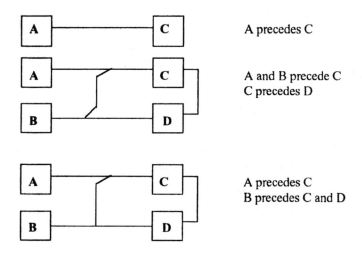

A precedes C

A and B precede C
C precedes D

A precedes C
B precedes C and D

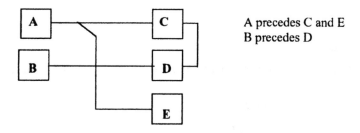

When we have to cross lines, we have to be careful not to interpret precedences wrongly. In general, crossed lines do not connect work items.

A precedes C and E
B precedes D

Crossing lines that connect work items are shown thus:

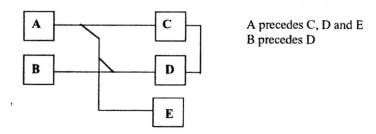

A precedes C, D and E
B precedes D

Lag Factors

To better visualize lag factors, they are graphically compared with a bar chart:

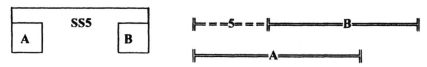

B may start five time units
after start of A.

B may not complete three time
units after completion of A.
Starts of A and B are independent.

B may not start three time units
after completion of A. This is
equivalent to the constraint dummy
used in an arrow diagram.

B cannot be finished five time units
after the start of A. The earliest start of B
may not be affected.

Calculating Early and Late Times

The calculations are not more difficult, but much more involving than those of the arrow diagram. We need to use start and finish times in addition to lag factors and durations. Similar to ADM, forward and backward paths are necessary to calculate earliest and latest times.

By definition,

- ◆ The latest start day is the latest day that a work item can start without affecting the final project duration assuming that it is completed in its expected time and all subsequent work items start as soon as they are able to and are completed in their expected times.
- ◆ The latest finish day is the latest day a work item can finish without affecting the project duration, assuming that all subsequent work items start as soon as they are able to and are completed in their expected time.

Those definitions are not easy to comprehend. Forward and backward paths are best demonstrated graphically (Figures 3-48 to 3-51):

- ◆ The forward path is shown on top of the diagram (earliest time).
- ◆ The backward path is shown on the bottom of the diagrams (latest time).
- ◆ Total float TF = LS – ES applies to all nodes.

Calculation without lags:
forward: ES1+D1=EF1=ES2+D2=EF2=ES3+D3=EF3 ...etc
backward: LF3–D3=LS3=LF2–D2=LS2=LF1–D1=LS1. TF = LS – ES

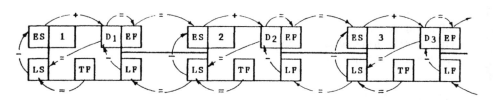

Figure 3-48
PDM without lags

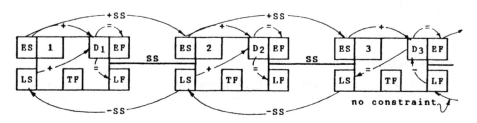

Figure 3-49
PDM with SS Lags

Calculations with SS lags:

forward: ES1+D1=EF1; ES1+SS1,2=ES2+D2=EF2; ES2+SS2,3=ES3+D3=EF3
backward: LF3−D3=LS3−SS3,2=LS2+D2=LF2; LS2−SS2,1=LS1+D1=LF1

Calculations with FF lag:

*forward:*ES$_1$+D1=EF1+FF1,2=EF2−D2=ES2; EF2+FF2,3=EF3−D3=ES3
backward: LF$_3$−D3=LS3; LF3−FF3,2=LF2−D2=LS2; LF2−FF2,1=LF1−D1=LS1

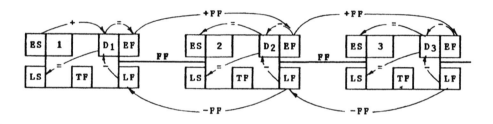

Figure 3-50
PDM with FF lag.

Calculations with FS lag:

forward: ES1+D1=EF1+FS1,2=ES2+D2=EF2+FS2,3=ES3+D3=EF3
backward: LF3−D3=LS3−FS3,2=LF2−D2=LS2−FS2,1=LF1−D1=LS1

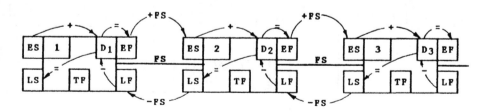

Figure 3-51
PDM with FS lag.

Applying this to our foundation example first shown in Figure 3-42, we obtain the PDM diagram shown in Figure 3-52:

Table 3-08
PDM foundation example

Work Item	Duration	Predecessor	Resources
A Prepare Site	5	-	4 Laborers
B Excavate	15	A	1 Operator, 1 Laborer
C Build Fence	10	A	1 Ironworkers, 2 Laborers
D Formwork	15	SS5, B	3 Carpenters
E Pour Concrete	15	C, B, D	4 Laborers, 1 Finisher

Zero float identifies the critical path.

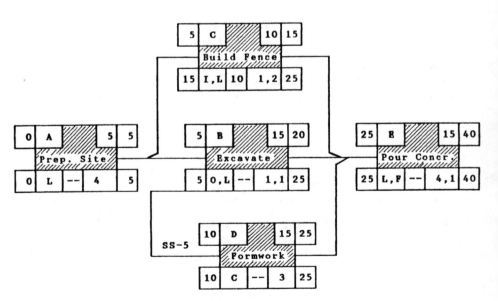

Figure 3-52
Applied sample of PDM.

Similar to the arrow diagram example, the durations must now be converted into dates. If the work starts at June 1, 1994 (1994-06-01), the earliest start dates for B and C are 1994-06-08 and 1994-07-08 respectively when weekends and statutory holidays have been taken into account. The job is finished by 1994-07-28.

3.3.6 Resource Planning

The initial planning is based on an unrestricted supply of resources. We have assumed, that the resources would be available whenever we needed them. We should ask ourselves

♦ In computing start times, when two or more activities are competing for a single resource, which one has priority?

♦ Can we change the sequence of some activities (soft logic) in order to have more flexibility in resource allocation?

If we had to schedule the installation of cable trays in several sections, the main constraint may be the availability of the area. The schedule may look like this:

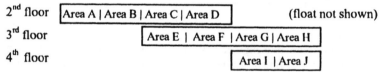

2nd floor [Area A | Area B | Area C | Area D]   (float not shown)

3rd floor [Area E | Area F | Area G | Area H]

4th floor [Area I | Area J]

There is usually no reason to be that specific. The scheduler should leave more flexibility to the installer. It may be better to apply soft logic:

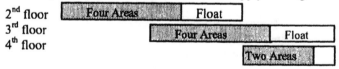

2nd floor [Four Areas | Float]

3rd floor [Four Areas | Float]

4th floor [Two Areas]

Consider the following activities (Table 3-09):

Table 3-09

Resource planning

Activity	Duration	Predecessor	Resource
A Prelim. Design	4	-	1 Designer
B Engineering Review	8	A	3 Engineers
C Production Review	6	A	2 Engineers
D Marketing Review	10	SS5, C	1 Economist
E Detailed Design	4	B, D	2 Designers

The above activities are depicted in the arrow diagram below (Figure 3-53).

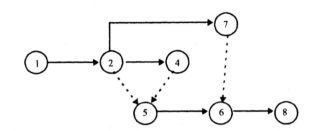

Figure 3 53
ADM resource planning.

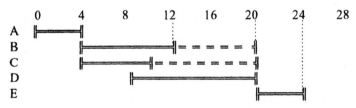

With the above initial plan, two engineers would be required for a period of seven days. Activity B has a late finish of nineteen days. The schedule could call for a start on the earliest finish of activity C.

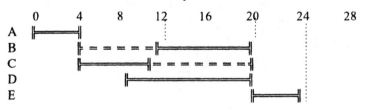

Furthermore, we only need one engineer without affecting the overall schedule. This is called *resource leveling*. Leveling with limited resources will be discussed in the next chapter.

Instead of

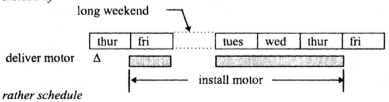

rather schedule

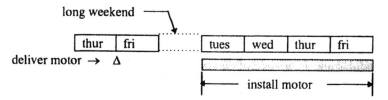

If there is float on a job to be done, we may want to look at the start and finish calendar dates. It is usually advantageous to start a new job after a long weekend for logistic reasons.

Here, we are going into a very detailed plan as produced at the working level. These are often called production level schedules. We will deal with the various schedule levels in more detail later in Chapter 4 - Cost Control.

All phases of the construction project concept, engineering, acquisition, construction, startup will have to be integrated into one project plan. This "master schedule" is the highest level time estimate. It is usually, but not necessarily, done in bar chart format. It shows milestones and durations that affect all areas of responsibility such as engineering, procurement and construction activities.

Depending on the size of a project, the next level could be a control or co-ordination schedule at the master package level of the work breakdown structure.

However we plan and estimate the time, activity durations must constantly be reviewed. Planning and scheduling is not a line function alone, all responsible members of the project team must contribute their expertise to maintain a realistic schedule.

The schedule will then be used as the base to estimate the project's cost flow and final cost in more detail.

YOU WANT IT WHEN??

3.4 CAPITAL COST ESTIMATING

As discussed under 2.1.2, we have two major capital investment periods, the capital expenditure and the capital recovery period. The first is a *fixed capital* investment to finance the development, engineering and construction of the physical facilities. The second is the earning period during plant operation, yielding a return used to recover the capital investment. The capital needed to operate the plant is the *working capital*.

Estimating and scheduling services are performed from the very beginning of the project life cycle until the plant is shut down or decommissioned. This chapter deals mainly with capital cost estimating of fixed capital requirements.

Relationship with Time Estimating

We need to put dollars against the time estimate of the project. Time estimating and cost estimating are very closely related. *We must not separate scheduling from costing!*

Planning Project Execution	$\Rightarrow$ Time Estimating $\Rightarrow$ Schedule Control $\Downarrow$	$\Rightarrow$ Cost Estimating $\Rightarrow$ Cost Control $\Downarrow$
Progress Reporting	Cost/Schedule Integration	

When we perform capital cost estimating we must be conscious of *when* the costs are going to be incurred. Only then can we meet the objective to be on time and within budget.

The reason why many texts on project management consider estimating completely separate from scheduling may have historical reasons. When cathedrals were built in the middle ages, people took their time.

Their only concerns seem to have been proper skills and adequate cheap resources. If it took ten weeks to shape one column, it did not matter, as long as it looked good.

 ne letter in a document may have taken several hours to produce. Even in the early 1950s a draftsman spent about four hours to draw a northpoint in ink on a map. We now use computer icons that do this in seconds.

Cost estimating and budgeting is as old as recorded history. There is plenty of literature on the subject. We will therefore only take a cursory look at capital cost estimating.

3.4.1 Overview of Estimating Processes

When preparing estimates, the estimator is continuously evaluating future project conditions about which only assumptions may be available. The *reliability* of the bottom line estimate depends on the accuracy and quality of the information available and also on the skill and experience of the estimator. There are many diversified *types of estimates*. We will try to summarize them into four broad categories:

The Quickie	also guesstimate, back-of-a-matchbox, while on-the-phone.
The Concept	also preliminary, study, order-of-magnitude, factored etc.
The Budget	also the official, the approval etc.
The Definitive	also final, control, firm, detailed, hard bid, take-off etc.

As we can see, different expressions are used for similar types of estimates. Those who prepare and contribute to cost estimates use various *techniques*. Some of those are

Factoring:	Empirical or "S" Curves
	Ratio Factors
	Parametric Estimating
	Turnover Ratio
	Average Unit Cost
Analytical:	Hirsch & Glazier
	Probabilistic (range) Estimating
Modular:	Average Unit Cost
	Guthrie Modules
	Standard Pattern
	Generic Models
Itemized:	Quantity Take-off
	Unit Rates
	Standard Allowances
	Data Base

Capital costs can be divided into two major *cost streams*:

Direct Cost	Indirect Cost
Permanent Equipment	Construction Facilities
Permanent Materials	Construction Services
Design Engineering*	Work Equipment and Tools
Installation Labor	Engineering Services
Supply Contracts	Corporate Administration*
Consultants*	Interests, Insurance, Taxes

(*) = These items may be categorized differently by some companies.

In addition to the normal cost estimate, the following *adjustment factors* may apply:

> Cost Location
> Escalation
> Contingency
> Productivity Adjustment

3.4.2 Estimating Qualities

Management must be aware that estimates have weaknesses and shortcomings. An assessment of their quality is very desirable. Some of the terms used are (32)

> Credibility
> Accuracy
> Exactness, Precision
> Reliability
> Tolerance
> Materiality
> Validity

Validity

An estimate can only be considered valid if the bases for preparing it are also valid. Assuming a quick estimate is required for a vacuum cleaning system in a power plant (B) to be built four years from now. A similar plant (A) has been built last year at a different location.

For the estimate to be valid, *all* differences between the two plants must be properly taken into account, i.e.

♦ *What are the technical differences?* Plant A has a centralized system, plant B does not. Plant A was smaller. Does plant B need a higher capacity system? Was the proper factor applied?
♦ Has the *correct source* been used for the escalation?
♦ Was a *location factor* applied? Was it valid?
♦ *When and where* will plant B be built? In winter up north? Has the different climate an effect on productivity? Was the proper adjustment made?
♦ Have any *known factors* been left out? This would make the estimate not as valid as it should be.

The purpose of the estimate must be fully understood and its bases documented in order to be judged valid. A quick estimate is based on a different scope definition than a detailed estimate. A valid estimate applies methods that adequately addresses both, scope and bases.

Reliability

"A good estimator produces reliable estimates." "It is the knowledge and experience of the estimator and his or her good performance record that inculates confidence in the estimate." The above statements indicate that the estimates have been produced by a reliable *estimator*.

"Good estimates are produced by consistently identifying, evaluating, using and recording all relevant information available."

This statement indicates that a reliable *estimate* has been produced. Here, the emphasis is on the *estimate*, not the *estimator*.

If the estimate for the vacuum cleaning system mentioned above states

Preliminary Estimate $ 100 000

Range (+50%, – 30%) $ 70 000 to $ 150 000

and nothing else, one has to have very great confidence in the *estimator*, who may or may not be a reliable person.

To judge the reliability of the *estimate*, it is good practice to document specifics, such as data used and their sources. Using the fictitious example above, the estimate may now look as shown in Table 3-10.

Table 3-10

Preliminary estimate

Estimate Date:.................... Produced by:....................................

PLANT A:	Vacuum Cleaning System: Recorded Cost	**$ 50 000**
PLANT B:	Capacity based on own previously constructed similar plant A @ 90% (source: Engineering Department).	$ 45 000
	Capacity factor = 0.7 (source: Cost Engineers' Notebook) $(45\ 000/50\ 000)^{0.7} = 0.9289$ (scale down).	
		$ 41 800
	Escalation rate 6% p.a. for 5 years (source: Statistics Canada) $41\ 800(1.06)^5$.	$ 55 938
	Location factor = 1.08 (source: Cost Engineers' Notebook).	$ 60 413
	Productivity loss = 20% (source: Company Files).	
	Point Estimate	**$ 72 496**
	Contingency method: Monte Carlo Simulation (see calculation printout attached) 38% =	$ 29 000
	Range: 50% , –30% , Confidence: 80% **TOTAL COST COMMITMENT** (rounded) .	**$100 000**

It is very important that risk assessment and contingency be added to the estimate to indicate the degree of reliability. Without it, even if the estimate is absolutely *accurate*, the degree of *reliability* cannot be established. The detailed breakdown will give the reviewer of the estimate the opportunity to check for validity and correctness of the data.

Materiality

The concept of materiality was originally used in accounting. It simply suggests that it is not worthwhile bothering with meticulous calculation of small amounts. The cost of the estimate may exceed the benefit derived from including too much detail. As a rule of thumb, 20% of items chosen in the order of magnitude will encompass 80% of the cost.

Accuracy

When we multiply "two" by "two", only the result "four" is correct. Any other figure such as 4.3 or 3.8 would not be *accurate*. Accuracy can be defined as "freedom from error." The example under reliability above shows the calculation to be accurate by

Point Estimate	= $ 72 496
38% Contingency	= $ 27 548
Total Cost	= $100 044
Cost Commitment	= $100 000
Accuracy	= – 1%

Inaccuracies can be caused not only by rounding, but also by simple mistakes. Those are usually eliminated when checking the calculations. *Systematic* errors can affect the accuracy of an estimate considerably. Assuming there are five items to be calculated by using average labor rates for a job:

1 Carpenter	$ 24.56 / Whr
1 Plumber	$ 31.16 / Whr
1 Electrician	$ 32.12 / Whr
1 Laborer	$ 18.62 / Whr
Total	$106.46 ,
Average =	**$ 26.615 / Whr**

If we round the average labor rate for the job *before* calculating individual cost, we incur inaccuracies in one direction. Even though the error is small for five items, it can lead to a noticeable inaccuracy when dealing with longer columns.

Estimated Workhours	Estim. Cost $ 26.615/Whr.	Estim. Cost Approx. 27	
34	909.91	918	
15	399.22	405	
45	1 197.68	1 215	
28	745.22	756	
31	825.06	837	
Total $ 4 072.10		$ 4 131	Difference = $ 58.90

> The 1.4% error is unidirectional.

Making errors when taking off quantities from a drawing or accidentally slipping a line when reading figures from a cost data book will also affect the accuracy of the estimate. Those errors are hard to detect.

> **The expression *accuracy* is sometimes mistakenly used instead of *reliability*. An estimate can be error-free and quite accurately prepared and yet the figures may be unreliable because of uncertainties, which are expressed in range form ($\pm$ x%).**

Precision

It is simply an expression for "how many significant digits we use in calculations." The precision is usually related to the reliability of the estimate. If the reliability ranges from 10% to 25%, a minimum precision of 10% may only be required. If we round the precise figure 1123 to 1000 we would have an error of $123/1123 = 11\%$, which is >10%. We would therefore express it in two significant figures, i.e. 1100, with an error of 2%.

We know, that the rounding of figures produces inaccuracies. The above rule only applies when manual calculations are made and when we need to save time. Most estimates are now produced with the help of the computer where rounding is not necessary. The more precise we calculate, the closer the result is to the truth.

> **It should be noted, however, that an estimate produced with precise figures can still be inaccurate, if an error was made.**

Fortunately, the addition of rounded figures does not introduce a *systematic* error when the proper rules of rounding are applied (see 1.1.1 Elementary Business Calculations).

The Reliability of an Estimate

At the very basis of project evaluation, which leads to the Go/No-Go decision, is the cost estimate. For whatever reason, virtually all projects are built with the expectation of maximum return for the money invested. The evaluation stage can therefore be the most critical stage during the life of a project. Unfortunately, it is during *this* stage that the *least* information is available to the estimator (Figure 3-54).

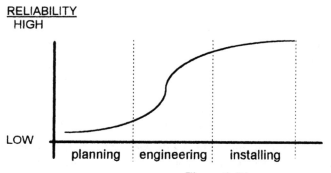

Figure 3-54
Reliability of an estimate.

RATIO OF UNCERTAINTY

Estimate Type:	-50 -40 -30 -20 -10 0 +10 +20 +30 +40 +50
Concept..........................	
Order-of-magnitude...........	
Factored..........................	
Parametric......................	
Definitive........................	
Control...........................	
Tender Check....................	

Figure 3-55
Ratios of estimating uncertainties.

3.4.3 Range Estimating

Capital cost estimating is concerned with the development of the cost of the project in relation to its *investment* cycle. It represents the anticipated final cost flow during the installation of project facilities.

The *reliability* of an estimate is dependent on the quantity and quality of information available, which in turn is a function of time. That means predicting final investment cost in the early stages is less reliable than at later stages. This can be presented graphically (Figure 3-54). This results in ratios of *uncertainty* ranging from over 50% down to about 5% variance from the cost-at-completion (Figure 3-55). It used to be general practice in industry to submit *single point* estimates to management. This way, most of the *intimate knowledge* the estimator has is lost in the process. Estimators who submit single point estimates keep most of their knowledge a secret (8). Of course, management is often at fault if they say: "Don't give me all that detail, just give me a bottom line figure!"

We *know*, that figure is wrong, the question is by how much!

Single point estimates have an entry called *contingency*. This amount is added to other costs to cover unforeseen events, errors and omissions. In most cases it is a subjective number often depending on the "feel" for the project. But how do we explain this "feel" to the client?

If the estimator has a close working relationship with the decision maker, at least they can discuss the confidence level of the estimate. In companies where the decision making executive is far removed from the working level, however, a hierarchy of documentation must be produced. The single point estimate is not good enough any more.

Range estimates should be produced on the basis of low and high boundaries. These boundaries are the lowest and highest cost which could be incurred barring any calamities. It does not force the estimators to be optimists or pessimists, in fact, they are both. It forces them to recognize the *uncertainties* inherent in the production of the estimate. Plotting minimum, most likely (ML) and maximum values against a scale, showing probabilities that actual cost will be as stated, we may obtain an absolute range with a most likely target (Figure 3-56).

The *target* will have the highest degree of confidence. This would be the single point estimate. But minimum and maximum have a confidence level that diminishes to *zero*. Since there is *no* confidence in the lowest or highest figure, it should not be included in the range. It means that a range of 100% is unrealistic, therefore, we stipulate a range between "o"% (optimistic or low) and "p"% (pessimistic or high) with a 50% probability that cost will be as stated for the most likely estimate (Figure 3-57). The estimate can now be stated as follows:

It is our prediction, that the most likely cost for the equipment is $500 000. There is a one in ten probability that it may only cost $400 000 but also a one in ten probability it may be as high as $900 000.

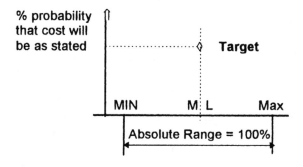

Figure 3-56
Estimating target.

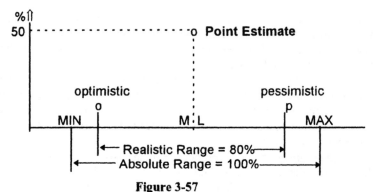

Figure 3-57
Realistic range estimate.

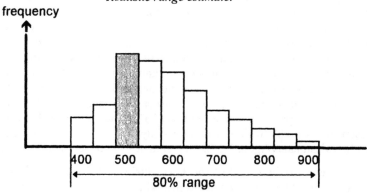

Figure 3-58
Range estimating histogram.

The Monte Carlo Method

Statistics are at the very base of risk analysis. We have looked at the decision tree (see Section 2.2.3) and established the various probabilities of cost overruns or underruns and the most likely cost. This will give us an estimate in range form. This can best be illustrated by use of a histogram (Figure 3-58).

The Monte Carlo Method is a computer simulation model that mimics the real world. This method is used for cases that contain great uncertainties or are too complex to be solved analytically. It is based on the statistical sampling theory using computer generated random numbers (stochastic model). The idea is similar to the flipping of a coin. We obtain a "head" or "tail" randomly, but get close to a 50/50 result the more often we flip the coin.

In regard to range estimating, each run of the model in the computer program is considered one trial within the area of uncertainty. By using mathematical simulation, we are generating a random distribution of numbers to be as free as possible from any pattern or bias. This mathematical model simulates a very large number of estimates, whereby the most frequently quoted estimate lies in the area of 500 k\$. The boundaries of 400 k\$ and 900 k\$ are specified as being exceeded only with a 1 in 10 chance (Please refer to 1.4.2 Probability Distribution) This, in statistical terms, would lead us to the Figure 3-59 frequency diagram. 500 is the point estimate with 80% confidence that there is a 10% chance of the cost exceeding 900.

The mean can be *approximated* by reducing the curve to a *triangle*. The total area would then be

$$(900-400)(0.1+0.4/2)=150 \text{ units}$$

The area to the left of the mode is

$$100(0.1+0.4/2)=30 \text{ units}$$

Assuming the mean to be at a frequency of 0.4, the mean can be calculated to be at 600. (Because of the extreme skewness, the BETA distribution

$$\mu =(400+2000+900)/6=550$$

may have been a better choice).

The unit area under the curve between μ and x is 40% of the total or 0.4. The distribution table shows z = 1.28 for a unit area of 0.4.

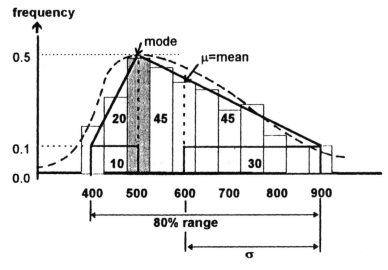

Figure 3-59
Range estimating frequency distribution.

Therefore

$$\sigma = \frac{x - \mu}{z} = (900 - 600)/1.28 = 234$$

If we have a large number of cost packages, each described by their low value, their target, and high value, then, the most likely value of the composite curve is *not* the value of each target, but the *sum* of the *means* of the constituent curves.

According to the *central limit theorem* the resulting new distribution is close to a bell-shaped normal curve.

As an example, we add three distributions with various shapes (Figure 3-60). Because the mean μ divides the area under the curve into two equal parts, we can look at it as the pivotal point balancing the two areas regardless of the shape of the curve above it.

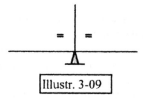

Illustr. 3-09

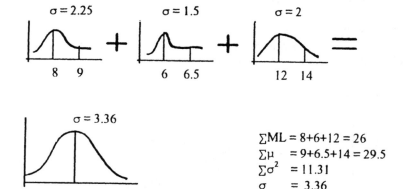

$\Sigma ML = 8+6+12 = 26$
$\Sigma \mu = 9+6.5+14 = 29.5$
$\Sigma \sigma^2 = 11.31$
$\sigma = 3.36$

Figure 3-60
Central limit theorem-1.

Some Practical Applications

A major project is broken down into various cost components first. We then tabulate upper and lower percent ranges for each component and the percent of the range itself. A computer calculates all mean and standard deviations to each and every input figure and prints out results in tabular or graphic form.

Table 3-11

Monte Carlo Simulation

NORMAL APPROXIMATION TABLE

PROBABILITY OF VALUE BEING GREATER THAN INDICATED

	90	80	70	60	50	40	30	20	10	
TOTAL COST TO DATE										
81	63.9	69.3	73.2	76.5	79.8	82.7	86.0	89.9	95.3	NORMAL APPROX.
82	16926	17457	17839	18166	18471	18776	19103	19485	20016	
83	77433	78978	80091	81043	81933	·82822	83774	84888	86433	
84	111061	113131	114624	115899	117091	118284	119559	121052	123122	
D1	111061	113131	114624	115899	117091	118284	119559	121052	123122	
85	139322	141975	143889	145524	(147052)	148580	150216	152129	154783	

FREQUENCY TABLE

PROBABILITY OF VALUE BEING GREATER THAN INDICATED

	90	80	70	60	50	40	30	20	10	
TOTAL COST TO DATE										
81	63.5	69.2	73.8	76.9	80.0	83.1	85.9	89.7	94.4	FREQUEN-CY
82	16934	17423	17741	18167	18483	18777	19099	19462	20085	
83	77254	78961	80239	81053	81936	82734	83790	85039	86321	
84	111096	113211	114692	116080	117014	117916	119442	121182	123135	
D1	111096	113211	114692	116080	117014	117916	119442	121182	123135	
85	139314	141815	144076	145754	(144997)	148511	150280	152027	155248	

SAMPLE STATISTICS

	MEAN	STD DEV	SKEWNESS ±<0.3	KURTOSIS 3.0+<0.3	10PC CONF MEAN	90PC	
TOTAL COST TO DATE				NORMAL			
81	79.58	12.25	-.1	3.1	78.88	80.28	
82	18471	1205	.1	2.8	18402	18540	
83	81933	3511	-.1	3.0	81732	82134	
84	117091	4706	.0	2.8	116822	117361	
D1	117091	4706	.0	2.8	116822	117361	
85	147052	6032	-.1	2.7	146707	147398	
ENTER POOL OR MODELING LANGUAGE COMMAND					± 0.2%	± 0.3% accuracy	

Example: For the sake of demonstrating how the estimate is prepared for input into a computerized range estimating program, let us assume we have a projected cost flow for a fictitious package, broken down by cost streams as shown in Table 3-11.

The printout shows a normal approximation table with an expected value (50%) of k$ 147 ± 5.3% with a 10% probability that the cost may be higher than k$ 154.78. It also gives probabilities for 20%, 30% and 40%.

In addition, a randomly generated frequency table and sample statistics will give the decision maker the background to assess the risks and assign the amount of contingency that is desired for the project (refer to 2.2.0 Decision Making).

If 90% of the area under the standard normal distribution is considered a realistic range, then the boundary limits are **Z = 1.645** standard deviations away from the mean. The *mean* for the estimate can be calculated as μ = the most likely estimate (ML or mode) multiplied by the median of the range. If we consider the example estimate by work package shown in Table 3-12, the following simplified calculation would apply:

The median is [−5% to +10%]/2 = 7.5% or, expressed as a ratio = 1.075

$$\therefore \mu = 1605 \times 1.075 = \mathbf{1725}$$

The upper boundary limit is

$$x = 1605 + 10\% = 1765.5$$

and from

$$\sigma = (x - \mu)/Z = (1765.5 - 1725)/1.645 = \mathbf{24.65}$$

Total project cost can also be predicted and presented as shown in the graph below (Figure 3-60):

The graph indicates that it was decided to accept a 10% chance of an overrun, and allowing a contingency of k$ 22, which is the difference between the most likely and the calculated expected cost in this case.

Table 3-12

Range estimating example

YEAR	19__	19__	19__	19__	19__	19__	TOTAL
Mechanical Equipment	35	120	210	220	40	20	645
Electrical Equipment			160	180	60	50	450
Construction Material			50	100	150	80	380
Construction Labor				50	25	25	100
Engineering	15	10	5				30
TOTAL	50	130	425	550	275	175	**1605**
					Confidence Range		−5% +10%

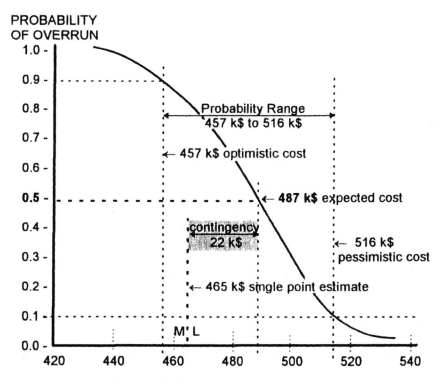

Figure 3-61 *Range estimating contingency.*

For smaller projects or for estimators who do not have access to a computer there is a manual model which can be used to analyze risks (11).

Graphic Model

All that is needed is a calculator and probability paper. The three range points (optimistic, most likely, pessimistic) are plotted on probability graphpaper. The curve is then analyzed as to skewness. Skewed right would be an upward deviation from a straight line. Adjustments are then made if necessary and the expected value is calculated.

Besides being a less expensive method, there is no "black box" and the estimator is on top of the work communicating closely with engineers in order to analyze the graph. The accuracy of this method is less than that of the Monte Carlo method but still a great improvement over the single point method.

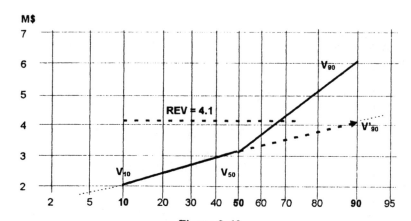

Figure 3-62
Range estimate - graphic method.

Using a most likely value of $V_{50} = 3$ M$, an optimistic value of $V_{10} = 2$ M$ and a pessimistic value of $V_{90} = 6$ M$, then V'_{90} is the standard distribution. V_{90} shows the skewness of the curve. Using proportions,

$$V'_{90} = 2V_{50} - V_{10} = 4$$

The approximate skewness is

$$Sk = (V_{90} - V'_{90}) / V'_{90} = 0.5$$

The expected value (EV) differs from V_{50} (50% probability) by the skewness value times adjustment factor $F_a = 0.4$

$$EV = [1 + (F_a \times Sk)] \times V_{50} = 3.6$$

This now needs some adjustment to obtain a resultant expected value (REV) based on the number of elements (N-value). Those calculations, not shown here, result in

REV = 4.1 M$

most likely to occur, when dealing with 15 elements. The decision to go ahead with the project is now based on this figure.

Simplified Triangular Distribution

Using the approximation of simple triangular distribution is another method of probabilistic estimating where the use of complicated computer programs is not necessary (33).

The range estimate for each work package is represented in the form of a triangle, with the base having cost intervals and the ordinate showing ten or more frequency divisions.

The area above the cost intervals is calculated or counted. The share each of those smaller areas have in regard to the total area is considered a probability of occurrence.

Work packages L, M, N, are depicted on squared paper and then tabulated as to area, probability and mean cost.

The dividing line for two equal areas of a rectangular triangle is projected at approx. 30% from the right angle.

For example, the area L1 has an average cost of $(3 - 0.3) = 2.7$ M$. Costs for the other areas are at the mid-point of each interval.

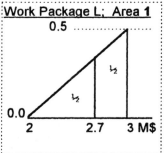

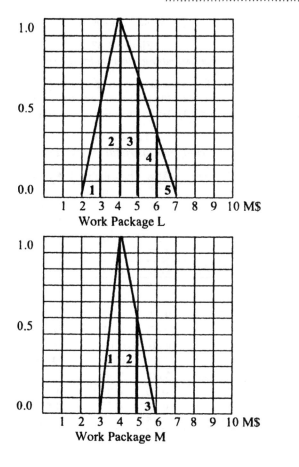

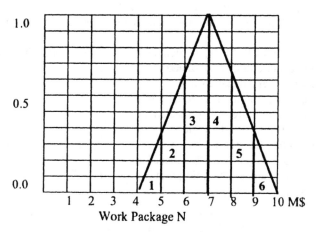

Figure 3-63
Range estimating - triangular distribution.

We can now tabulate the areas and corresponding probabilities (Table 3-13).
From this, we need to calculate the total cost and combined probabilities for all 90
combinations (5 × 3 × 6 = 90). To do this manually would be quite a bit of work.

Table 3-13
Triangular distribution - averages.

Package	Area	Probab.	Aver.M$
L1	0.25	0.10	2.70
L2	0.75	0.30	3.50
L3	0.83	0.33	4.50
L4	0.50	0.20	5.50
L5	0.17	0.07	6.30
TOTAL	2.50	1.00	
N1	0.17	0.06	4.70
N2	0.50	0.17	5.50
N3	0.83	0.27	6.50
N4	0.83	0.27	7.50
N5	0.50	0.17	8.50
N6	0.17	0.06	9.30
TOTAL	3.00	1.00	
M1	0.50	0.33	3.70
M2	0.75	0.50	4.50
M3	0.25	0.17	5.30
TOTAL	1.50	1.00	

		n1-M$ 4.70	n1-p 0.06	n2-M$ 5.50	n2-p 0.16	n3-M$ 6.50	n3-p 0.27	n4-M$ 7.50	n4-p 0.27	n5-M$ 8.50	n5-p 0.17	n6-M$ 9.30	n6-p 0.06	TOTAL 1.000000
L1-M1- 6.40	p 0.03	11.10	0.00	11.90	0.01	12.90	0.01	13.90	0.01	14.90	0.01	15.70	0.00	0.033265
L1-M2- 7.20	p 0.05	11.90	0.00	12.70	0.00	13.70	0.00	14.70	0.00	15.70	0.00	16.50	0.00	0.050401
L1-M3- 8.00	p 0.02	12.70	0.00	13.50	0.00	14.50	0.00	15.50	0.00	16.50	0.00	17.30	0.00	0.017136
L2-M1- 7.20	p 0.05	11.90	0.01	12.70	0.02	13.70	0.03	14.70	0.03	15.70	0.02	16.50	0.01	0.099795
L2-M2- 8.00	p 0.05	12.70	0.01	13.50	0.03	14.50	0.04	15.50	0.04	16.50	0.03	17.30	0.01	0.151204
L2-M3- 8.80	p 0.05	13.50	0.00	14.30	0.01	15.30	0.01	16.30	0.01	17.30	0.01	18.10	0.00	0.051409
L3-M1- 8.20	p 0.10	12.90	0.01	13.70	0.02	14.70	0.03	15.70	0.03	16.70	0.02	17.50	0.01	0.109774
L3-M2- 9.00	p 0.17	13.70	0.01	14.50	0.03	15.50	0.04	16.50	0.04	17.50	0.03	18.30	0.01	0.166324
L3-M3- 9.80	p 0.06	14.50	0.00	15.30	0.00	16.30	0.02	17.30	0.02	18.30	0.03	19.10	0.01	0.048523
L4-M1- 9.00	p 0.07	13.90	0.00	14.70	0.01	15.70	0.02	16.70	0.02	17.70	0.01	18.50	0.00	0.066530
L4-M2-10.00	p 0.10	14.70	0.01	15.50	0.02	16.50	0.03	17.50	0.03	18.50	0.02	19.30	0.01	0.100803
L4-M3-10.80	p 0.03	15.50	0.00	16.30	0.01	17.30	0.01	18.30	0.01	19.30	0.01	20.10	0.00	0.034273
L5-M1-10.00	p 0.02	14.70	0.00	15.50	0.00	16.50	0.01	17.50	0.01	18.50	0.00	19.30	0.00	0.023285
L5-M2-10.80	p 0.04	15.50	0.00	16.30	0.01	17.30	0.01	18.30	0.01	19.30	0.01	20.10	0.00	0.035281
L5-M3-11.60	p 0.01	16.30	0.00	17.10	0.00	18.10	0.00	19.10	0.00	20.10	0.00	20.90	0.00	0.011996
TOTAL			0.06		0.16		0.27		0.27		0.17		0.06	1.000000

Figure 3-64

Range estimating spreadsheet.

Using the electronic spreadsheet will require only two input columns for each package, average cost and probability.

Combining cost is an *addition;* combining probabilities is a *multiplication.* For example, the combination L2, M3, N2 would yield

$$3.5+5.3+5.5=14.3 \text{ M\$}.$$

The corresponding combined probability is

$$0.30 \times 0.17 \times 0.17 = 0.00867.$$

The initial spreadsheet calculation is shown in Figure 3-64. Please note, that probability p = 0.~ on the printout is truncated to two decimals which does not affect the final calculation.

There are many M$ duplications which we need to sort and summarize. This is simply done by listing M$ in ascending order and adding the corresponding probability fractions (summarization by descending order may be preferred by some cost engineers/consultants).

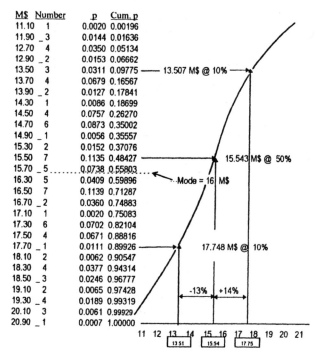

Figure 3-65
Range estimating - cumulative probabilities.

The calculation shows that the amount which has a 50% chance of *not* being exceeded is $ 15 543 000. With an 80% confidence level, the optimistic/pessimistic range is −13% to +14%. The mode of the frequency distribution is at 16 M$ (Figure 3-65).

A contractor may use this procedure to establish a policy *not* to place a bid on a price exceeding a certain percentage.

3.4.4 Estimating Techniques

The techniques used depend very much on the cost data available to the estimator. Those techniques range from individual or collective judgment to detailed quantity take-offs from design sketches or drawings.

Judgment

The phone rings in the contractor's office: "I want to build a shopping center with approximately 20 small stores, 4 stores medium size, and one standard department store. Can you give me the approximate building cost?"

How can an estimator even consider to give an answer with so little information? It needs an experienced estimator to come up with a "while-on-the-phone" estimate. Using the "rule-of-thumb" method, the estimator would compare this project with other similar shopping centers that were built recently and use good judgment. The answer may be: "You can expect to spend 10 to 15 M$ on this project."

If there were a little more time available, some pointed questions could have been asked. The estimator would then consult others who are experts on similar projects to reach a consensus on the capital cost. Needless to say, judgment type estimates carry a high risk and are very much dependent on the confidence one has in the estimator who produces the estimate.

Project Cost Curves

When historical data is available on several *similar* projects, a cumulative cost curve (S-curve) can be plotted on a time series graph (Figure 3-66). Those curves can give a surprisingly accurate cost flow for repetitive projects such as apartment buildings, slip form type construction, shore protection and even some processing plants. The empirical curve can be smoothed for general application and a mathematical formula developed that will fit this empirical curve. This formula is usually a modified version of continuous compounding (see 1.3.1) such as

$$Y = A + \frac{B}{1 + D(e)^{-ct}}$$

"Y" is the cumulative cost at time interval "t".

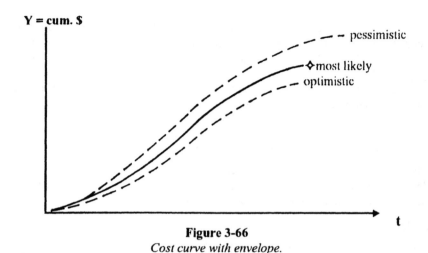

Figure 3-66
Cost curve with envelope.

Cost Curves are often expressed in percent time versus percent cost, where

$$0 < t < 1.0$$

Standard curves should only be used as a preliminary cost projection, not for cost control.

Factoring

When preparing order-of-magnitude estimates or when time does not permit detailed estimating (fast tracking a project), *factor estimating* is a quick and inexpensive way of predicting future costs of a project. It is beyond the scope of this text to provide more than just a cursory overview of factor estimates. The reader should refer to the AACE *Cost Engineers' Notebook* or other literature on the subject.

If the cost of a given unit (C_1) is known at one capacity (Q_1) and it is desired at another capacity (Q_2) "R" times as great, the known cost (C_1) multiplied by R^x will give the cost at the second capacity (C_2), i.e.

$$C_2 = C1(R)^x$$

where the capacity ratio $R = Q_2/Q1$ and x is the exponential factor. This factor can vary anywhere from 0.2 to > 1.0.

Considerable care should be taken when using this formula. Similar configuration and scope must exist when comparing Q_2 with Q_1. Correlation of equipment cost data has been done for many years by using the equation

$$y = ax^b$$

where "y" is the cost, "a" is a constant for the type of equipment, "x" is the capacity of the unit, and "b" is also a constant.

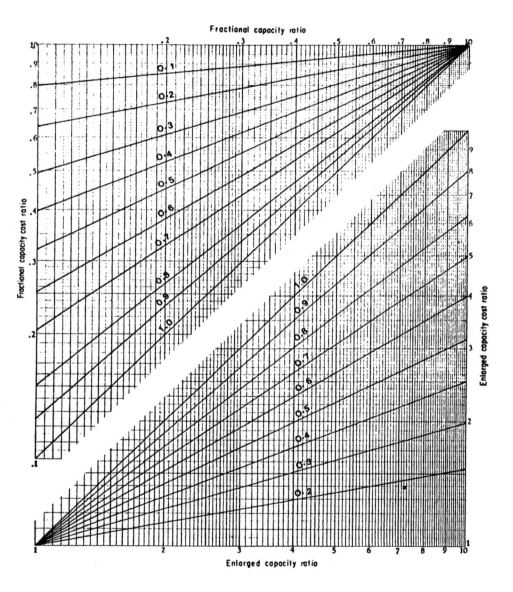

Figure 3-67
Capacity curves.

Plotting y against x on logarithmic graph paper will produce straight lines for various values of "b" (Figure 3-67).

Assume a 10 Mg (metric ton) package air conditioner costs $ 18 000. To get the approximate cost of a 15 Mg air conditioner of a similar type (exponent 0.6), we find the value by moving at 1.5 up on the lower scale, intersecting with line 0.6. The capacity ratio reads 1.2. Moving to the upper fractional capacity ratio of 0.5 projects at 0.67 of line 0.6. Therefore, the upscaling by 1.5 has a capacity cost ratio of 1.267. The estimated cost for the 15 Mg unit is

$$1.267 \times 18\ 000 = \$\ 22\ 800.$$

A good estimator will take the cost reliability of the 10 Mg air-conditioner into account: Is it based on *one* other estimate or on a data base averaging many estimates, or was actual cost data used of a recently installed similar unit?

The reliability of an estimate increases as follows:

Even though the exponent 0.6 is typical for a large range of equipment such as piping, pumps and vessels, it should only be used when more specific statistical data is not available. A centrifugal fan with a capacity of 150 m^3/min may have an exponent of 0.4, whereas for a fan with a large capacity of 1 500 m^3/min the exponent may be 1.1.

A central data bank that includes factors to be used by estimators should be established and continually updated.

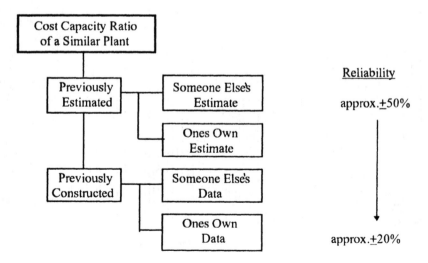

Figure 3-68
Increasing estimate reliability.

Another factor widely used for order-of-magnitude estimates and for "quickies" for process plants are the LANG factors. Those factors are based on a statistical relationship between delivered process equipment cost and the total installed cost.

Because it is an *average* of a multitude of process plant costs, it does not allow for specific plant configuration. (A non-swimmer was told that the river had an average depth of 60 cm. Unfortunately he drowned at a spot where the river was two meters deep). It is very important to recognize, that the LANG factor is based on *averages*.

Example: The ratio for a fluid process plant is 4.74 : 1.

If the estimated cost of all process equipment F.O.B. site is 15 M$, then the estimate for total installed cost is

$$15 \times 4.74 = 71 \text{ M\$}$$

before adjustments for escalation, location, cost of money etc. Factors can also be used for different types of equipment. This will make the estimate more reliable.

To install 6 M$ worth of equipment will cost approx. 16.6 M$. Depending on the definition of the multiplier, this may include the cost of site development, buildings and structures, electrical installations, instrumentation, civil work, heating and ventilating, piping, insulation, engineering and supervision (Table 3-14).

To derive a multiplier, either published data can be used (Wroth, Chilton, Peters-Timmerhouse, etc.)(5) or preferably the data bank of the company can be accessed, e.g.

Table 3-14
Estimating multipliers

Equipment	Cost in k$	Multiplier	Installed k$
Fans	120	2.4	288
Compressors	600	2.2	1 320
Furnaces	1 200	1.9	2 280
Heat Exchangers	1 000	4.8	4 800
Motors	800	8.3	6 640
Pumps	280	6.6	1 848
Tanks	1 000	3.6	3 600
Other Equipment	1 000	3.0	3 000
TOTAL	6 000	3.76	22 588

Table 3-15
Estimating cost components

Summary Cost Component		Permanent Facilities Material	Labor
Process Equipment	27%	45%	28%
Work Items: Materials	25%	55%	-
Field Labor	20%		72%
Constr. First Cost	6%	100%	100%
Constr. Operating Cost	5%	Engineering Workhours	
Constr. Supervision	4%	Design Services Admin.	
Engineering	13%	55% 25%	20%
TOTAL	100%		

Components are made up of more detailed items, for example

	Materials	Labor
Work Items:	55%	72%
Civil	3%	8%
Buildings	4%	4%
Structures	7%	6%
Piping	25%	42%
Electrical	8%	9%
Instrumentation	8%	3%

Labor cost for electrical work would be

$$22\ 588 \times 0.72 \times 0.09 = 1\ 464\ k\$$$

If an electrician costs \$26/Whr, working about 2 000 Whrs/y, 1 464 000/26/2 000 = 28 WY (worker years). If the job duration is nine months, $28 \times 4 / 3 = 37$ electricians are required on the average. Assuming that the peak requirement is twice the average, up to 74 electricians need to be hired to meet the schedule.

To estimate *engineering cost*, historical records from similar projects will give an average percent of workhours for major categories such as Design, Services, Administration. Each of those can be further broken down, i.e. Design into Flow Sheets, Specifications, Drafting, Proposals. The engineering costs are then made up of average salaries, payroll burden, supplies & services, expenses, overhead. Payroll burden is usually a percentage of salaries. This will vary, depending how many benefits (sick leave, vacation pay, pension plan, health & accident insurance) the employer is willing to provide. The multiplier varies from 1.2 to 1.6.

Expenses include computer charges, reproduction and printing cost, travel expenses, supplies and misc. services. They are normally 25% to 30% of salary cost.

Overheads are indirect costs for such things as office rent, utilities, corporate management, some functional support (standards development, training). Overheads are partially prorated to various projects, usually on the basis of cost flows. This varies with accounting practices of individual companies. The multiplier can have a range of 1.2 to 2.0.

If the average engineering salary is $25/Whr, the engineering cost to the project could be $25 \times 1.4 \times 1.27 \times 1.75 = \$78$/Whr. 10 to 15% profit may be added for proposal estimates. We had factored 13% of total project cost in our example above, or $0.13 \times 22\ 588 = 2\ 936$ k$. Design work would cost $0.55 \times 2\ 936 = 1\ 650$ k$. This, divided by the cost per Whr is $1\ 615/78 = 20\ 705$ Whrs. or $20\ 705/2\ 000 = 10.35$ work-years.

For a design job duration of 9 months, the number of designers and drafting staff is $10.35(4/3) = 14$ on the average. If the peak is twice the average, a maximum of 28 will be required to meet the schedule.

For process equipment factored for *total installed cost*, the reliability increases with the quality of data available (Figure 3-69).

Cost/Price Indexes

The term *index* number or index for short is used for a single ratio often expressed in percentage which measures the averaged change of several variables between two different times, places or conditions.

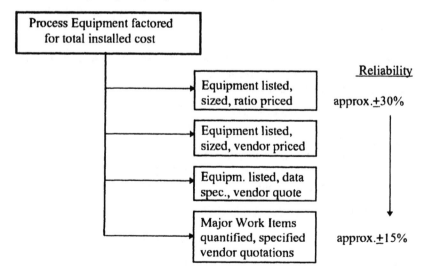

Figure 3-69
Capacity ratio - similar plant.

The term *escalation* refers to the change in prices for various goods and services over time periods. There are many escalation indexes such as

Consumer prices
Classes of equipment
Materials (steel, copper)
Building construction
Complete processing plants
Cost of labor

When applying escalation to an estimate one must be careful to select the appropriate index (AACE Cost Engineers' Notebook - Cost Index Committee). Many companies employ economists to derive at indexes that pertain specifically to the company's projects. The production of an index is a very complex process, however, the application is simple. To convert costs from one point in time to another

$$C_F = C_P \times \frac{I_F}{I_P}$$

C_P and C_F are present and future costs and I_P and I_F their respective indexes.

For example: A company published the following escalation table for large boilers

1991	1992	1993	1994	1995	1996	1997	1998	....
571	628	657	708	748	799	851	902	....

The cost of a boiler delivered for a generating station in 1992 was 100 k$. Another plant is to be built with an identical boiler to be delivered in 1996. What is the escalated cost of that boiler?

$$C_F = 100 \ (799/628) = 127 \ k\$$$

Some historical indexes have very high numbers because their base may go as far back as 1926 (Marshall & Swift). To change the base year of the above escalation table for large boilers to the base year 1991 = 100

$$I_F(new) = I_P(new) \times \frac{I_F \ (.old.)}{I_P (.old. \ )}$$

For 1996 and 1992 it would be

$$(799/571)100 = 140 \ \text{and} \ (628/571)100 = 110 \ \text{and}$$
$$C_F = (140/110)100 = 127 \ k\$.$$

The new table compared to the old:

1991	1992	1993	1994	1995	1996	1997	1998	....	
571	628	657	708	748	799	851	902	....	(old)
100	110	115	124	131	140	149	158	....	(new)

Variables between *places* instead of time are usually called *location factors*. The MEANS Construction Cost catalog publishes "City Cost Indexes" based on average cost of 162 major USA and Canadian Cities (1992 = 100). The calculation for cost in city A as compared to average national cost is

$$\text{Cost, City A} = \frac{\text{(National Average Cost)(Index for City A)}}{100}$$

For Example: A contractor in Birmingham, AL pays $60/m³ of concrete. The city cost index is 92. He needs to do a job in Anaheim, CA where the index is 106.2. How much does he have to expect to pay in Anaheim ?

$$\text{Cost in Anaheim} = (106.2 \times \$60)/92 = \$69/m^3$$

International cost/price indexes must include location factors that reflect the country's

- infrastructure
- climate
- government legislation
- safety regulations
- taxation and insurance
- productivity

It must also include the buying power of the currency.

For example: A plant was estimated to cost 10 M$ in 1992. The same type of plant is to be built in Germany. The location cost factor for Germany at that time was 112 as compared to the US base factor of 100. The currency exchange rate was DM 1.35 per Dollar, or $0.74/DM.
The location cost factor is

$$\text{LCF(G/US)} = (112/100)0.74 = 0.83$$

To build the plant in Germany would cost 8.3 M$ in 1992.

In most cases, the estimator will have to use past data that needs to be escalated to the present or future, e.g.

	1992		1996	
	USA	Germany	USA	Germany
Escalation Index	110	283	140	312
Exchange Rate	0.74$/DM	1.35DM/$	0.90$/DM	1.11DM/$

From the above data, the escalated location cost factor for 1996 is

$$\text{LCF(G/US)} = 0.74 \ \frac{312/283}{140/110} \times \frac{0.90/1.11}{0.74/1.35} = 0.95$$

Therefore, the escalated cost to build the 10 M$ plant in Germany would be 9.5 M$ in 1996.

Table 3-16 shows an example of plant cost indexes (base 1980 = 100) - *Excerpt from Process Economics International - Promotional Pamphlet:*

Table 3-16

International Cost Indexes

	1986 →				1987 →				1988 →	
	1 →	2 →	3 →	4 →	1 →	2 →	3 →	4 →	1 →	2 →
Western Europe										
Belgium	131	130	130	130	130	130	130	130	131	131
Denmark	146	149	149	150	153	159	159	162	167	167
Finland	157	159	160	162	165	169	172	174	176	173
France	165	165	165	166	165	166	168	170	172	173
West Germany	123	124	126	127	128	129	129	130	131	132
Greece	319	324	330	340	347	355	363	370	378	387
Italy	208	210	212	215	219	224	226	230	232	234
Netherlands	121	121	122	122	122	122	123	123	124	124
Norway	148	153	159	162	171	175	177	180	183	186
Portugal	246	263	304	302	277	296	344	344	356	372
Spain	207	204	208	212	218	221	225	230	234	238
Sweden	153	154	154	156	157	161	160	162	163	167
Switzerland	122	122	123	124	126	126	126	127	128	129
United Kingdom	158	162	162	167	168	173	173	175	176	178

Estimate Components

With an increased scope definition, the *order-of-magnitude* estimates are further refined, thereby increasing the reliability of predicting future cost. Those "semi-detailed" estimates are based on flow sheets, design sketches, layout drawings, equipment lists, and master schedules. They are called *budget estimates* by the AACE, having a reliability of +30% to −15%.

With further design work, increased drawing production, information on quotations etc., Definitive Estimates are produced with a reliability of better than 15%. There is no distinct dividing line between those types of estimates.

To categorize the estimate components, we use the work breakdown structure (see 3.2.1 - The Work Package Concept). Major components are

| DIRECT COST | | INDIRECT COST |

Direct Cost

The WBS will help an estimator to organize direct cost data. "Directs" are the cost of equipment, material and labor that can be directly attributed to the physical facility. This may or may not include engineering cost.

An engineer responsible for the design of a chlorination system for example will charge the time to the appropriate direct work package, and so will the draftsperson producing the drawing for that system.

The division between direct and indirect cost must be well defined because it varies with company policy. When the owner provides its own design and construction labor, the breakdown will differ from a contracted construction job.

A *typical example* of an owner's estimate for the work package Q-25 = Chlorination System of a large project is shown in Figure 3-70. This manual estimate has three sub-components:

> Cost of Labor
> Equipment Rental
> Permanent Material

Construction equipment that is not rented but owned is not considered direct cost in this case. The owner has an equipment pool from which various projects draw the bulldozers, cranes etc. that are needed. Those costs are general project cost and are not included in specific work packages.

estimate	work order no. ABC-15		property no. DE-5	estimate no. 3	date		page 5 of 20
	project XYZ-28 - WBS # Q25 - Chlorination				prepared by b.		
description	account no.	quantity	man-hours	labour	equipment rental	material	total
Pumps 60 kW	71652	2	220	6,000	–	14,000	50,500
120 kW	71652	1	150	4,500	–	26,000	
Valves 50 mm check	71653	4	5	200	–	2,000	
100 mm gate	71653	2	12	300		2,000	
" 100 mm globe	71653	1	6	200		3,000	7,700
Tanks 6 kL	71654	2	17	500	–	4,000	10,500
19 kL	71654	1	32	1000	–	5,000	
Chlorinator	71656	1	28	900	100	3,000	4,000
Pipe Supports	71658	included	in 71659 - Piping				
Piping SS40 - 25mm	71659	100 m	60	1,700	–	2000	
SS40 - 20mm	71659	200 m	130	3,000	–	4000	
" Epoxy - 50 mm	71659	150 m	150	4,000	–	2500	
Fittings	71659	200	300	8,000		5000	30,200
SUB-TOTAL			1,110	30,300	100	72,500	102,900
Engineering	71650	15%	200	15,400			
TOTAL			1,310	45,700			118,300

Figure 3-70
Estimate worksheet.

For a *turnkey project* that is contracted out, the engineering costs are usually not considered direct cost. The cost for operating the equipment to install permanent materials will be a direct cost component to the contractor. This includes ownership cost as well as leasing cost.

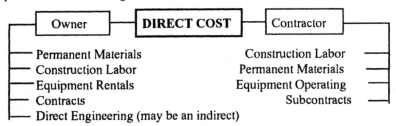

Owner	**DIRECT COST**	Contractor

Permanent Materials   Construction Labor
Construction Labor   Permanent Materials
Equipment Rentals   Equipment Operating
Contracts   Subcontracts
Direct Engineering (may be an indirect)

When estimates are produced from completed drawings, which is often the case for small to mid-size building construction (apartments, office buildings, housing developments), quantities are priced in detail. These *take-off* estimates have a high degree of reliability.

This is specialized estimating.

For example, an electrical material check list may show over 300 items to be evaluated with headings such as

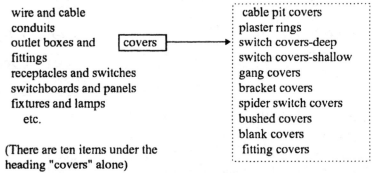

wire and cable   cable pit covers
conduits   plaster rings
outlet boxes and   covers ───────▶   switch covers-deep
fittings   switch covers-shallow
receptacles and switches   gang covers
switchboards and panels   bracket covers
fixtures and lamps   spider switch covers
     etc.   bushed covers
                                                  blank covers
(There are ten items under the   fitting covers
heading "covers" alone)

Applicable quantities are entered into the checklist and then priced. For buildings, the estimate would be summarized by system, i.e.

- ◆ wiring (outlets, junction boxes, switches, conduits, wire)
- ◆ fixtures (lamps, lights, supports, range hoods)
- ◆ panels (switchboards, cabinets, meters, transformers)
- ◆ feeders, services (conduits, wire, fittings - runs & risers)
- ◆ motors (list price or cost per kW)
- ◆ special systems (instruments, alarms, detection devices)
- ◆ at-cost-items (tools, scaffolding, cellular phone, travel)

Labor is estimated on the basis of unit cost, such as cost per outlet or per number of motors. The cost of tools can be added to the estimate on a pro-rated basis or as a separate item, but it must be clearly stated as such.

Indirect Cost

AACE defines indirect construction cost as those cost that do not become a final part of the (permanent) installation but are required for the orderly completion of the installation. In our take-off example above, the indirect cost to the electrical contractor are such incidentals as

> insurance, office expenses, office and stockroom maintenance, rent, bond, drawings, license, inspection, storage; salaries for job superintendent, estimator and clerk; allowance for delays & conditions, material handling.

Those indirects are usually expressed as overheads, listed as a percentage of total direct cost. Also listed as separate indirect cost items are

- ♦ *Sales taxes* that have to be paid for the purchase of materials,
- ♦ *Interest* on money borrowed to purchase materials and to pay wages,
- ♦ *Profit* is an incentive for an owner or contractor to stay in business. It may also be considered compensation for the risks involved when undertaking the work.

Indirect cost components will differ somewhat according to the type and size of project and corporate accounting practice.

For large corporations, the following items may be added to the list of indirects:

- ♦ research and development
- ♦ corporate policy and procedures
- ♦ health and safety
- ♦ public and employees' relations
- ♦ general projects' management
- ♦ systems development (main frame)
- ♦ data processing and corporate data base
- ♦ government liaison
- ♦ quality engineering (ISO 9000)
- ♦ environmental studies and investigations
- ♦ corporate finance and accounting
- ♦ laws and legal matters
- ♦ personnel planning and development

Projects make use of those corporate services. Each project will carry a prorated corporate overhead on the basis of a selected cash flow to pay for those services.

Contingency

The previous section 3.4.3 (Range Estimating) dealt with the reliability of an estimate and the use of the Monte Carlo method as a tool to express uncertainties in terms of ranges. The decision maker can determine from this information the amount of contingency to be assigned to the total estimated cost of the project.

This works very well, *provided the contingency assignment does not become a routine calculation of the difference between most likely and expected cost.*

It would defeat the intention of risk assessment and may lure us into a false security. The decision-maker's attitude toward risk based on experience need always be a factor for contingency provisions. Obviously, there is more to contingency than meets the eye.

Even the *definition* of contingency varies widely with the nature of industry, the type of project and its financing. The Oxford Dictionary calls it "Chance Occurrence" and Random House calls it "An Uncertain Event". The AACE Standard No. 10S-90 defines contingency as

"An amount added to an estimate to allow for changes that experience shows will likely be required. May be derived either through statistical analysis of past project costs or by applying experience from similar projects. Usually excludes changes in scope or unforeseeable major events such as strikes, earthquakes, etc."

The PMI PMBOK Glossary says

"Specific provision for unforeseeable elements of COST within the defined project SCOPE; particularly important where previous experience relating estimates and actual costs has shown that unforeseeable events which will increase costs are likely to occur. If an allowance for escalation is included in the contingency it should be a separate item, determined to fit expected escalation conditions of the project."

There are various other definitions in cost engineering literature such as:

An allowance added to an estimate to cover unforeseen costs for items such as strikes and delays due to weather, accidents, changes, etc.

Obviously, there is no "standard" on the application of contingencies, but those definitions express the general idea of allocating some dollars not specified in the estimate to cover events that cannot be foreseen. It is an indeterminable amount (stochastic) that must not be buried in the estimate. One thing is quite definite:

Contingency is not a cost item that can be *used* to reduce overspending.

If the forecast shows 11 M$ vs. a budget of 10 M$, we must make an effort to reduce future costs and get back on track. This is what cost control is all about! Staying within the budget and finishing the project on time is the main objective of project management.

The contingency is a *variable* amount, an "insurance" against unforeseen events. It may or may not be used. The contingency *allowance* is not routinely *used up* in the common sense but reduced as the project progresses because the risk decreases.

We should also make a distinction between "unforeseen" and "unforeseeable" events. We may included the following in the category of *unforeseen* events:

- ♦ escalation rate increases
- ♦ scope details not identified (extras)
- ♦ schedule elements missing
- ♦ equipment breakdowns
- ♦ interruptions due to weather (floods, storms)
- ♦ labor productivity
- ♦ constructability (rework, claims)
- ♦ contract cancellation

In general, we could say that the estimate may well have contained most of the items above, but not to the extent anticipated at the time. Escalation is *always* calculated on a long duration job. Inflation may be more severe than the estimator predicted. This amount would be drawn from the contingency reserve.

The time estimate (schedule) should be assigned a "time contingency" in the form of float in addition to the normal float that is part of the typical scheduling process, covering *unforeseen* changes.

Contingencies should *not* include *unforeseeable* events that are impossible to predict such as

- ♦ earthquakes   ♦ new government regulations
- ♦ riots, sabotage   ♦ economic collapse
- ♦ acts of war

Even though the policies on contingency will differ from one company to another, it should definitely be based on an experienced review of uncertainties of estimate details (packages). An arbitrary bottom line percentage is not good enough.

Range estimates produced by the Monte Carlo method are built up from packages each of which shows already the reliability (most likely, pessimistic, optimistic) of the estimate. The bottom line here will be based on the judgment of the expert that reviews the estimate.

Interest During Construction (IDC)

Interest charges can be a considerable portion of the total capital expenditure. The financial cost for a nuclear power station for example may be as high as 30% of the total cost.

Being two months behind the scheduled in-service date may cost the owner over 40 M$. The calculation of interest is based on the cash flow of the project (not cost flow) and depends on the financial arrangements made with the lender. Common arrangements are monthly payments with an annual compounding or an annual interest rate with monthly compounding.

The financial cost is an unallocated indirect cost entered as a single line on the project estimate document. It is usually capitalized with total final cost on the in-service date in order to separate major plant facilities from the current operations of the business.

However, a company may decide to treat interest as a current operating expense to lower the net earnings. Total interest cost for the company is then allocated to specific projects. The company therefore acts as the "lender" to the project manager (indirect capitalization).

There are other forms of interest charges to the project such as imputed interest which is an opportunity cost based on the expected value of the rate of return.

Cost Flows

Cost Flows are simply incurred costs spread over the project's scheduled duration and expressed within constant time periods. To cost flow the items below, we would do "spread sheet" calculations:

Table 3-17
Cost flow spread

ITEM	DELIVERY		INSTALLATION	
	Date	Cost k$	Duration	Cost k$
Feed Pump	1 May	200	Jun - Sept.	100
Storage Tank	1 June	50	Jul - Aug.	50
Large Piping	1 July	100	Aug - Nov.	200
Piping	1 July	50	Aug - Dec.	100
		400		450

ITEM	Apr	May	Jun	Jul	Aug	Sep	Oct	Nov	Dec
Feed Pump		Δ 200	25	25	25	25			
Storage Tank			Δ 50	25	25				
Large Piping				Δ100	50	50	50	50	
Piping				Δ50	20	20	20	20	20
Monthly Cost		200	75	200	120	95	70	70	20
Cumulative Cost		200	275	475	595	690	760	830	850

Figure 3-71
Cost flow spread.

Q #3.08: The reliability of an estimate is expressed by a ± % range. If 50 laborers are needed on a job (+ 5% − 6% confidence in the estimate) and their present rate is $11.89 per hour (exact), what is the most likely estimate if the duration of the job is 10 weeks (± 10% accurate) for a 38 hour week?

A #3.08: The bottom line estimate is

$$50 \times 10 \times 38 \times 11.89 = \$\,225\,910.$$

This figure is equivalent to the mode of a frequency distribution. A chain is only as strong as its weakest link, which is ± 11% or approximately $24 850. We must also look at the variance in schedule, which lies between 9 and 11 weeks (± 10%). Combining the two will give us an area of most likelihood.

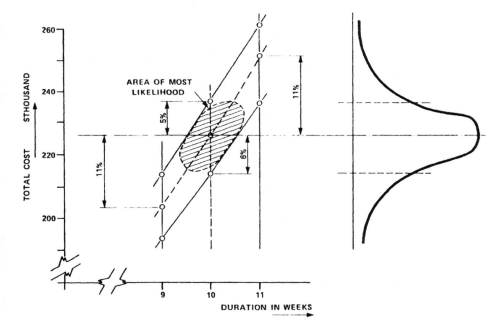

Q #3.09: We must accept that ten estimators will produce ten different estimates based on the same information supplied to them.

You have lunch with a prospective client whose firm intends to start construction of a process plant in Idaho late next year. This plant will have a capacity about one-third larger than a plant your firm completed for $ 355 000 three years ago. Of this cost, 65% was for equipment, about 15% was for erection and 20% for engineering and administration. This plant was constructed in the Golf Coast area.

A #3.09: It will be left up to the reader to estimate a dollar value. Consideration should be given to
- A productivity factor for adjusting the Gulf Coast labor productivity to Idaho.
- An escalation factor over the time span for labor, material and engineering.
- The cost of money (financing) and corporate overheads.
- Proper contingency calculations based on reliability of information.
- Other factors such as size of plant, winter conditions, plant configuration, Government regulations (safety, environmental, permits), transportation, utilities, etc.

Q #3.10: *(Based on lecture notes by Mr. Gord Zwaigenbaum, CCE, Ryerson Polytechnic University; AACE sponsored program).* The purchase cost of a 5600 liter steel tank in 1986 was $ 4300. The tank is cylindrical with a 2 m diameter flat top and bottom. The entire outer surface of the tank is to be covered with 5 cm thick rigid insulation which cost $ 20.00 per square meter in January 1991 while the labor for installing it was $ 40.00 per square meter. Estimate the present (Dec. 1996) total cost for the installed and insulated tank.

End-of-year escalation indexes are: 1986 = 303,   1990 = 398
                                      1991 = 444,   1996 = 688

Use Chilton multiplication factor of 1.43 for tank installation (see also Jelen (5)). The reliability of the cost estimate is considered to be ± 5%.

A #3.10:   Flat top area: $r^2\pi = 3.14$ m^2
                Height (h):   Volume ÷ 3.14 = 5.6 m^3/3.14 m^2 = 1.78 m
                Circumference $2\pi h = 11.18$ m^2

Insulation cost (Dec. 1996 $)   ($ 20+$ 40)× $\dfrac{688}{398 \times \left(444 / 398\right)^{0.5}}$ = $ 1 093

Installed equipment cost (Dec. 1996 $) $ 4300 × 1.43 × $\dfrac{688}{303}$ = $ 13 962

Subtotal in 1996 dollars = $ 15 055
Contingency allowance = $ 753

Total estimated cost $ 15 808

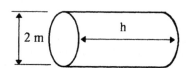

(Indirect cost are not included)

If quoted as a separate item, the estimated cost would probably be reported as

$15 800.

CHAPTER 4
COST CONTROL

The objectives of this chapter are:

1. to explain the methods by which a project manager uses estimates and schedules to control the cost of a project.

2. to describe conventional cost control and how it evolves into an integrated cost control system.

3. to explain the difference between financial and cost control.

4. to provide practical applications with regard to cost collection, cost reporting, forecasting and subsequent actions to keep a project within budget.

Control means taking charge and actively affecting changes to meet the project objective.

Cost control is often confused with cost monitoring. Monitoring is a passive investigative activity that should lead to control.

Planning without action is futile; there is no control without authority. The project manager must have the means to regulate expenditures within the scope of the project.

Introduction

Knowledge is a treasure; practice is the key to it.
                                                                      Old proverb

Chapter 3 has given us a plan, a *path* to follow in order to reach a goal. We now know where we are going. We have to trace this path by constantly evaluating and adjusting our steps in order to reach our predetermined destination in time and within budget.

To control cost, we must be able to make a valid comparison between the elements of the plan and corresponding elements of actual performance.

This chapter deals with the collection of costs, how those costs flow through a control system, and the effect of schedule changes on project cost. A systematic approach to cost control will be emphasized.

4.1. COST COLLECTION

An effective cost control system requires a hierarchy of periodic status reports. What do we mean by *status*? The dictionary defines status as position, situation, state, condition. That means: Where do we stand at this point in time ? How many dollars have we expended up to now?

It should be a simple matter, all we have to do is *ask the accountant*. It is the accountant who records the money spent during the execution of the capital project. The accountant must also produce the corporate financial statements which include capital expenditures. Accounting policies can vary from company to company, however, the basic concepts have a standard application.

It will be shown later, that there is a marked difference between financial control and cost control. This in turn affects the collection of cost incurred vs. cash disbursements made.

Cost data frequently require allocation of accounts to different products and processes. To do this properly requires a considerable technical familiarity with materiel installation and operations. This technical familiarity becomes more and more important with increasing technology. When technology is changing and operations are getting more complex, it is important that cooperation and communication between engineers and accountants be nurtured and their expertise fully used.

The accountant is not an engineer or architect, neither is the engineer an accountant. Both must work together when establishing the rules by which these allocations are made. The cost engineer or technician must have a knowledge of the language and methods of accounting if he/she is to help set up and use accounting data properly.

Classification of accounts must be decided *before* recording is performed. This is done in conjunction with engineering and construction and other functions by means of the Work Breakdown Structure (WBS). Whoever is responsible for cost control must make arrangements with the accounting staff to have pertinent data included in the

Code of Accounts

Accountants and engineers are not only physically separated within an organization, their training and experience are quite different, often causing conflicting points of view. Project engineering is involved in the acquisition of fixed assets and related feasibility studies and economic evaluations and later optimizing the use of resources of all kinds, human, materiel and money. The engineer is concerned with designing the best and most efficient product or process at the best price.

Every activity that takes place in an enterprise has a cost attached that is eventually reflected in the company's earnings statement and balance sheet. In the complex setting of a corporation, it is the accountant who determines the availability of *cash* for projects. In this context, the accountant is the engineer's best partner in translating products and processes into the profits and losses, figures the owner need to make vital decisions. This includes incremental costing, tax savings through depreciation allowances, capitalizing or expensing of cost, the present value of future cost and so on.

Accountants and engineers must appreciate the talents and virtues of each other.

4.1.1 Basic Accounting

Accounting is a technique of measuring economic events in terms of money (5). Accounting uses a double-entry system for every transaction, namely credit = debit, or

$$\text{EQUITY (capital, net worth)} = \text{ASSETS} - \text{LIABILITIES}$$
$$\text{ASSETS} = \text{EQUITY} + \text{LIABILITY}$$

This is reported on the *balance sheet* in a form such that the two sides of the equation, the debits and credits, do always balance.

A company is in business to make a profit. This increases its equity. Revenues increase the equity and are therefore entered as a *credit*. Expenses decrease the equity and are entered as a *debit*. Similarly, to increase a liability requires a credit entry. To decrease a liability requires a debit entry.

DEBIT	CREDIT		HEADING
ASSETS $100	LIABILITIES $ 25		ASSETS $100
	EQUITY $ 75	OR	DR **$100**
$100	**$100**		
			LIABILITIES $ 25
			EQUITY $ 75
			CR **$100**

However, an increase to an asset requires a debit entry and to decrease an asset requires a credit entry. Table 4-01 summarizes these rules.

Table 4-01

Debit/credit entries

TYPE OF ACCOUNT	DEBIT COLUMN ENTRY	CREDIT COLUMN ENTRY
Asset	increase	decrease
Expense	increase	decrease
Income	decrease	increase
Liability	decrease	increase
Equity	decrease	increase

Those unfamiliar with accounting may have a hard time to interpret the meaning of debit and credit especially in the asset account, where any increase in assets is entered in the debit column.

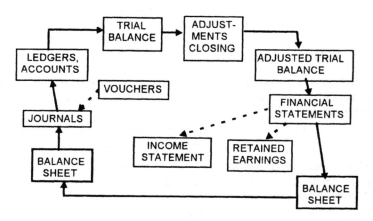

Figure 4-01

Bookkeeping cycle.

Instead of using those terms, we may call it the "left" and the "right" column. A diagram of the total bookkeeping cycle is shown in Figure 4-01.

The *general ledger* consists of a group of accounts. All the asset, liability, expense, and revenue transactions are recorded in the ledger. The general ledger may become too cumbersome for large corporations. It is customary to remove certain accounts from the general ledger and bring them together in a *subsidiary ledger.*

Recording and classifying is known as *journalizing.* It is the final classification of data within the accounting system. All transactions are entered *daily.*

Those new transactions will affect the ledger, which was in balance at the beginning of the period. Since transactions should affect both the debits and credits equally, the ledger should still be in balance at the end of the next reporting period.

To make sure this is the case, a *trial balance* is taken off the ledger (total debit = total credit). Necessary adjustment is made and temporary accounts are *closed* at the end of the fiscal year in order to prepare *financial statements*. This, in a nutshell, outlines the bookkeeping cycle.

Journals and Vouchers

Most companies use computerized Journal Processing Systems (JPS) where all cost information are collected in accordance with proper accounting codes. The classifying and numbering of accounts does not have a uniform application and can vary from company to company. There are, however, generally accepted pro forma charts of summary level *control accounts* issued by associations of chartered accountants (AICPA, CICA, etc.). The basis for the coding of *subaccounts* was discussed in Chapter 3.2 "The Work Breakdown Structure."

Transactions are recorded every day as they occur in documents of some form, usually specific to the company. There are vouchers, requisitions, time sheets, expense reports, delivery slips and many others. Transactions do not necessarily mean payments. They could be transfers from one account to another or allowances and specific charges to a project.

For example, if a pump is bought for cash, the cash asset decreases. But at the same time, the equipment asset increases. This is a simple two-column, double entry (debit/credit) general journal. In practice, journals are designed to meet the financial need of the company and have multiple entries such as sundries, sales credits, accounts receivable, cash and others.

Projects may be assigned one or more general ledger accounts with specialized journals for engineering, purchasing, cash disbursement, office supplies and other accounts.

Posting to the Ledger Accounts

After transactions have been *journalized*, they need to be transferred or *posted* to the various *ledger accounts*. This is done on a daily and monthly basis (Table 4-02).

Example:	Diagnosis:	Journal Debit	Entry Credit
Bought pump $ 800	Asset increased	800	
$200 cash down	Cash asset decreased		200
$600 still to pay	Liability increases		600
(Credit obtained)	(Accounts payable)	-----	-----
		800	800

The last column posts the total of all transactions under each account during the day. If several transactions had taken place under the category "pump", it would show the total dollar amount and the balance (debit minus credit), if applicable, at the end of each month (Table 4-03).

Table 4-02
Posting to ledger accounts

General Journal (J):　　　　　　　　　　　　　　　　　　　　　　Page 1

Date	Particulars	Acc't #	Debit	Credit
1996	Pump	135	800	
04-13	Cash	101		200
	Credit	201		600

General Ledger:

PUMP　　　　　　　　　　　　　　　　　　　　　　　　　　　#135

Date	Particulars	Journal	Debit	Credit	TOTAL	
1996		J1	800		D	800
04-13						

CASH　　　　　　　　　　　　　　　　　　　　　　　　　　　#101

Date	Particulars	Journal	Debit	Credit	TOTAL	
1996		J1		200	C	200
04-13						

CREDIT　　　　　　　　　　　　　　　　　　　　　　　　　　#201

Date	Particulars	Journal	Debit	Credit	TOTAL	
1996		J1		200	C	200
04-13						

Table 4-03
Month-end journal

CASH   #101

Date	Particulars	Journal	Debit	Credit	TOTAL	
1996		J1	1800		D	1800
04-13		J1		1000	D	800

If an account were kept in the general ledger for each sub- account, the ledger would become too bulky. It is therefore the practice to categorize and remove specific accounts from the ledger and separate them in a *subsidiary ledger*.

There may be hundred customers listed under accounts receivable for example. In this case those customer subaccounts would be removed from the general ledger and posted in a separate subsidiary ledger. The total of all accounts receivable would then become the

Accounts Receivable Control Account.

The subsidiary ledger is therefore only an extension of the control account. Since each transaction is included in both, the individual accounts in the subsidiary ledger *and* the control account of the general ledger, the summary of the individual customer accounts must equal the amount in the control account at the end of each month (if this is corporate policy) and at the end of each fiscal year.

Accounts payable is handled in the same way as the accounts receivable. In most cases there are three journals involved, the *Purchase* Journal, the *Cash Disbursements* Journal and the *General* Journal.

The purchase journal usually has multiple columns such as sales tax, sundries, office supplies and others. For simplification, the examples in table 4-04 below show only the accounts payable column.

The total accounts payable entered into the journals will show in the general ledger control account as shown in Table 4-05.

This control account has subsidiary ledgers (sometimes called subledgers for short) shown in Table 4-06.

All subsidiary ledgers that have a balance account are summarized at the end of each month.

Accounts Payable Subsidiary Ledger Summary
as at 1996-04-30 *

Pump Supply Co.	$ 2 500
Clean Pumping Ltd.	$ 3 600
	$ 6 100

* *Please note, that the heading of the date column "1996" is put in front of month and day. This is international practice (ISO).*

Table 4-04
Journal - accounts payable

Purchase Journal (PJ)   Page 1

Date 1996	Supplier Description	Fo	Acct. Payable Credit	Purchase Debit
04-13	Pump Supply Co.		800	800
04-15	Clean Pump Ltd.		1 000	1 000
04-20	Better Pump Co.		200	200
04-30	Total Acc't 135		**2 000**	2 000

Cash Disbursement Journal (CD)   Page 8

Date 1996	Supplier Description	Fo	Acct. Payable Credit	Cash Credit
04-13	Pump Supply Co.		200	200
04-15	Clean Pump Ltd.		400	400
04-20	Better Pump Co.		200	200
04-30	Total Acc't 135		**800**	800

General Journal (PJ)   Page 2

Date 1996	Particulars Supplier	Fo	Acct. Payable Debit	Credit
04-18	Pump Supply Co. (Damaged part returned)	5	100	100
	Total Acc't 135		**100**	

Table 4-05
General ledger - accounts payable

GENERAL LEDGER   **Accounts Payable**   Page 5

1996	Journal	Pa	Debit	Credit	Balance	
04-01	Balance				C	5 000
04-13	PJ	1		2 000	C	7 000
04-15	CD	8	800		C	6 200
04-20	J	2	100		C	6 100
04-30					C	6 100

Table 4-06

Subsidiary ledgers

Pump Supply Co.

1996	Journal	Pa	Debit	Credit	Balance	
04-01	Balance				C	2 000
04-13	PJ	1		800	C	2 800
04-15	CD	8	200		C	2 600
04-20	J	2	100		C	2 500
04-30					C	2 500

Clean Pump Ltd.

1996	Journal	Pa	Debit	Credit	Balance	
04-01	Balance				C	3 000
04-13	PJ	1		1 000	C	4 000
04-15	CD	8	400		C	3 600
04-30					C	3 600

Better Pump Co.

1996	Journal	Pa	Debit	Credit	Balance	
04-01	Balance					0
04-13	PJ	1		200	C	200
04-15	CD	8	200			0
04-30						0

The total amount of $ 2 500 + $ 3 600 = $ 6 100 must agree with the month of April balance of the *Accounts Payable* control account in the general ledger.

The Trial Balance

When all postings have been completed at the end of the accounting period and before the financial statements are prepared, it must be checked if the ledger accounts are in balance. This does *not ensure posting accuracy*, but only that debits still equal credits since the previous accounting period. Posting errors are corrected through the journal for account distribution, which is a separate procedure. The trial balance simply shows debits and credits, as shown in the example below (Table 4-07).

Table 4-07
Trial balance

THE XYZ COMPANY LIMITED
TRIAL BALANCE
as at 1996-12-31

Acc't #	Account Description	Debit	Credit
101	Cash	10 456	
106	Accounts Receivable	180 468	
107	Allowance (bad debts)		323
135	Inventory- Jan. 1	5 999	
183	Land	300 000	
173	Building	500 000	
600	Accumulated. Depreciation		90 000
201	Accounts Payable		456 000
307	Share Capital		360 000
318	Retained Earnings		80 600
413	Sales		60 000
502	Cost of Sales	30 000	
200	Other Expenses	20 000	
		1 046 923	**1 046 923**

Account Distributions

Assuming a particular project is identified in the General Ledger of the company as code 123. Charges from all cost locations to account 123 are posted in specific subsidiary ledgers. This information is sent to the Project Manager's office (31).

THE
MEEK
SHALL
INHERIT
THE
COST

Each transaction should show the source of the charge as well as the distribution. It gives the project the chance to check the legitimacy of an input. Errors and unacceptable charges should be corrected by journal transfers. One must not be *meek* because *not* transferring irrelevant charges means accepting the unrelated cost to the project.

Below is a fictitious example of a transfer done for a major project of a large corporation (Figures 4-02 and 4-03).

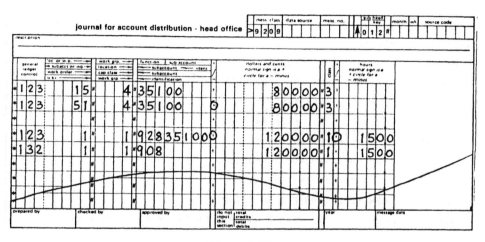

Figure 4-02

Account distribution.

This project is identified by the general ledger control account #123. Out of 15 work orders, the fuel handling system is work order # 15. The subaccount is based on the work breakdown structure's fuel transfer and storage system 351. The project was charged $800.00 under class 3 (equipment rental) for work order 51. The journal therefore shows the following input:

Ledger Control	Work Order	Unit	WBS #	Dollar	Class	Hours
123	51	4	35100	+800.00	3	-

The work order number is obviously in error. It should have been 15, not 51. The charges must be transferred back to work order 15 to avoid cost distortions. The circle indicates a removal (credit) from the wrong work order (see second line entry) to work order 15 entered as a debit under a plus sign (see first line entry).

In some cases corrections to erroneous input can be automatically flagged or corrected. Since there are only 15 work orders in control account 123 and 51 is shown, either an error report is issued or the input may not be accepted by the computer.

Furthermore, since work order 15 deals with the fuel handling contract of the project, the subsystem number must have a 35... for the first two digits. The computer would match 15 with 35 as a default in the system and not accept any other combination.

```
        REACTOR-BOILER & AUXILIARIES                              FILING

                                                            COST ACCOUNT

35000       FUEL HANDLING                                          E    F

  35010       MANUALS & PROCEDURES                                      F

  35100       NEW FUEL TRANSFER & STORAGE   <===              D,E  F

  35110       NEW FUEL TRANSFER                                         F
  35111         Transfer Port Mechanisms
  35117         Tools & Accessories
  35118         Fluid Systems
  35119         Electrical Equipment

  35120       NEW FUEL STORAGE & HANDLING                              F
  35121         Storage & Transport
  35122         Inspection & Loading

  35130       NEW FUEL TRANSFER AUXILIARIES                            F

  35170       NEW FUEL TRANSFER & STORAGE SAFEGUARDS & EQUIPMENT  D,E  F

  35200       FUEL CHANGING                                       D,E  F
```

Figure 4-03
Cost account breakdown.

The other example is similar except that another project 132 has journalized 150 hours of work to project 123. A three-digit function number starting with 9 is entered in front of the five-digit subaccount (92835100). This 9 identifies an engineering charge. Assuming that department 928 is the project department for project 132. We can be almost certain, that project 123 was wrongly charged. By consulting the engineering journal, showing all time sheet inputs including employee numbers, the error has now been corrected on this "journal for account distribution".

There are many other ways of identifying incorrect data inputs, one of which is the "sum-of-digits" method. Here, all digits in a number must add up to 9. An extra digit is added to the base account number, making the total sum-of-digits add up to 9. For example:

Base Acc't.	Sum of Digits	Add Number	Input Account	New Sum-of-Digits
1234	1+2+3+4=10; 1+0=1	8	12348=18;	1+8=9
6744	6+7+4+4=21; 2+1=3	6	67446=27;	2+7=9
9018	9+0+1+8=18; 1+8=9	0	90180=18;	1+8=9

That means an input such as 67546 would either be rejected or posted as an error for correction because its sum of digits add up to 1 instead of 9 (6+7+5+4+6=28; 2+8=10; 1+0=1).

With all the care we take, we will probably never reach a *perfect* cost collection system. The higher the quality of input data, the more costly are preparation and checking. But if we neglect to monitor the accuracy of charges to our project, the true cost picture is distorted, causing management to make wrong decisions which can also be very costly. Some trade-off is necessary.

> *"No project is ever a complete failure; it can always serve as a negative example."*
>
> *Murphy's Facility Factor*

Accrual Accounting

The financial statements produced by the accountant's book-keeping procedures will not give a project manager the ability to judge the value of the progress made and thereby is not good enough for *controlling* purposes. Expenses must be recognized when incurred, whether or not paid in cash. The accounting done to meet that need is called *accrual accounting*.

There is a difference between financial control and cost control.

Just as an example, the Canadian Institute of Chartered Accountants in the publication "Accounting Principles - A Canadian Viewpoint" (R.M.Skinner) describes accrual accounting as
"accounting based on recording the effect of transactions on financial condition and income when the transactions take place, not merely when they are settled in cash." Furthermore, "From the standpoint of the purchaser, if he is buying services, he normally should record the cost and liability in step with performance, i.e. as the work progresses. If he is buying goods, the normal test of performance will be delivery. A few exceptional types of costs, however, are considered to relate to time periods and are accrued in the accounts according to the passage of time rather than in accordance with a performance test."

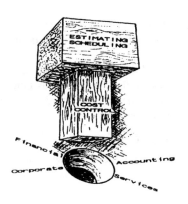

It is never an easy task to fully implement a system where accrual accounting is completely integrated with the cost control requirement of collecting "incurred" cost.

There are situations when the end account cannot be identified. This is especially the case for bulk orders or general material supplies.

For example: According to the flow diagram for a processing plant, there are 50 valves of a common make and size required for the work package K7.

The same type of valves are also needed for some other work packages. Total order is 660 valves which are ordered in bulk from one supplier.

Even though the valves are not needed at once and for different systems, delivery will take place at three scheduled dates with 220 valves each. Those valves will be put in storage at the construction site and withdrawn by the various foremen when needed.

The cost are incurred to the project at the delivery date and entered into an inventory holding account under the category "valves". The distribution to an end account will take place when withdrawn from storage for installation.

The financial module of a cost collection system should be programmed to take care of those common problems. This needs close cooperation between cost accounting and project cost management.

The main criterion is the compatibility between the estimated cost flow based on the schedule and the collection of incurred cost based on accrual accounting.

4.1.2 Cost vs. Cash

Capital cost control requires that we are able to measure progress made and the related incurred cost at any point in time.

The estimate is based on the schedule, and the schedule indicates *when* the work has to be performed and on the timing of materials supplied, equipment delivered and installed.

Project management monitors engineering and construction efforts as they relate to the plan. We must therefore devise a system that recognizes as much as possible the cost related to schedule dates and durations. In other words, we want to express in dollars the value of work done and equipment on site.

This is documented in a hierarchy of cost reports.

Those reports are different from financial statements which deal with cash disbursements. Periodic statements of cash disbursements are called *cash flows*.

They are essential for a corporation to plan project funding and regulate each month's cash draw-down. Interest charges to the project are based on cash flows, not cost flows. There is another type of cost that has a somewhat different flow. This is the *installed* cost.

It is defined as the value of work completed on site and represents the rate at which physical progress is made. In summary, there are basically three types of project expenditures:

CASH FLOW — needed to plan project funding, to set fiscal budgets, to calculate interest (IDC) during the construction period and to evaluate tenders.

COST FLOW — identifies incurred cost, measures contract performance against the approved time phased plan and is used for cost control.

INSTALLED COST — is the value of work constructed and is used to calculate insurance premiums.

The three expenditure flows are somewhat out of phase. The rate at which costs are incurred and ultimately funds are disbursed depends on the nature of expenditure. Cash flow delays such as late invoicing, payment approvals, escalation and holdback recognition leave gaps between cost flows and cash flows (Figure 4-04).

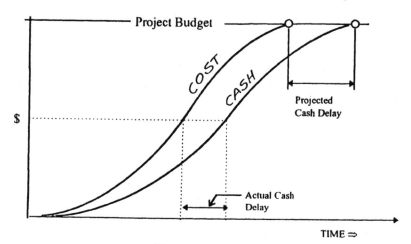

Figure 4-04
Projected cash delay.

Analysis of typical cash flow delays for various types of contracts for a power station in Alberta indicated the following (21):

Site Contracts:	8 weeks from certification
Supply Contracts:	5 weeks from delivery date
Consulting Contracts:	5 weeks from the end of the month

Prestipulated Progress Payments

The incurred cost forecast is developed on a uniform monthly basis and is refined in accordance with the expediters' progress or percent complete reports. The disbursement cash flow depends solely on the contractual pre-stipulated progress payment schedule. *For example:* A *supply contract* has the following payments prestipulated:

A   60%   of contract amount if progress is on schedule. (The owner is paying part of the fabrication cost).
B   25%   of contract amount on delivery.
C   10%   of contract amount as *estimated* by the owner to be paid *extra* to the contract for escalation.
D   10%   of contract amount when the equipment is placed in service.
E   ?   *Actual* escalation payment based on STATCAN* index to be made to supplier.
F   5%   Holdback.

*STATCAN = Official Government Statistics of Canada.

The total contract amount is 1000 k$. The published STATCAN index shows actual escalation over the period to be 8%. The chart shows the dollar values for cash payments and cost incurred in k$ (Table 4-08).

Obviously, the cash flow is lagging behind the cost flow in this case. Progress payments considerably in excess of the value of work done is generally cause for contract scrutiny.

This comparison is one area of effective cost control.

Table 4-08
Cost flow - progress payments

	A	B	C	D	E	F
Cash Payments	600	250	-	100	80	50
Cost Incurred	600	400	100	-	(20)	-
Cum. Cash Flow	600	850	850	950	1030	1080
Cum. Cost Flow	600	1000	1100	1100	1080	

Monthly Progress Payments

Example: A large unit price civil contract requires that 50 000 m³ concrete be poured for $ 20.00/m³. The scheduled start is on April 1ˢᵗ with completion on June 31st. Twice the quantity is scheduled to be poured in July than in June and in August.

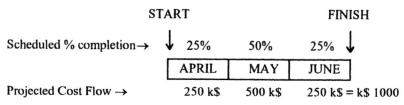

The projected cost flow is the budgeted cost of work scheduled. The actual quantities installed are then measured at the end of April.

The quantity surveyor measures the amount of concrete poured by the end of April and finds that only 20% was put in place instead of 25%. The cost incurred is now 200 k$. Assuming there was a 10% holdback imposed for 90 days after certification, what is the *cash* flow in this example?

The progress certificate for April is prepared in early May and sent to the contractor who prepares the invoice. There is mailing time, management and client approval time, check preparation time and bank processing. This delay in cash disbursement for this type of contract is at least two months after the cost have been incurred. The contractor will receive a check for 200–20 = 180 k$ by the end of June or in early July. Holdback will be paid 90 days after progress is certified (Figure 4-05).

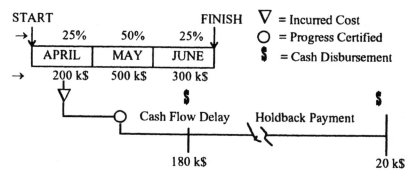

Figure 4-05
Cash flow delay - progress payments.

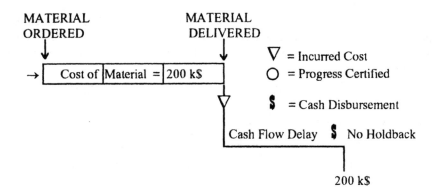

Figure 4-6

Cash flow delay -payment on completion.

The cash flow delay due to holdback may even be longer for major process equipment, where holdback may be retained until the in-service date (or plant commissioning).

Payment on Completion

Contractual terms for smaller purchase contracts or purchase orders generally stipulate payment certain days after delivery. The recognition of incurred cost (cost flow) is on the delivery date. Cash payments are delayed at least one month (Figure 4-06).

Direct Hire Labor

Pay is handed out weekly or monthly as the work proceeds. There is no cash flow delay for labor cost.

Consulting Contracts

Those contracts can exhibit a non-linear rate at which costs are incurred and payments result. The spread between cost incurred and cash disbursements can vary according to the cumulative cost profile.

General Application

The difference between cost flow and cash flow is not always recognized in practice. The term *cash flow* is used when it should be *cost flow*. We cannot even find a distinction between the two terms in some literature dealing with project management!

One of the reasons may be that accountants prepare the cost reports for project management based on cash disbursements. Labor intensive small projects have little cash flow delay and management may feel that there is no need for separate incurred cost monitoring.

The expression *actual cost to date* (or in short "actuals") is often used on cost reports. The definition of "actuals" must be fully understood.

> **If the actual costs are not collected on an accrual basis, the budget comparison is distorted and will look more favorable than it really is.**

Look at Figure 4-04 again.

4.1.3 Account Numbers and Codes

When we discussed journals and account distributions, we put numbers on each item. *(It would be a good idea at this time to review "The Cost Plan" Section 3.2 - The Work Breakdown Structure (WBS)).*

The WBS is the foundation for project control. There are many similarities within the various types of construction projects. Can we somehow devise uniform codes that would allow us to compare cost data between similar projects and programs?

Codul resurselor	Denumirea resurselor	U/M	Cantități				
			G 07 00	H 08 00	I 09 00	J 10 00	K 11 00
	Materiale						
563413390006	Ciment M 30	kg	22,05	22,05	14,595	14,595	—
563711100004	Ciment alb PA 300	kg	—	—	10,000	10,000	—
563313300006	Ciment PA 35	kg	—	—	—	—	21,000
561821122000	Piatră din mozaic calcar	kg	—	—	—	—	—
561811122007	Piatră din mozaic marm.	kg	26,300	26,300	26,300	26,300	26,300
218345120001	Nisip 0...3 mm	m³	0,037	0,037	0,037	0,037	0,037
487000140006	Piatră de frecat	kg	0,253	0,253	0,253	0,253	—
529859120001	Rumeguș de lemn	kg	0,050	0,050	0,050	0,050	—
511116122000	Sipci de rășinoase geluite	m	0,360	0,250	1,50	—	—
210121200003	Apă	m³	0,014	0,014	0,014	0,014	0,014
591351111000	Geam tras simplu 5 mm grosime	m²	—	—	—	0,090	—

Figure 4-07
A national standard code.

A great amount of time and effort has been expended by AACE Committees (2) and others to develop a standard code. In spite of considerable progress made, we still do not have one national standard. Can it be done? Should it be done? Is it even necessary? The size and diversity of free enterprise in North America may not lend itself to one national standard that *must* be adhered to.

It could be possible and has been accomplished in some countries. The standard DIN, first developed in Germany, is widely excepted in Europe and other countries in the world. As an example, Figure 4-07 is an excerpt from the national standard code for the construction industry in Rumania.

Since Work Breakdown Structures are wholly project oriented and projects vary from one type to another (see 3.1.2-Types of Projects), standardization is very difficult to obtain. There is also resistance to change a numbering system that has been in use for years. What seems adequate at the time may have become a cluttered unwieldy assortment of numbers. But since then it became completely entrenched in the daily business routine of the company. Internal consistency can become a bottle neck for an efficient project cost control program. Whatever the structure, a project's numbering system has to be integrated with a viable corporate financial accounting scheme.

The Subject Classification Index (SCI)

Only five fields of the corporate accounting structure were allowed for the SCI system described in Section 3.2 to identify a huge number of capital construction elements. Five digits from 00000 to 99999 should theoretically give us over 6 million combinations.

This is drastically reduced to approximately 10000 when some desirable logic is added to the numbering system. If boilers and preheaters are 33110, the instrumentation for those would have a 6 as a first digit, i.e. 63311 by dropping the last digit zero. Furthermore, if we want the capability to summarize certain types of equipment, we may restrict the fifth digit to specific items:

> 2 = pumps, fans
> 3 = valves, dampers
> 4 = tanks, drums and other containers
> 5 = filters, strainers
> etc.

It was found that for *controlling* purposes, the 5-digit SCI system was adequate enough.

Other Systems

Because of their similarities, it is easier to derive at standardization for building construction projects.

The *Cost Engineer's Notebook* (2) suggests a 4-level division with an 8-digit code for processing plants:

Level	Description
1	2-digit components
2	2-digit functions
3	3-digit elements
4	1-digit element detail

Similar to the SCI system, a left digit always has priority over a digit to its right. The first four digits describe major components and their functions. The other four are cost elements.

Whereas the SCI system uses the IN → PROCESSES → OUT concept, those codes use the "area" concept. There are four major areas, the process area (battery limit) and three off-site areas, storage and handling, facilities, utilities, and services.

A breakdown by discipline is also quite common:

100 Civil	200 Mechanical	300 Electrical
110 Concrete Work	210 Equipment	310 Batteries
111 Substructure	211 Pumps	311 Battery Charger
etc.	etc.	etc.
120 Structural	220 HVAC	
121 Columns	221 Ductwork	
etc.	etc.	

Ahuja (12) refers to a modified Uniform Coding Index (UCI) with a seven-level breakdown. This index is alphanumeric:

Level	Description
1	Functional Elements A to S
2	Functional Subelements A to M
3 to 7	Work Items (numeric)

Example:

- Mechanical
- Heating
- Power Heat Generation

K B 15600

Kerzner (35) describes a six-level structure:

Level	Description
1	Total Program
2	Project
3	Task
4	Subtask
5	Work Package
6	Level of Effort

Example:

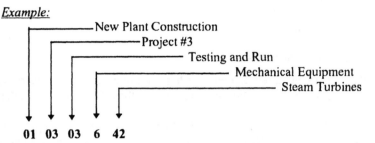

01 03 03 6 42

Tompkins (36) suggests Tertiary Account Codes:

Section	Description	
A-1	X.0-0000	Primary
A-2	0.X-0000	Secondary
A-3	0.0-XY00	Tertiary
other	0.0-00XY	Detailed

Example:

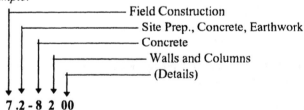

7 .2 - 8 2 00

The Construction Specification Institute (CSI) has a "Masterformat" with four levels:

Level	Description	Example
1	Division	02 = Sitework
2	Section	1 = Site Preparation
3	Activity	4 = Dewatering
4	Estimate	0 = (detail)

Some publishers of estimating reference data such as R.S. Means use the CSI code as a base to categorize estimates. They expand cost data into 16 divisions.

Giammalvo (45) considers an application of the Masterformat a tool for Total Cost Management. He describes in his paper how a modified CSI code can be used as a WBS for a project.

It is obvious from the few examples above that the coding of accounts is based on individual preferences. However, they all have something in common:

♦   In data collection, all elements of a lower level are summarized and become an element at the next higher level (rolling up).

- Some flexibility is maintained to allow for expansion due to scope changes and different project configuration.
- The lowest level element is of manageable size and value, i.e. the benefit must be greater than the cost for excessive details.
- Not all work needs to be subdivided to the lowest level.

4.1.4 Quantitative Cost Collection

Actual costs are not only incurred in dollars but also in terms of time, materials, products, processes, activities, and borrowed money. This leads us to the fact that

Much of the expenditures for a capital project are *measured* in terms of time and quantities and then *converted* into dollars for cost reporting.

Labor

For cost conversion, *time* is normally measured in hours duration. If the work day in construction is 8 *hours* (8 hrs), a worker on the job for 8 hours will have performed 8 *workhours* (8 Whrs.).

It is important for cost control that all workhours are categorized under the proper direct and indirect subaccount. Engineering workhours are usually collected on a weekly time sheet. The method of collection and the format vary with each company. Direct computer input has become so common, that we forget how a collection system works. The examples below may be a helpful explanation, but they do not advocate to be a standard. Some companies consider engineering time an "indirect" activity. Especially on larger jobs, design work on capital accounts could be classified as a "direct" time expenditure. The time sheet (or electronic time input) would then include a space for entering the WBS subaccount. The approved time sheet is also used for payroll purposes. It will show time off such as sick leave, vacation, paid and unpaid leave of absence. It is also used for the preparation of employees' paychecks.

Time sheet information is journalized as a *primary* distribution to the control account of the originating department. To collect engineering cost for a project, a *secondary* distribution is established.

Figure 4-08 is an example of a secondary account distribution as applied to a capital project. In this case engineering expenditures are essentially time (Whrs.) and expenses (dollars) documented on time sheets and expense reports.

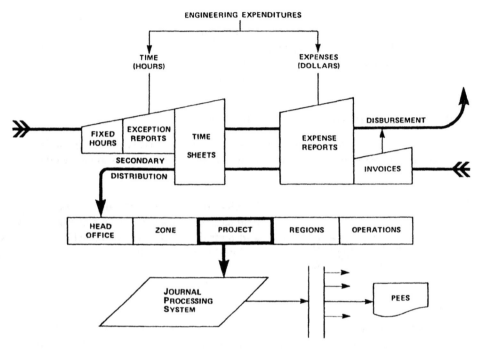

Figure 4-08
Secondary account distribution.

This information goes to the Payroll Department for disbursement to the employee *and* as a *secondary* distribution into the company's Journal Processing System (JPS). The JPS is accessed by another computer system dedicated to the reporting needs of a specific project. In this example we will call it the Project Engineering Expenditure System (PEES). PEES is a data base that selects pertinent information from JPS, combines and summarizes it, and then generates reports needed to monitor engineering cost. In reference to section 4.1.2 we can say that primary distribution produces a cash flow and the secondary distribution gives the project the necessary cost flow. Those costs generally consist of:

PAYROLL	+ PAYROLL BURDEN	+ OVERHEADS
Employee Salary	Vacation Pay	Office space
Expenses	Sick Leave Grant	maintenance
Bonus	Paid Absence	equipm supply
	Insurance Taxes	Cafeteria
	etc	Stand-by time
		etc.

In some cases *profit* is added. This depends on the policy of the company and on the type of contract.

Overhead cost are very difficult to enumerate. They are in fact estimates. Because of many changes in staff and office requirements, overhead rates will have to be adjusted periodically.

Assuming an owner does his own engineering for one of his projects. The engineering cost to the project would be as follows:

Department	:	Mechanical Design
Staff Level	:	50
Overhead Cost	:	$ 30 200/week
Payroll	:	$ 45 000/week
10% Burden	:	$ 4 500/week
Add to Payroll	:	34.7/45 = 77%
Average Payroll per Employee	:	45/50 = $ 900/week
Add 77% Charge to Project	:	900 * 1.77 = $ 1 593/week
With a 35-hr week	:	1 593/35 = $ 45.51/Whr.

This is the amount the Mechanical Design Department would charge the project per hour of work. Instead of average payroll, a department may categorize the staff by salary levels:

Level	Title	Average Payroll	Total $ per week	Hourly Rate	Gross Rate(+77%)
1	10 Engineers	1 500	15 000	42.86	75.86
2	8 Designers	1 000	8 000	28.57	50.57
3	10 Technicians	800	8 000	22.86	40.46
4	12 Draftsmen	667	8 000	19.06	33.73
5	10 Clerks	600	6 000	17.14	30.34
	50		45 000	Average =	46.19

This will increase the validity of departmental gross rates. *Actual* overhead cost are accumulated in a holding account and negative or positive balances cleared at year-end or at budget review time. A new rate is then established.

Caution! Those *departmental* overheads must not be confused with *corporate* overheads charged to the project for corporate general services such as legal, medical, advisory and other support. Those costs are not distributed to system accounts.

In order to gather all information necessary to control the job for a large project and also satisfy payroll requirements, and the corporate accounting structure, the layout of a time sheet can be quite a challenge. Smaller companies will need less complicated time sheets. It is always easier to reduce requirements than the other way around. The layout shown below can be simplified.

When engineering time is considered direct *and* indirect work, a system classification number (SCI) must be included in addition to the general ledger control account, work order and subaccount. The layout may look somewhat like this:

Design Department	1	2	3	4	5

XYZ Company
Name of Employee:

2	3	7	6	0	1	2	3	4	5	3	Y	Y	M	M	D	D

Joe Doe

This is the *heading* of the time sheet, whereby
1. = the pay center's payroll number
2. = the check digit for the payroll number (2+3+7+6+0=18; 1+8=9, see section 4.1.1).
3. = employee number
4. = employee number check digit
5. = year, month, day end of time period (Wednesday night)

Now for the account distribution:

DESCRIPTION	6	7	8	9	10
Fuel Handling	6 5 1	3 5	1 9	9 2 8	0 3
Reliability Study	6 5 1	0 6 5	1 9	9 2 8	0 3
Sick Leave				7 2 8	0 3

6. General Ledger Control account. This identifies the project.
7. System Classification Index (SCI) number. First digit 1 to 7 is direct work pertaining to permanent facilities. First digit 0 (zero) is used for indirect engineering such as estimating, scheduling, procurement, special studies and other work that benefits the overall project. This field is left-justified. The trailing zeros for 35000 need not be shown.
8. The Work Order number is a breakdown of the project into units. 19 is head office work. This field is right justified. The leading zeros need not be shown.
9. Department numbers. They always start with 9 which must coincide with the last digit of the work order. This helps to avoid errors. Time charged to the department overhead is coded with 7 instead of 9.
10. (A possible 11) can be used for special coding such as department section or discipline (engineer, draftsman, clerk etc.)

The hours will now have to be filled in for each description:

DESCRIPTION	Total	Th	Fr	Sa	Su	Mo	Tu	We	O.T
Fuel Handling	20.0	7	9	-	-	-	4	-	2
Reliability Study	13.0	-	-	-	-	-	6	7	3
Sick Leave	7.0	-	-	-	-	7	-	-	-
Totals	40.0	7	9	-	-	7	10	7	
Overtime (O.T.)		-	2	-	-	-	3	-	5
Travel Time									

Checked by _Joe Doe_ _____ Certified by _Doe All_ _____
     Cost Technician   Supervisor

The reason for overtime total (O.T.) in the right column is that *all* direct hours are monitored under system accounts. The trend is for less manual prepared time sheets in favor of electronically transmitted time distributions which still must be checked and certified.

Cost collection for *Construction Forces* is similar to that for engineering except that *daily* time cards are required. This is needed because of the variety of jobs and worker turnover. In addition to hours, quantities installed (meter of pipe, number of joints) will also be collected for each work item. This is necessary in order to monitor productivity of the work force. Figure 4.09 is an example of a time card:

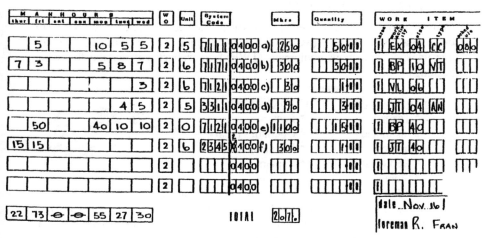

Figure 4-09
Daily time card.

Those time cards are filled in by the foreman for his crew, not by the individual worker.

Permanent Equipment and Materials

Major equipment and materials will be ordered by the head office' purchasing department. The bulk of routine purchases are made by the field office. Whatever, all purchase orders must show the proper account distribution so that delivered material can be identified physically and on the invoice. Cost Accounting will then be able to enter quantities and cost incurred into the Project Material Management System (PMMS) when equipment and materials are delivered (see Figure 4-10).

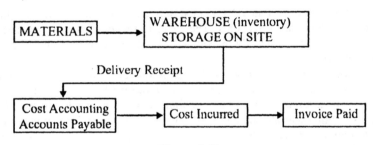

Figure 4-10
Material cost accounting.

Transport and Work Equipment:

Construction equipment can either be owned or rented. Those costs are not distributed to a system account (SCI) but to a special equipment account. The cost for this transport and work equipment (TWES) are collected under the following headings:

Owned:

Cost of Ownership	Operating Cost
* Taxes	* Fuel
* Insurance	* Operator's Salary
* Interest	* Maintenance
* Depreciation	* Repairs
* Storage	

The average cost per hour is calculated under both categories.

Ownership costs are incurred whether the equipment is used or not. The average cost per hour is therefore calculated on calendar time (years of working life). Average operating cost per hour is calculated on the number of hours the equipment is operating (quantitative cost collection). Both combined are the total ownership cost to the project.

<u>Rented:</u>

<u>Rental Cost:</u>	<u>Operating Cost:</u>
* Arrival Time	* Operator's Salary
* Period (from....to)	* Fuel
* Rate per time unit	
* Cost per Period	
* Cost to Date	
* Equipment Release (return)	

The hours of possession of the rented equipment are recorded and converted into dollars. Operating cost may not apply if the operator comes with the rented equipment. This is often the case with transportation equipment.

Financing Cost

They could be categorized under

- ♦ Interest During Construction (IDC)
- ♦ Corporate Overheads

Interest are calculated on the owner's total *cash* outflow to the project. Each trade performs a certain amount of work within each increment of time, and is paid reasonably soon after completing that work. The money used to pay each contractor or sub-contractor each month is provided by a construction loan. Monthly "draws" or "take-outs" are made against the loan. Once a draw is made, interest charges start accumulating on that money. Because the construction loan debt is not capitalized until the completion of construction, all amounts actually disbursed must be financed from the time of disbursement until the loan is retired.

There are many ways of doing those calculations depending on who does the financing. The interest cost are usually charged monthly and compounded annually on long duration projects.

It is quite common on projects with two or less years duration to compound the interest rate on a monthly basis. Interest charges for a particular month are added to the principal (take-out) of the following month. The interest charges for the following month are then calculated on the new total which consists of principal and the previous month's interest. This is repeated to the end of construction.

In case of a contract, the progress payments received and the funding required by the contractor may leave a gap, resulting in interest cost to the contractor. Here again, the calculation of those costs depend on the type of financing available.

The share of corporate overheads is distributed to each project on a cash flow percentage basis in accordance with corporate policy. It is based on the owner's involvement in the design and construction of the project. Those overhead charges are under constant review. Projects built under a management contract will draw very low corporate overheads or none at all.

Final Words

Collected data on materials, labor and
overheads must not only satisfy the needs
of financial accounting but must also be
structured to serve as a base for the control of project costs.

Q #4-01:

Why is it often that the difference between cost incurred and cash flow is not recognized?

A #4-01:

There may be two main reasons:

1.   The project is small and very labor intensive. Since there is no cash flow
      delay for incurred labor cost and the material invoices are paid within a
      month of their receipt, the difference between cost and cash is not apparent.
      The manager for small projects may also be responsible for the accounting
      function on the project, fusing both, financial control *and* cost control.

2.   A company's accounting function is not set up to recognize incurred cost
      and report on an accrual basis. A system has not yet been implemented
      where accrual accounting is fully integrated with the project manager's cost
      control requirements.

4.2. COST CONTROL SYSTEMS

Cost Control is an important part of total cost management. Whatever action we take, whatever we do or fail to do, there are costs involved.

Due to the complex interrelationships between the many activities and responsibilities required to carry out project work, a *systematic* approach to project management is required. One aspect of this approach is *cost management* (14).

The word "system" is used in its broadest sense, i.e. it encompasses the total process of cost management.

4.2.1 System Definition and Concepts

What is meant by "system"? The word comes from the Greek sun (together) + histanai (to make stand, setup) which literally means "a set of interacting elements." Kerzner (35) defines a system as:

"A group of elements, either human or non-human, that is organized and arranged in such a way that the elements can act as a whole toward achieving some common goal, objective, or end".

SYSTEM

♦   is programmed actions leading to desired ends. It is the product of intelligence designed to save steps and work and money.

SYSTEM

♦   provides a sense of direction, poise and preparedness.

SYSTEM

♦   *and planning and orderliness* are not different words for the same thing, but they fit very well together as a guide to a more efficient business and personal life (16).

SYSTEM

♦   and order are means toward carrying out *plans*.

Much of the expensive waste in industry and government is due to ineffective planning and lack of system. These lead to inefficient use of resources such as materials, equipment, time and labor. The best chance of success is a plan where everybody is involved as a team, working toward a common goal.

> **A system is a plan for getting work done, under control and
> by using reliable data.**

The essence of a *Total System Concept* (17) is, that a business exists to serve certain *objectives* and that the overall system (or interrelated factors of people, equipment, materials and money) can be divided into subsystems having specific functions, which in turn can themselves be subdivided just as the components of the human being can be identified and then studied.

Clear grasp of this concept gives a *task* oriented view of problems and solutions, which is superior to attacking symptoms rather than causes (fire fighting).

Closed and Open Systems

Closed systems are relatively static over time and can be regarded as being independent of environment. Outcomes are determined by internal conditions, such as a closed chemical reaction system or memos to others within a department.

Open systems have a continual input/output exchange with their environment. They may appear static in form, but this is in reality a "dynamic" stability. Open systems behave in an adaptive manner in line with "survival" objectives. Living organism, social systems and the *project environment* are characteristically open systems.

The fundamental representation of an open system is:

In an open system, the transformation of the "throughput" (process) can be of different kinds. For example:

INPUT		PROCESS		OUTPUT
Existing Skills	⇒	Learning	⇒	Enhanced Skills
Powder	⇒	Molding	⇒	Plastic Article
Fuel	⇒	Furnace	⇒	Heat
Cost Data	⇒	Analyzing	⇒	Cost Report
One Location	⇒	Transport	⇒	Next Location

Taking the simple example of a heat producing furnace, we know, that a feedback is necessary to maintain a level of heat required. We also know, that some losses are incurred during the burning (processing) of the fuel.

It is also important to understand the processes leading to the *transformation* of the input into the desired output. In case of the furnace, the input (fuel) is *converted* into heat. The fuel is used up.

In case of cost reports, the cost data is *used* but does not necessarily change. It still exists in a summarized form after processing. Cost reports may also require feed-back in the form of a review of input data that enters the process. Not all data is useful, errors in coding could occur or some data may not enter the processing in time for specific reports etc.

The learning process will only *change* certain characteristics (operation) but not the structure and the functioning of the object (student).

When we move an object, a person or information from one point to another, the only change is the *location*. Within the environment of the open system, we have a feed-back, some inefficiencies of processing (drain-to-waste), and an inter-action of resources with the processes (Figure 4-11). A well designed system should work in a smooth balanced way. All the processes should act together in harmony to perform one of the main functions of an operation.

When we specify the output of a system within a certain environment we like to *minimize the losses* incurred while we process incoming data and information.

Elimination of loss is as necessary to a successful project as the making of profit.

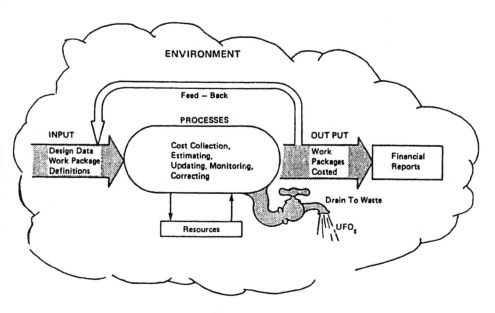

Figure 4-11
Input/output model.

Some of the losses, depicted as "drain-to-waste" in the input/output model can be called UFOs, UnFortunate Oversights (or Undetected Fatal Oversights). They are polluting the environment and affect the credibility of cost reporting to management. We should strive to minimize these losses by improving the processes and the input data.

Other System Concepts

Systems thinking has permeated most branches of knowledge. It leads us to the thought that the nature of an entity is discovered not alone by study of its separate elements, but also by observing how these elements interact with each other and with the environment.

The property by which open systems can adjust their inputs and processes (operations) while still achieving the same end-result is called *equifinality*. Equifinality is the essence of cost control. The project objective is achieved through flexibility when managing resources.

The word *holism* has been given by J.C. Smuts to the concept that the whole system cannot be looked at as just the sum of its parts. A kind of holism is the much publicized *chaos* theory. Rather than reducing the system to its component parts, chaos examines the system as a whole, tracking repetitions and variations that occur on different levels within the system (market trends, rate of exchange, weather patterns).

When input/output transactions take place across boundaries within their environment, they *interface* with each other. This can affect the operation or even the survival of systems. Interfacing underscores the importance of interdependence of systems rather than their independence.

When systems are composed of multilevel-level sub-systems and components, interacting in a logical arrangement, this is called a *hierarchy* of systems. The work breakdown structure is such a system.

4.2.2 Systems Overview

The PMI Standards Committee (PMBOK Exposure Draft Aug. 1994) has identified control within cost management as follows (3):

> "The processes of comparing actual performance with planned performance, analyzing variances, evaluating possible alternatives, and taking appropriate corrective action".

AACE's Cost Engineer's Notebook (2) defines cost control as

> "...the application of procedures to monitor expenditures and performance against progress of projects and manufacturing operations; to measure variances from authorized budgets and allow effective action to be taken to achieve minimum costs."

There seem to be as many definitions as there are organizations or books written about cost management, but this is basic to all definitions:

Costs shall be known or be foreseen in sufficient time to enable remedial action to be taken if any variation from the target is observed.

Controls must be effected *before* a commitment is made. After a contract is placed, little cost reduction is possible. That means, as the project develops, the manager will have less influence on project control.

One important function of cost control is the recording of actual cost performance during the various construction phases. This includes durations, suppliers' and construction performance, design criteria etc.

Cost management on capital projects is a dynamic process. During the initial or concept stages of a construction project the expenditures are minimal, but at this time, the *commitments* for future expenditures are very crucial. Decisions made then have a large influence on cost prevention. Further on as we specify in more detail what goes into the plant and after we requisition some of the major equipment, commitments rise sharply. At the same time, the influence, to reduce total cost diminishes.

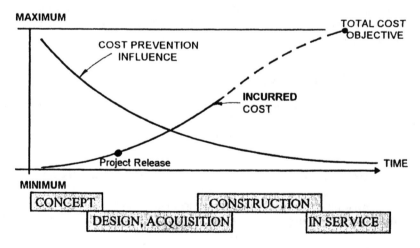

Figure 4-12

Cost prevention influence.

By the time a project task force is established, major design features have already been defined. Even though we are still at the very beginning of the expenditure curve, the cost prevention influence at this time is already reduced drastically in regard to total project cost (Figure 4-12).

As we go along, we do not want to move the TARGET, but we certainly want to improve our AIM.

This requires control over our expenditures even beyond the end of the design period. We must, therefore, be acutely aware of *cost flows* and the means of establishing them. We must know the portion already expended as well as the timing of future expenditure in relation to the planned program. This encompasses:

♦ Predicting the final capital cost level at any point in time.
♦ Knowing the proper expenditure "path" to reach this level.
♦ Identifying the magnitude and causes for deviation from the planned cost flow.

Many attempts have been made to establish standard action plans and procedures for a successful cost management. The diverse complexity and uniqueness of each type of project makes standard applications very difficult. There are, however, tools available that can greatly reduce project failures. Those are discussed in this and other chapters.

4.2.3 The Control Cycle

The elements of cost control follow a fairly simple *sequence*. This is similar to any other engineering control cycle (5).

Lets assume an engineer wants to control the temperature of a room. There are several logical steps necessary:

1. <u>Have an objective:</u>   Maintain room at an even temperature.
2. <u>Have a plan:</u>   i.e. install a thermostat and set at 22°C.
3. <u>Measure performance:</u>   Read temperature, which may be 20°C.
4. <u>Compare with plan:</u>   Analyze the variance from the objective (-2°C).
5. <u>Forecast future performance:</u> Thermostat may lose accuracy 10%/year.
6. <u>Take action to improve performance:</u> Calibrate the thermostat.
7. <u>Revise the plan:</u>   Install the calibrated thermostat.
8. <u>Report on new status:</u>   Continue to measure performance etc.

Those elements, when applied to capital cost control become the Cost Control Cycle (Figure 4-13). Whatever the size or type of project may be, the same control cycle is applicable.

This control cycle is a multiple input system with feed-back.

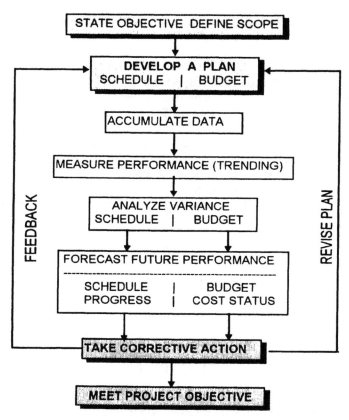

Figure 4-13
The control cycle.

Assuming we want to build a cottage:

<u>Objective:</u>	Build a cottage on Rice Lake before the end of this year, with less than $ 80 000.
<u>Scope:</u>	3 bedrooms, double garage, close to the lake. Prepare scope statement.
<u>Develop a Plan:</u>	**Schedule:** Start April 15, finish November 1st **Budget:** $ 70 000 (cost flow). Prepare a statement of work incl. drawings and material list.
<u>Accum. Data:</u>	**Measure** physical progress and incurred cost. When did work start? Expended Whrs., performance, etc.
<u>Trending:</u>	**Compare** progress with schedule (behind by one month). Compare cost with budget (below budget).

<u>Variance:</u>	**Schedule:** Carpenters work overtime to make up for lost time. Electricians and plumbers on standby. Actual percent complete: 20%, planned: 30%.
	Budget: Cost Variance: $ 2 000. Labor below budget by $ 5 000, excess material on hand ($ 3 000).
<u>Forecast:</u>	Revised completion date is Dec.1st.(November work indoors). Design changes require $ 6 000 additional funds. Forecast: $ 80 000.
<u>Corrective Action:</u>	Temporarily release electricians and plumbers, hire more carpenters, streamline procurement activities, cut some corners on "fringe" designs (cheaper doors and trimmings). Re-estimate cost for single garage.
<u>Revise:</u>	Implement revisions to the plan. Continue to move through cost control cycle.

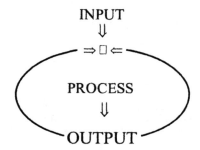

4.2.4 Systems Objectives and Requirements

The input into the cost control cycle (Fig. 4-13) shows

STATE OBJECTIVE DEFINE SCOPE

Project objective and scope definition are closely related. The scope is defined by expressing the objective in terms of outputs, required resources and timing. To design a costing system it is essential, that company policies and objectives be used as a framework. Below is an example of one company's stated project objective for a large construction project (19):

> **Complete the project on time and within budget, consistent with safety, quality and due regard for the environment.**

This overall summary objective could have further been subdivided with statements such as

Generate, improve and maintain cost consciousness throughout the organization through information, education, delegation and accountability.

or

Supervising design engineers must accept the cost responsibility for their work packages and report the cost impact of changes in these work packages to their superiors for approval.

The project objective is further sub-divided into cost system objectives and cost system requirements:

Cost System Objectives

I.   All significant design, procurement and scheduling decisions are to be made by reducing economic cost to a feasible minimum.
II.   Maintain a cost data base and keep it current for changes that have an impact on the project estimate and the budget.
    a   Monitor the impact of the changes and initiate corrective action in order to reduce variances.
    b   Minimize variance due to errors in cost distribution
III. Commitment and accountability for controlling the cost of work at various levels in the organization hierarchy.
    a   An understanding among project staff of the total flow of financial and cost information from the production of estimates to the collection of actual cost.
    b   Acceptance of the approved scope and related cost estimates by those responsible for the work. Management's full support for project cost control.
IV.   Be consistent with Corporate and Division procedures such as the Work Program Budget, the Signing Authority Register, and the Financial Management Guide.

Cost System Requirements

From those basic objectives, system requirements are derived. For example:

I.
    a) Cost information must be provided to allow individuals to assess the impact of their decisions on life cycle project cost.
    b) All significant design decisions must be supportable as the "least cost" option to meet the technical requirements.
    c) All technical requirements must be supportable by regulations and economic benefit.
    d) All purchase awards must be made on the basis of the lowest evaluated tender while meeting technical specifications.

II.
    a) Integrate the project estimate and the work program budget.

b) Maintain and apply an accurate and well defined cost coding system for true cost collection.
c) Provide a valid comparison between the elements of the estimate (or budget) and corresponding elements of actual cost incurred.
d) Introduce a reporting process which keeps others in formed of cost status and changes in cost in a timely manner and consistent with their responsibility
e) Incorporate and monitor changes in design, schedule, contracts, economic factors etc. without delay.
f) Predict the final capital cost level at any point in time and know the proper expenditure "path" to reach this level.
g) Analyze the impact of costs to-date on the project estimate and, when necessary, provide management with alternative courses of action to minimize the impact.

III

a) Establish a hierarchy of work packages and assign responsibility for managing each work package.
b) Provide a hierarchy of reports which easily allows each successive level of management to keep abreast of changes and problems at lower levels.
c) Document adequately the impact of decisions on project cost.
d) Induce cost consciousness among project team members by providing documentation and reference information on project performance and cost status.
e) Provide system familiarization and educational courses to staff members from time to time as required.

IV

a) Follow corporate requirements for cost optimization and purchasing.
b) Document adequately the impact of changes on project cost.
b) Identify quantitatively significant causes for variances in regard to scope, estimate, schedule, economic.

The above project cost control system was designed on a *conventional* basis. The state-of-the-art in cost engineering requires that cost and schedule be fully integrated. This does not change the objectives. Only some requirements are different in wording, not in basic content.

Implementation

Project Objectives and Project Requirements outline expected results. We now have to organize those expected results by levels of the organization and by functional layers. Project control is implemented in two major phases, the project planning phase and the project execution phase. The implementation plan includes the following:

Scheduling and Control Functions:
Organize data such as work breakdown structures, sequencing of tasks and their interdependencies, critical path scheduling, resource allocation and leveling.

Cost planning, i.e. schedule based estimates, resource based estimates, budgets.

Implement cost control by integrating collected cost data and budgets with performance, linking other corporate data systems into it.

Levels of the Organization
Initially select the most skilled, the most experienced and likely to succeed group, able to use project management software. Expand resources once this group is stabilized. Get everyone informed and working as a team working as a team.

4.2.5 Cost Information Flow

Behavioral Implications

What is *information?*
Information is a set of facts which when examined by an individual, under circumstances which surround the individual, adds to that person's body of knowledge. The purpose of information is to "inform", not to confuse. To inform, we should ask ourselves

♦   Whom are we trying to reach?
♦   What is the real message behind the massive technical data and computer printouts?

There once was a blacksmith who told his apprentice: "I will hold the horseshoe and when I nod my head, you hit it with a hammer."

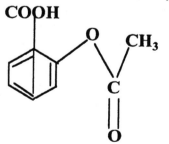

(Famous Last Words)

Words and symbols are used as signals that flow from mind to mind. For example: If you don't speak "engineering," you can't read "engineering." You have to have the words - the vocabulary - of the discipline. Engineers speak in formulas all the time. It's a valuable way of compressing large amounts of symbolization. Chemists will understand the symbol to the left. So may the rest of us - if we knew the simple way to say it: ASPIRIN.

The sculptor Rodin once explained the secret of his art: *"One simply gets a block of marble and chisels away the unwanted material concealing the masterpiece."*

Cost Engineers chip away at mountains of data in the hope that a work with logic, structure and clarity will emerge. And that work is called *information*.

The Random House Dictionary defines information

In·for·ma·tion (in/fər mā/shən), *n.* **1.** knowledge communicated or received concerning a particular fact. **2.** any knowledge gained through communication, research, etc. **3.** any data that can be coded for processing by a computer.

It could be argued that 3. Information should not have been defined as input, but *output*. Data is *input*, information is processed data. It would have been more appropriate to say: "3. Any data that can be coded, after it is processed by a computer."

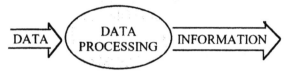

Data is the *input* of system processes. Raw data should not be used as an end product for decision making. The data should be converted into information by being analyzed, corrected, summarized and formatted for easy interpretation.

 It is important that we distinguish between cost data and cost information.

Information Overload

Managers have frequently complained that they are not able to act on cost information that is available to them. One reason why the manager may not act on this is, that he/she do not trust it. That means that either the reporting of the cost is inadequate or not in the form conducive to action. The amount and detail of data may be overwhelming, its content may be ill-defined or may contain outmoded historical costs instead of current values. It may be misleading or too complicated. In order to communicate cost information effectively we must consider the following:

Managers must not receive more information than they can possibly absorb. Up to a certain point, more information reduces uncertainty, but beyond that point information overload sets in. The usual reaction is to select certain portions from the mass available and base action entirely on it, ignoring the rest.

There are, however, managers that request large amounts of information. They want to screen *all* data coming into their department, then passing on routine matters to the areas concerned, delegating non-routine matters to supervisors and then looking after the important items themselves. Screening all data will not increase the quality of their decisions, but it may increase their *confidence* in those decisions (Figure 4-14):

The other mode in managing information is obtained by structuring multiple channels of information flow. Routine data is processed in accordance with established procedures and then passed on to the manager for decision making.

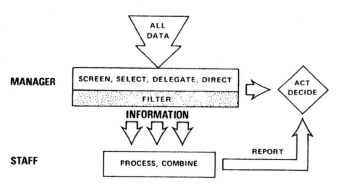

Figure 4-14
Information overload -1.

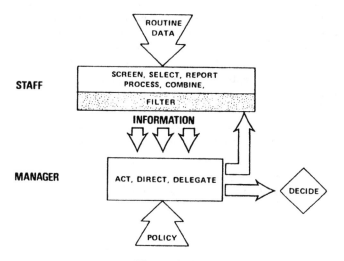

Figure 4-15.
Information overload - 2.

This will certainly decrease the amount of information, but it is a function of the CONFIDENCE the manager has in the decisions being made by lower level staff. (Figure 4-15).

Quality of Information

All items of information are *contaminated* with inaccuracies thereby causing uncertainties. The quality of data required for input into the system must be governed by the cost obtaining it and the risk in using it.

Contamination of information with inaccuracies causes uncertainties. This in turn affects the decision making process. It is a constant struggle to maintain reliable data.

GIGO

Obviously, the coding of accounts is a major consideration in controlling cost. A proper numbering system will permit the costs of like or related items to be collected together, and analyzed in relation to budgets so that trends can be examined.

> **To control costs, a valid comparison between the elements of the budget and corresponding elements of actual expenditures is essential.**

Unless the reporting of costs against established codes is closely adhered to, the information produced is of little value, and effective control and forecasting of costs is not possible. Using a wrong code for cost data is like dialing a wrong telephone number. On drawings, bills of material, time sheets, procurement documents, sightings of UFO's have been reported (see Section 4.2.1).

Those who cause the input of a wrong code should not always be blamed. It is often skimpy instructions, lack of follow-up, inadequate screening processes, complicated procedures etc., which may be at fault.

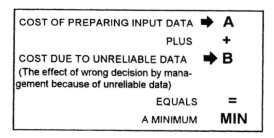

Figure 4-16
Optimum cost of data processing.

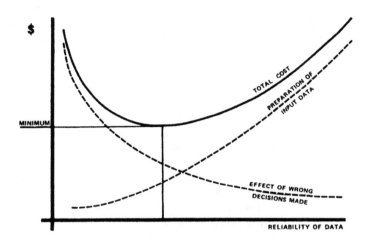

Figure 4-17
Reliability of data -minimum cost.

Unfortunately, the ultimate in identification conflicts somewhat with simplicity of input. If a system becomes too complex and detailed in its breakdown, it will increase the cost of obtaining reliable data. It is a constant struggle to maintain reliable data, but there has to be a trade-off somewhere. On the one hand, unreliable data caused by careless preparation of input can have an expensive outcome when management makes wrong decisions. On the other hand, the more reliable we require data to be, the more it will cost us to prepare the input.

Minimum total cost can be depicted in graphic form (Figure 4-17). Assuming the curves are exponential curves, then

$$Y = KX^s$$
$$\log Y = s \log X + \log K$$

which is the equation of a straight line.
The sum of the ordinates produce a total cost curve:

$$Y_A + Y_B = \text{minimum}$$

To determine the optimum point on the total cost curve, we equate the first derivative of the resulting curve equation to zero.

Information requirements

When a project is initiated, the manager appointed should be very much concerned about the type and level of information that is flowing through the task group organization. Very often, too much (junk mail), or poorly presented information is provided to management. The author was a member of a study team engaged to improve communication and to reduce paperwork for a Design and Construction Unit (D & C) of a large corporation (19).

The study team looked at the organizational units and their main functions first. As a second step, the primary functions of each unit were examined in detail. Secondary and intrinsic functions, which support the primary functions, were then analyzed.

The objectives were:

♦ Identify the key information required to support the overall goals and objectives of the D & C organization.

♦ Assess the relative importance of the information, identify opportunities and problem areas, and establish priorities.

♦ Promote user (head office and project staff) understanding and commitment to information management.

To manage each of these broad functions, there are common activities in each organizational unit of planning, resourcing, performing, monitoring and controlling.

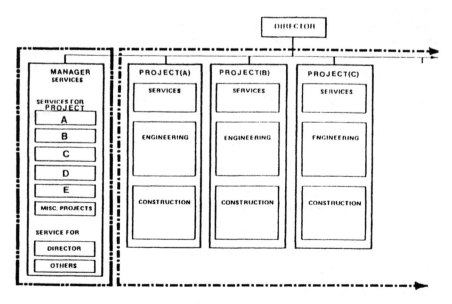

Figure 4-18
Information study organization.

To carry out those activities, specific information are needed. The classes or kinds of information used or produced in carrying out functional activities are then identified. An information requirements model is then produced, which includes two matrices (computerized data base):

♦ A Function/Organizational Unit Matrix
♦ A Function/Information Class Matrix

Even though the Information Requirements Model pertains to a specific Branch of a large corporation, it could be used as an example or framework for a modified information flow to meet the needs of a different organization. The model applied to the organizational unit shown in Figure 4-18.

As a first step, all functions were grouped into major categories and broadly defined:

Planning Function:

Establish divisional, project, department or section goals, functions and work program.

Resourcing Function:

Obtain resources required to reach unit goals.

Engineering, Construction and Services Function:

Provide engineering and construction of capital projects and other facilities. provide services in support of engineering, construction, and operations.

Monitor and Control Function:

Monitor and report performance of all unit and project activities and redirect the work if necessary.

Special Functions:

Provide personal expertise to internal (corporate) and external (national, international) bodies, representing corporate interests.

These groups are further broken down into Functions and Sub-Functions as shown below (Figures 4-23 to 4-27)

PLANNING FUNCTIONS

ESTABLISH ORGANIZATION STRUCTURE AND RESPONSIBILITIES
DEVELOP POLICIES AND PROCEDURES
ESTABLISH UNIT OBJECTIVES
ESTABLISH AN INTEGRATED WORK PROGRAM
      Establish the WBS and accounting codes
      Determine the work to be done
      Forecast resource requirements
      Prepare time estimates and schedules
      Prepare cost estimates and financial plans
      Establish contingency plans
      Establish record- and data management systems
ESTABLISH PROJECT MANAGEMENT INFORMATION SYSTEMS (PMIS)
      Establish materiel management systems
      Establish scheduling system
      Establish financial management and costing systems
ESTABLISH A QUALITY ENGINEERING PROGRAM
ESTABLISH A CONSTRUCTION LABOR RELATIONS PROGRAM
ESTABLISH A HEALTH AND SAFETY PROGRAM
ESTABLISH A SITE SECURITY PROGRAM
INITIATE PROJECT INSURANCE
ESTABLISH A CORPORATE CITIZENSHIP PROGRAM

Figure 4-19
Information requirement functions - 1.

RESOURCE FUNCTIONS

ACQUIRE AND RELEASE HUMAN RESOURCES
      Acquire senior staff
      Acquire head office engineering staff
      Acquire construction field staff
      Acquire construction trades labor
      Acquire head office service staff
ACQUIRE HEAD OFFICE SPACE AND FACILITIES
ACQUIRE CONSTRUCTION FACILITIES
ACQUIRE CONSTRUCTION MATERIAL
ACQUIRE CONSTRUCTION EQUIPMENT AND TOOLS
TRAIN AND DEVELOP STAFF AND TRADES

Figure 4-20
Information requirement functions - 2.

ENGINEERING, CONSTRUCTION AND SERVICES FUNCTIONS

PROVIDE ENGINEERING
    PREPARE DETAILED DESIGN
        Integrate design requirements
        Provide design standards and procedures
        Co-ordinate design activities
        Provide design packages
        Review design and drawings documentation
        Participate in design reviews
    REQUISITION 'E' MATERIAL AND SERVICES
        Provide technical specifications
        Co-ordinate preparation of tendering documents
        Evaluate tenders and make recommendations
        Provide procurement engineering services
        Provide contract administration
    PROVIDE ENGINEERING SUPPORT
    MANAGE REGULATORY ACTIVITIES
    ADMINISTER Q. E. PROGRAM
PROVIDE CONSTRUCTION
    PROVIDE CONSTRUCTION ENGINEERING
    CO-ORDINATE SITE ACTIVITIES
    CONSTRUCT SYSTEMS AND FACILITIES
    ADMINISTER SITE CONTRACTS
    PROVIDE CONSTRUCTION QUALITY ASSURANCE
    ADMINISTER AND MAINTAIN CONSTRUCTION FACILITIES
    MAINTAIN HEALTH AND SAFETY PROGRAMS
    ADMINISTER CONSTRUCTION TRADES LABOR
    MAINTAIN LABOR RELATIONS PROGRAMS
    REQUISITION 'C' MATERIEL AND SERVICES
        Provide technical and commercial specifications
        Co-ordinate preparation of tendering documents
        Evaluate tenders and make recommendations
        Provide procurement engineering services
    TURNOVER SYSTEMS AND PERMANENT FACILITY COMPONENTS
    PROVIDE MATERIAL MANAGEMENT
    PROVIDE CONSTRUCTION SERVICES TO OPERATIONS
    MAINTAIN SITE SECURITY PROGRAM
PROVIDE ADMINISTRATIVE, PROJECT AND OTHER SERVICES
    PROVIDE INTEGRATED MANAGEMENT SUPPORT SERVICES
        Provide scheduling services
        Provide estimating services
        Provide financial services
        Provide administrative services
        Assess productivity improvement opportunities

(continued next page)

```
PROVIDE REQUISITIONING SERVICES
    Provide a requisitioning service
    Provide liaison procurement service
PROVIDE TECHNICAL SUPPORT SERVICES
    Contribute to the quality engineering program development
    Provide construction equipment service
    Assess constructibility
PROVIDE OTHER PROJECT MANAGEMENT FUNCTIONS
ADMINISTER CITIZENSHIP
ADMINISTER INSURANCE CLAIMS
TURNOVER COMPLETED FACILITY
PROVIDE OTHER STANDARD FUNCTIONS
    ADMINISTER STAFF
```

Figure 4-21
Information requirement functions - 3.

```
MONITORING AND CONTROL FUNCTIONS

MONITOR AND CONTROL PRODUCT QUALITY
MONITOR AND CONTROL RESOURCES
MONITOR AND CONTROL COSTS
    MONITOR AND CONTROL ENGINEERING COSTS
        Monitor and control permanent materiel cost
        Monitor and control consultants' cost
        Monitor and control other H.O. engineering cost
    MONITOR AND CONTROL CONSTRUCTION COST
    MONITOR AND CONTROL SERVICES COST
    MONITOR AND CONTROL PROJECT TEAM COST
    MONITOR AND CONTROL DIVISIONAL COST
MONITOR AND CONTROL SCHEDULES
MONITOR AND CONTROL GENERAL PROJECT ACTIVITIES
    MONITOR AND CONTROL CITIZENSHIP PROGRAM
    MONITOR AND CONTROL SITE SECURITY PROGRAM
    MONITOR AND CONTROL HEALTH AND SAFETY PROGRAM
    MONITOR AND CONTROL LABOR RELATIONS PROGRAM
```

Figure 4-22
Information requirement functions - 4.

```
SPECIAL FUNCTIONS

PROVIDE REPRESENTATION EXTERNAL TO THE DIVISION
PROVIDE SPECIAL PRESENTATIONS
```

Figure 4-23
Information requirement functions - 5.

The next step, probably the most difficult one, is the listing of *Information Classes* and their generalized definitions.

Each function and sub-function is then given a code. Here again, each function's performance criteria will be described. In our case, code numbers go from 1 to 29:

        1 - Establish organization structure & responsibilities

to   29 - Provide special presentations.

```
                    INFORMATION CLASS DEFINITIONS
                    -----------------------------

                    GENERATION PROJECTS DIVISION
                    -----------------------------


ACCOUNT INFORMATION
        INFORMATION RELATING TO ACCOUNT DESCRIPTIONS, CODES AND CLASSES INCLUDING
        SYSTEMS CLASSIFICATION (SCI), WORK ORDERS, WORK ITEMS, AND CORPORATE
        ACCOUNTS REGISTER.

ACTIVITY COMPLETION INFORMATION
        INFORMATION ON WORK COMPLETED INCLUDING DESCRIPTION OF ACTIVITY, DATE
        COMPLETED OR DEGREE OF COMPLETION.

ACTUAL COST INFORMATION
        DATA ON COST INCURRED (INCLUDING MAN-HOURS EXPENDED).

ACTUAL DESIGN INFORMATION
        A TECHNICAL DESCRIPTION OF A SYSTEM OR FEATURE AS DESIGNED. INCLUDING
        DESIGN NOTES, DESIGN DESCRIPTIONS, SPECIFICATIONS, DRAWINGS, MATERIEL
        LISTS, ETC.

CAPITAL EXPENDITURE GUIDELINES
        GOVERNMENT, CORPORATE OR BRANCH DIRECTIVES RELATING TO CONSTRAINING
        EXPENDITURES IN SPECIFIED TIME PERIODS, INCLUDING BUDGET PARAMETERS
        ESTABLISHED AT VARIOUS ORGANIZATIONAL LEVELS.

CITIZENSHIP INFORMATION
        INFORMATION RELATING TO THE COMMUNITY AND THE ENVIRONMENT.

COLLECTIVE AGREEMENTS INFORMATION
        AGREEMENTS BETWEEN ONTARIO HYDRO AND TRADE UNIONS OR OTHER COLLECTIVE
        BARGANING GROUPS SUCH AS SOPHEA.  INCLUDED IN SUCH AGREEMENTS ARE
        INFORMATION ON CONDITIONS OF EMPLOYMENT, HOURS OF WORK, WAGE RATES, TRAVEL
        AND MOVING ALLOWANCES ETC.

CONCEPTUAL DESIGN INFORMATION
        INFORMATION SPECIFYING  FUNCTIONS A GIVEN SYSTEM MUST PERFORM, AND THE
        INTERFACE OR BOUNDARY BETWEEN THAT SYSTEM AND OTHER ASSOCIATED SYSTEMS.
        INCLUDED ARE QUALITY ENGINEERING PLANS, PRELIMINARY DESIGN REQUIREMENTS AND
        PRELIMINARY DESIGN DESCRIPTIONS.

CONSTRUCTION ACTIVITY INFORMATION
        TECHNICAL INFORMATION ON SITE ACTIVITIES, CONDITIONS AND INCIDENTS.

CONSTRUCTION TECH. STDS. & PROCEDURES
        INFORMATION ON STANDARD CONSTRUCTION DESIGNS AND METHODS.

CONSULTANT PERFORMANCE INFORMATION
        HISTORICAL DATA AND FEEDBACK ON CONSULTANTS' ACTUAL PERFORMANCE.
```

CONTRACT INFORMATION
>   THE TERMS AND CONDITIONS OF ACTUAL OR POTENTIAL CONTRACTS BETWEEN ONTARIO
>   HYDRO AND CONTRACTORS FOR SUPPLY OF EQUIPMENT, PRODUCTS, MATERIAL AND
>   SERVICES.

CONTRACTOR INFORMATION
>   INFORMATION ABOUT WORK EXPERIENCE, PAST AND PRESENT PERFORMANCE,
>   CAPABILITY, AND RESOURCE LEVELS, OF CONTRACTORS.

CORPORATE INFORMATION
>   INFORMATION ABOUT THE OPERATION OF ONTARIO HYDRO.

COST ESTIMATES INFORMATION
>   INFORMATION ON ESTIMATED QUANTITIES AND COSTS RELATED TO A WORK PROGRAM.

COST ESTIMATING FACTORS INFORMATION
>   INFORMATION SUCH AS UNIT RATES, PRODUCTION RATES, BURDENS, OVERHEADS AND
>   ESTIMATING PARAMETERS.

CURRENT TECHNOLOGY INFORMATION
>   STATE-OF-THE-ART INFORMATION RELATED TO TECHNICAL AND MANAGERIAL ASPECTS OF
>   DIVISIONAL ACTIVITIES.

DESIGN REFERENCE INFORMATION
>   ALL REFERENCE TEXTS, MANUFACTURERS' DATA SHEETS, AND OTHER DESIGN DATA
>   REQUIRED FOR THE DESIGN OF A SYSTEM, A PIECE OF EQUIPMENT, OR STRUCTURE.

ECONOMIC INFORMATION
>   ECONOMIC FACTORS SUCH AS, DISCOUNT AND INTEREST RATES, FOREIGN EXCHANGE,
>   INCLUDING THEIR PROBABLE RANGES.

EMPLOYEE DEVELOPMENT INFORMATION
>   INFORMATION ON INDIVIDUAL EMPLOYEE EDUCATION, EXPERIENCE,CAREER PLANNING
>   AND DEVELOPMENT, AND TRAINING NEEDS.

EMPLOYEE PERFORMANCE INFORMATION
>   INFORMATION ON EMPLOYEE'S CONDUCT, ACHIEVEMENTS, AND LIMITATIONS.

EQUIPMENT AVAILABILITY INFORMATION
>   INFORMATION ON THE AVAILABILITY OF EQUIPMENT INSIDE OR OUTSIDE THE
>   CORPORATION.

EQUIPMENT INFORMATION
>   DETAILED DATA ON CAPABILITY AND CHARACTERISTICS OF EQUIPMENT.

ESTIMATED QUANTITY INFORMATION
>   QUANTITATIVE INFORMATION ON MATERIEL AND ACTIVITIES REQUIRED FOR COMPONENTS
>   OF A WORK PROGRAM.

FINANCIAL INFORMATION
>   INFORMATION CONCERNING BUDGETS, CASH FLOWS.

GOAL AND FUNCTIONS INFORMATION
INFORMATION ON THE GOALS AND PRIMARY DELIVERY FUNCTIONS OF AN
ORGANIZATIONAL UNIT AND ITS SUB UNITS.

GOVERNMENT ORGANIZATIONAL STRUCTURE INFORMATION
INFORMATION ABOUT THE ORGANIZATION AND AREAS OF RESPONSIBILITY OF FEDERAL,
PROVINCIAL AND MUNICIPAL GOVERNMENTS AS THEY AFFECT THE REGULATION OF THE
ACTIVITIES OF GENERATION PROJECTS DIVISION.

HEALTH AND SAFETY INFORMATION
INFORMATION ON CONDITIONS AND STATISTICS RELATING TO SAFETY, HEALTH, FIRST
AID AND FIRE PREVENTION AND FIRE FIGHTING.

INSURANCE INFORMATION
INFORMATION ON TERMS AND CONDITIONS OF INSURANCE TO COVER LIABILITY AND
PROPERTY DAMAGE AFFECTING GENERATION PROJECTS DIVISION'S OPERATION.

INVENTORY INFORMATION
INFORMATION ON RECEIPT, DEMAND, STORAGE AND ISSUE OF EQUIPMENT, MATERIAL
AND TOOLS.

LABOUR WORKING CONDITIONS INFORMATION
COMPILATION OF PRACTICES PERTAINING TO CONSTRUCTION WORKING CONDITIONS,
SUCH AS JOB TRAINING, APPRENTICESHIP, JOB RATES, WORK ENVIRONMENT, TOOLS
SUPPLIED, ETC.

LAWS, REGULATIONS AND GUIDELINES
INFORMATION ABOUT LAWS, REGULATORY REQUIREMENTS AND GUIDELINES AND THEIR
INTERPRETATION AS APPLIED TO THE DIVISION'S ACTIVITIES.

MANAGEMENT SYSTEMS INFORMATION
INFORMATION CONCERNING SYSTEMS AND PROCEDURES FOR ESTIMATING, ENGINEERING
'COSTING, PROJECT COSTING, ACCOUNTING, BUDGETTING, SCHEDULING, PROGRESS
REPORTING, MATERIAL AND RECORD MANAGEMENT.

MANPOWER INFORMATION
INFORMATION CONCERNING NUMBERS OF PEOPLE INCLUDING THEIR SKILLS AND
CATEGORIES, IN RELATION TO THE WORK PROGRAM OR UNIT.

MATERIEL INFORMATION
ALL DATA ON MATERIEL INCLUDING DEMAND, SUPPLY, INVENTORY MANAGEMENT,
QUALITY, STORAGE CONDITIONS, RE-USABILITY AND COST.

OBJECTIVES INFORMATION
THE ACCOMPLISHMENTS REQUIRED TO ACHIEVE THE GOALS OF A UNIT EXPRESSED IN
SPECIFIC AND MEASURABLE TERMS.

PERMITS, APPROVALS & LICENCES INFORMATION
INFORMATION SPECIFIC TO A PROJECT THAT RELATE TO PERMITS, APPROVAL AND
LICENCES ISSUED BY REGULATORY AUTHORITIES.

PERSONNEL INFORMATION
INFORMATION ON INDIVIDUAL EMPLOYEES SUCH AS PERSONAL AND PAYROLL DATA.

POLICIES & PROCEDURES INFORMATION
  THE INFORMATION SETTING OUT THE CONSTRAINTS, SELECTED PRINCIPLES AND
  GENERAL COURSES OF ACTION SPECIFIED BY SUPERIOR UNITS AND THE SPECIFIED
  METHODS AND SEQUENCE FOR PERFORMING ACTIVITIES.

PROCUREMENT INFORMATION
  INFORMATION RELATED TO TENDERING DOCUMENTS, TENDERS, REQUISITIONS, PURCHASE
  ORDERS, ETC., NECESSARY TO PURCHASE EQUIPMENT, MATERIALS, PRODUCTS AND
  SERVICES.

PROJECT DEFINITION INFORMATION
  INFORMATION CONCERNING THE SCOPE OF ALL COMPONENTS OF A GENERATION PROJECT,
  SUCH AS ARE INCLUDED IN SYSTEM PLANNING SPECIFICATIONS, PROJECT
  REQUIREMENTS, WORK ORDERS, PRELIMINARY DESIGN REQUIREMENTS AND PRELIMINARY
  DESIGN DESCRIPTIONS.

PROJECT HISTORY INFORMATION
  A DETAILED RECORD OF PAST WORK PERFORMANCE INCLUDNG SCOPE,SCHEDULES, COSTS
  AND SIGNIFICANT UNPLANNED HAPPENINGS.

QUALITY ENGINEERING INFORMATION
  ALL INFORMATION AND DOCUMENTATION RELATIVE TO THE Q.E. PROGRAMS.

REGULATORY CODES INFORMATION
  INFORMATION INCLUDED IN CODES OF PRACTICE WHICH HAVE BEEN ADOPTED BY THE
  REGULATORY AGENCIES.

RESOURCE MARKET INFORMATION
  INFORMATION ABOUT THE AVAILABILITY OF MANPOWER, MATERIAL, AND EQUIPMENT
  RESOURCES.

SCHEDULE INFORMATION
  INFORMATION CONCERNING HISTORIC, CURRENT AND FUTURE SCHEDULED ACTIVITIES AT
  AN APPROPRIATE LEVEL OF DETAIL. THEIR DURATION, SEQUENCE AND RESOURCE
  REQUIREMENTS, START DATES AND REQUIRED COMPLETION DATES.

SITE INFORMATION
  ALL SITE SPECIFIC INFORMATION, INCLUDING GEOLOGY, TOPOGRAPHY, LAYOUT OF
  FACILITIES, ACCESS, CLIMATE, SERVICES, PROPERTY AGREEMENTS, SECURITY
  SYSTEMS, ETC.

STRATEGIES INFORMATION
  INFORMATION DOCUMENTED IN THE FORM OF INTEGRATED SCHEDULES, ESTIMATES,
  BUDGETS, RESOURCE FORECASTS, 'MAKE OR BUY' DECISIONS, ETC. ON HOW TO
  ACHIEVE OBJECTIVES.

TECHNICAL PERFORMANCE INFORMATION
  CONSTRUCTION, MANUFACTURING AND OPERATIONS FEEDBACK ON EQUIPMENT, MATERIAL
  AND SYSTEMS PERFORMANCE.

TECHNICAL SPECIFICATIONS INFORMATION
  ALL TECHNICAL INFORMATION RELATIVE TO THE SCOPE DEFINITION, AND CONSTRAINTS
  OF A GIVEN STRUCTURE, PROCESS OR CONTROL SYSTEM OR OTHER FACILITY.

TECHNICAL STANDARDS INFORMATION
ALL DESIGN AND DRAFTING STANDARDS DEVELOPED INTERNALLY AND EXTERNALLY IN
THE FORM OF TABLES, CHARTS, GRAPHS, DRAWINGS AND WRITTEN DIRECTIONS.

TRAINING PROGRAM INFORMATION
INFORMATION ON O.H. AND EXTRAMURAL COURSES AND SEMINARS AND STAFF
DEVELOPMENT PROGRAMS AND ON TRAINING FACILITIES.

UNIT ORGANIZATIONAL STRUCTURE INFORMATION
INFORMATION ABOUT THE RELATIONSHIP OF AN ORGANIZATIONAL UNIT WITH ITS
SUPERIOR, PEER, AND SUB-UNITS.

UNIT PERFORMANCE INFORMATION
INFORMATION RELATING TO THE ACHIEVEMENTS OF A UNIT IN THE AREAS OF QUALITY,
QUANTITY, SCHEDULE, COST, RESOURCES, SAFETY AND CITIZENSHIP.

VARIANCE INFORMATION
A COMPARISON BETWEEN ACTUAL WORK ACCOMPLISHED AND THE APPROVED WORK PROGRAM
INCLUDING AN EXPLANATION OF DEPARTURES.

VENDOR INFORMATION
INFORMATION ABOUT SUPPLIERS OF PRODUCTS OR SERVICES, INCLUDING COMMERCIAL
CONDITIONS, RELIABILITY, FINANCIAL STATUS, QUALITY OF GOODS AND SERVICES.

WORK PACKAGE INFORMATION
INFORMATION PERTAINING TO A COMPONENT OF A GENERATION PROJECT AS DEFINED BY
THE WORK BREAKDOWN STRUCTURE.

WORK PROGRAM INFORMATION
INFORMATION CONCERNING THE TITLES, SCOPE, COST ESTIMATES AND REQUIRED
COMPLETION DATES FOR PROJECTS AND OTHER WORK.

WORK TRADE ASSIGNMENTS INFORMATION
INFORMATION CONCERNING THE ASSIGNMENT OF CONSTRUCTION TRADES TO SPECIFIC
TYPES OF WORK.

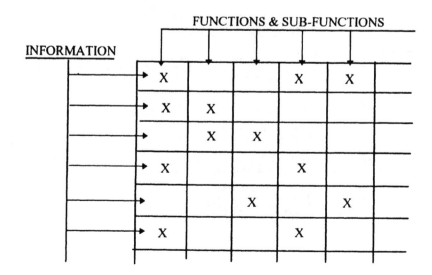

Information Class Definitions can be considered the

Dictionary of Information Flow.

Each function and sub-function obviously will require specific information to operate properly.

The type of information needed is then taken from the dictionary and mated with each function and sub-function. In addition, the information can have three specific purposes, it could be

- ◆ Used (U)
- ◆ Produced (P) or
- ◆ Both, used and produced (U/P).

The function definitions have now been enhanced by information requirements. The data base is now expanded to the equivalence of 30 printed pages.

Below are a few examples of the expanded data base covering some of the planning and control functions:

Planning Functions

```
4     ESTABLISH AN INTEGRATED WORK PROGRAM
            DETERMINE THE CONTENT AND TIME FRAME OF THE WORK FOR WHICH THE
            MANAGER IS ACCOUNTABLE.

4.1   ESTABLISH WORK BREAKDOWN & ACCOUNT STRUCTURES
            ADAPT BRANCH SYSTEMS CLASSIFICATION INDEX AND COST ACCOUNTS TO UNIT
            NEEDS. PREPARE SYSTEM AND WORK PACKAGE DESCRIPTIONS.

            U/P      ACCOUNT INFORMATION
            U        POLICIES & PROCEDURES INFORMATION
            U        PROJECT DEFINITION INFORMATION
            U/P      WORK PACKAGE INFORMATION
            U        WORK PROGRAM INFORMATION

4.2   DETERMINE THE WORK TO BE DONE
            COLLECT INFORMATION SUCH AS DESIGN REQUIREMENTS, DESIGN
            DESCRIPTIONS, RELEASE ESTIMATES, AND MASTER SCHEDULES TO DETERMINE
            THE SCOPE, SEQUENCE AND TIMING OF WORK TO BE CARRIED OUT.

            U        CONCEPTUAL DESIGN INFORMATION
            U        COST ESTIMATES INFORMATION
            U        LAWS, REGULATIONS AND GUIDELINES
            U        SCHEDULE INFORMATION
            U/P      STRATEGIES INFORMATION
            U        WORK PROGRAM INFORMATION

4.3   FORECAST RESOURCE REQUIREMENTS
            PREPARE FORECASTS FOR FUTURE HUMAN AND OTHER RESOURCE REQUIREMENTS
            TAKING INTO ACCOUNT THE UNCOMMITTED PROGRAM.

            U/P      EQUIPMENT INFORMATION
            U/P      MANPOWER INFORMATION
            U        RESOURCE MARKET INFORMATION
            U/P      STRATEGIES INFORMATION
            U        WORK PROGRAM INFORMATION
```

4.4   PREPARE SCHEDULES
              LIST ACTIVITIES AT APPROPRIATE SCALE DETERMINE DURATION AND SEQUENCE
              AND PREPARE SCHEDULES. IDENTIFY MILESTONES AND INTERFACES.

              U   CONCEPTUAL DESIGN INFORMATION
              U/P   MANPOWER INFORMATION
              U   PROJECT HISTORY INFORMATION
              U   RESOURCE MARKET INFORMATION
              U/P   SCHEDULE INFORMATION
              U   WORK PACKAGE INFORMATION
              U   WORK PROGRAM INFORMATION

4.5   PREPARE COST ESTIMATES AND FINANCIAL PLANS
              THE PREPARATION OF REQUIRED COST ESTIMATES AND FINANCIAL PLANS FOR
              DIVISIONS, PROJECTS, DEPARTMENTS OR SECTIONS.

              U   ACCOUNT INFORMATION
              U   ACTUAL COST INFORMATION
              U   CITIZENSHIP INFORMATION
              U   COLLECTIVE AGREEMENTS INFORMATION
              U   CONTRACT INFORMATION
              U/P   COST ESTIMATES INFORMATION
              U   COST ESTIMATING FACTORS INFORMATION
              U   ECONOMIC INFORMATION
              U/P   FINANCIAL INFORMATION
              U   GOAL AND FUNCTIONS INFORMATION

Control Functions

24.3   MONITOR & CONTROL MANAGEMENT SYSTEMS PRODUCTIVITY
              EVALUATE THE EFFECTIVENESS OF THE VARIOUS SYSTEMS USED TO COLLECT,
              DESCRIBE AND SUMMARIZE DATA AND REPORT PLANNING AND CONTROL
              INFORMATION AT VARIOUS FUNCTIONAL LEVELS. IDENTIFY PROBLEM AREAS
              WHERE IMPROVEMENTS ARE ACHIEVABLE. INITIATE SYSTEM REVISIONS IN
              THOSE AREAS. ASSESS VALIDITY OF SYSTEM CHANGES AND ADDITIONS AND
              RE-DIRECT ACTIVITIES.

              U   ACTIVITY COMPLETION INFORMATION
              U/P   INVENTORY INFORMATION
              U   MANAGEMENT SYSTEMS INFORMATION
              U   OBJECTIVES INFORMATION
              U   POLICIES & PROCEDURES INFORMATION
              U   SCHEDULE INFORMATION
              U/P   STRATEGIES INFORMATION
              P   VARIANCE INFORMATION

25   MONITOR & CONTROL COSTS
              MONITOR COSTS FOR VARIOUS COST STREAMS WITHIN THE APPROVED WORK
              BREAKDOWN STRUCTURE AND COMPARE WITH ESTIMATES AND BUDGETS.
              IDENTIFY AND ANALYZE CAUSES FOR THE VARIANCE FROM ESTABLISHED
              PARAMETERS. RE-DIRECT WORK PROGRAM IF NECESSARY.

25.1   MONITOR AND CONTROL ENGINEERING COSTS
              MONITOR COSTS FOR THE ENGINEERING COST STREAM WITHIN THE APPROVED
              WORK BREAKDOWN STRUCTURE AND COMPARE WITH ESTIMATES AND BUDGETS.
              IDENTIFY AND ANALYZE CAUSES FOR THE VARIANCE FROM ESTABLISHED
              PARAMETERS. RE-DIRECT WORK PROGRAM IF NECESSARY.

25.1.1   MONITOR & CONTROL PERMANENT MATERIEL COSTS
          MONITOR COSTS FOR THE PERMANENT MATERIEL COST STREAM WITHIN THE
          APPROVED WORK BREAKDOWN STRUCTURE AND COMPARE WITH ESTIMATES AND
          BUDGETS. IDENTIFY AND ANALYZE CAUSES FOR THE VARIANCE FROM
          ESTABLISHED PARAMETERS. RE-DIRECT WORK PROGRAM IF NECESSARY.

               U   ACTUAL COST INFORMATION
               U   CAPITAL EXPENDITURE GUIDELINES
               U   COST ESTIMATES INFORMATION
               U   COST ESTIMATING FACTORS INFORMATION
               U   MATERIEL INFORMATION
               U   OBJECTIVES INFORMATION
               U   STRATEGIES INFORMATION
               P   VARIANCE INFORMATION
               U   WORK PACKAGE INFORMATION
               U   WORK PROGRAM INFORMATION

25.1.2   MONITOR & CONTROL CONSULTANT COSTS
          MONITOR THE CONSULTANT'S COST REPORTS AND COMPARE WITH ESTIMATES AND
          BUDGETS. IDENTIFY AND ANALYZE CAUSES FOR THE VARIANCE FROM
          ESTABLISHED PARAMETERS. RE-DIRECT WORK PROGRAM IF NECESSARY.

               U   ACTUAL COST INFORMATION
               U   CONSULTANT PERFORMANCE INFORMATION
               U   COST ESTIMATING FACTORS INFORMATION
               U   MANPOWER INFORMATION
               U   STRATEGIES INFORMATION
               U   VARIANCE INFORMATION
               U   WORK PROGRAM INFORMATION

25.1.3   MONITOR & CONTROL OTHER O.H. ENGINEERING COSTS
          MONITOR COSTS FOR THE ENGINEERING COST STREAM WITHIN THE APPROVED
          WORK BREAKDOWN STRUCTURE AND COMPARE WITH ESTIMATES AND BUDGETS.
          IDENTIFY AND ANALYZE CAUSES FOR THE VARIANCE FROM ESTABLISHED
          PARAMETERS. RE-DIRECT WORK PROGRAM IF NECESSARY.

               U   ACTUAL COST INFORMATION
               U   COST ESTIMATING FACTORS INFORMATION
               U   MANPOWER INFORMATION
               U   STRATEGIES INFORMATION
               P   VARIANCE INFORMATION
               U   WORK PACKAGE INFORMATION
               U   WORK PROGRAM INFORMATION

25.2   MONITOR & CONTROL CONSTRUCTION COSTS
          MONITOR COSTS FOR THE CONSTRUCTION INDIRECT, DIRECT AND CONTRACT
          COST STREAMS WITHIN THE APPROVED WORK BREAKDOWN STRUCTURE AND
          COMPARE WITH ESTIMATES AND BUDGETS. IDENTIFY AND ANALYZE CAUSES FOR
          VARIANCE FROM ESTABLISHED PARAMETERS. RE-DIRECT WORK PROGRAM IF
          NECESSARY.

               U   ACTUAL COST INFORMATION
               U   CONSTRUCTION ACTIVITY INFORMATION
               U   COST ESTIMATING FACTORS INFORMATION
               U   MANPOWER INFORMATION
               P   VARIANCE INFORMATION
               U   WORK PACKAGE INFORMATION

25.3   MONITOR & CONTROL SERVICES COSTS (EXCLUDING CONSTRUCTION)
          MONITOR COSTS FOR HEAD OFFICE SUPPORT GROUPS SUCH AS PROJECT AND
          DIVISIONAL SERVICES AND SUPPLY PROCUREMENT DIVISION. COMPARE WITH
          ESTIMATES AND BUDGETS. IDENTIFY AND ANALYZE CAUSES FOR VARIANCE
          FROM ESTABLISHED PARAMETERS. RE-DIRECT WORK PROGRAM IF NECESSARY.

               U   ACCOUNT INFORMATION
               U   ACTIVITY COMPLETION INFORMATION
               U   ACTUAL COST INFORMATION

```
        U           CONSULTANT PERFORMANCE INFORMATION
        U           COST ESTIMATES INFORMATION
        U           COST ESTIMATING FACTORS INFORMATION
        U           FINANCIAL INFORMATION
        U           MANAGEMENT SYSTEMS INFORMATION
        U/P         OBJECTIVES INFORMATION
        U/P         UNIT PERFORMANCE INFORMATION
        P           VARIANCE INFORMATION
```

25.4   MONITOR & CONTROL PROJECT AND UNIT COSTS
        MONITOR TOTAL PROJECT AND UNIT COSTS INCLUDING INTEREST, OVERHEAD
        AND CONTINGENCY AND COMPARE WITH ESTIMATES AND BUDGETS. IDENTIFY
        AND ANALYZE CAUSES FOR THE VARIANCE FROM ESTABLISHED PARAMETERS.
        RE-DIRECT WORK PROGRAM IF NECESSARY.

```
        U           ACTUAL COST INFORMATION
        U           CAPITAL EXPENDITURE GUIDELINES
        P           VARIANCE INFORMATION
        U           WORK PACKAGE INFORMATION
        U           WORK PROGRAM INFORMATION
```

25.5   MONITOR & CONTROL DIVISIONAL COSTS
        MONITOR DIVISIONAL EXPENDITURES AND COST ESTIMATES. COMPARE WITH
        APPROVED BUDGETS AND ESTIMATES. IDENTIFY AND ANALYZE CAUSES FOR THE
        VARIANCE FROM ESTABLISHED PARAMETERS. RE-DIRECT WORK PROGRAM IF
        NECESSARY.

```
        U           ACTUAL COST INFORMATION
        U           CAPITAL EXPENDITURE GUIDELINES
        P           VARIANCE INFORMATION
        U           WORK PROGRAM INFORMATION
```

26   MONITOR & CONTROL SCHEDULES
        MONITOR THE TIMING OF ACTIVITIES AND COMPARE WITH DETAILED WORK
        PLANS AND SCHEDULES. ENSURE THAT INTERFERENCES ARE RESOLVED AND
        SCHEDULES ARE UPDATED AND VARIANCES ARE EXPLAINED. MONITOR
        CONSULTANT'S SCHEDULING SYSTEMS AND ACTIVITIES.

```
        U           ACTIVITY COMPLETION INFORMATION
        U/P         SCHEDULE INFORMATION
        U           TECHNICAL PERFORMANCE INFORMATION
        P           VARIANCE INFORMATION
```

27   MONITOR GENERAL PROJECT ACTIVITIES
        MONITOR, DESCRIBE AND MAINTAIN RECORDS OF GENERAL ACTIVITIES SUCH AS
        SITE SECURITY, LABOUR RELATIONS, SAFETY, AND CITIZENSHIP AND TAKE
        CORRECTIVE ACTION IF NECESSARY.

27.1   MONITOR & CONTROL CITIZENSHIP PROGRAM
        MONITOR THE "MAKE" VS. "BUY" PORTIONS OF THE PROJECT. MONITOR
        IMPACTS ON THE ENVIRONMENT AND THE COMMUNITY, AND TAKE NECESSARY
        ACTION.

```
        U/P         CITIZENSHIP INFORMATION
        U           COLLECTIVE AGREEMENTS INFORMATION
        U           GOVERNMENT ORGANIZATIONAL STRUCTURE INFORMATION
        U           LAWS, REGULATIONS AND GUIDELINES
        U           PERMITS, APPROVALS & LICENCES INFORMATION
        U           POLICIES & PROCEDURES INFORMATION
        U           REGULATORY CODES INFORMATION
        U/P         SITE INFORMATION
        U           WORK PROGRAM INFORMATION
```

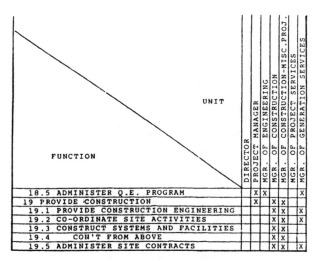

FUNCTION	UNIT	DIRECTOR	PROJECT MANAGER	MGR. OF ENGINEERING	MGR. OF CONSTRUCTION	MGR. OF CONSTRUCTION-MISC.PROJ.	MGR. OF PROJECT SERVICES	MGR. OF GENERATION SERVICES
18.5 ADMINISTER Q.E. PROGRAM			X	X				X
19 PROVIDE CONSTRUCTION			X		X	X		
19.1 PROVIDE CONSTRUCTION ENGINEERING					X	X		X
19.2 CO-ORDINATE SITE ACTIVITIES					X	X		X
19.3 CONSTRUCT SYSTEMS AND FACILITIES					X	X		
19.4 CON'T FROM ABOVE					X	X		
19.5 ADMINISTER SITE CONTRACTS					X	X		X

Figure 4-24
Information requirements - functions vs. unit

One requirement is still missing to complete the model. Where does the required information go? The individual organizational unit that has to deal with the information in order to perform its function needs to be identified. This is shown in a matrix which lists functions vs. unit (Figure 4-24).

It should be noted that this matrix is neither quantitative nor is it qualitative, i.e. it does not show the frequency or level of performance of the functions. It only indicates that the function is performed within the unit listed. A standard responsibility code for the positions listed is later added to the system separately.

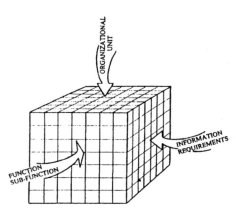

The final picture of the model is a three-dimensional one. The implementation has resulted in a considerable reduction in paperwork, improved productivity and increased quality of technical information A periodic review must be done to keep the data base up to date. When reviewing, the *user* must tell *why* the information is needed and *how* he or she is going to use it. This, in conjunction with a linear responsibility chart can be of great help in the planning and implementation of project management.

Q #4.02:

The manager of a project task force has to identify the key information required by his/her organization. A study was made to compare the cost of increasing sophistication in preparing input data vs. the effects unavailable information would have on decision-making. The result of the study was plotted on a graph to obtain optimal requirements.

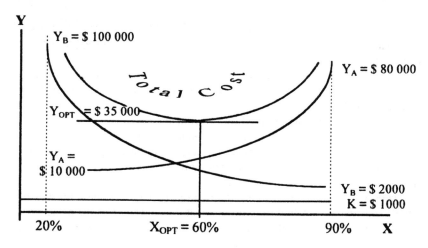

By spending only $ 10 000 (Y_A) on data entry (DE), the confidence in decision-making is at its lowest ($X_A = 20\%$). This could result in an estimated penalty of $ 100 000 (Y_B).

If a maximum available amount of $ 80 000 (Y_A) is spent on DE, the captured data (DC) results in very reliable information ($X_B = 90\%$) with little adverse effect on decisions ($Y_B = $ 2000$). The marginal cost of setting up the data base is $ 1000.

What is the percentage reliability of data for minimum cost?

A #4.02:

$Y_A + Y_B$ = total cost. In many cases, YA and YB are exponential curves and can be fitted by calculating $Y = KX^S$. If the curves can not be fitted, these total cost can be plotted on the graph, the horizontal tangent on the curve gives us the minimum cost = $ 35 000 with a reliability of 60%. Any increase in reliability will be at premium cost.

Q #4.03:

Shortly describe the purpose of the information flow model.

Cost System Flow Diagram

Having acquired a basic understanding of the "systems" approach
(input-processes-output), combined
with objectives and requirements of
a project, we should now be able to
design a cost system flow diagram.
This will encompass the total cost
and financial information flow for a
specific project.

It is not often that this will be produced from scratch. Processes are in most
cases existing, but they may need modification. The main purpose of the flow
diagram is twofold:

- Design, redesign or modify cost system
- Use flow diagram to inform and educate project staff

The reason for modification is usually a change in project requirements, dissatisfaction with the present system, corporate policy revisions, or updated computer
processes.

There is often a lack of awareness among project staff of the *total flow* of financial and cost information from the estimate stage through to the actual cost. A
flow diagram will identify the major systems that are in use to process cost information. Giving all members of the project team a view of the overall picture and
how *they* fit in, will instill a sense of belonging and more interest in the work itself. As mentioned at the start of this Section, learning the "language" of the cost
system will enhance communication with others that contribute to the work.

Project management appears to entail independent separate steps. Each contractor plans and controls the work. Designers and draftspersons establish the
configuration, purchasing agents order the materials and the construction team
installs the facilities. And yet, for optimum effectiveness they all are interrelated.
There is one common denominator:

COST

The cost flow diagram will depict the cost flow in an integrated fashion. As a first
step we may want to look at the system as it relates to permanent materials only.

For example: As the project evolves through the various stages, definitions of
package contents are improved. Configuration is described in detail and recorded
in the Design Manual. Specifications, mechanical flow diagrams and drawings
with pertinent Bills of Material (BMs) are produced.

BM information is keyed into a computerized Project Material Management
System (PMMS) that monitors inventories. Purchase requisition data also goes
into PMMS.

Major purchases that originate in head office are processed by a system called PURCHAS which updates PMMS with delivery dates, unit prices, taxes, and discount data. Another program called Project Cost Management System collects cost information such as invoices, priced material receipts. To compare estimated material cost with actual cost from invoices, an Estimating Purchase Requisition (EPR) computerized processing system is used (Figure 4-25).

The above condensed description of interrelated cost processing systems may sound complicated. We can greatly simplify it by identifying

Input: System design, specifications, drawings, estimates, schedules, BMs, purchasing documents, etc.

Processes: Computerized and manual processing systems.

Output: Cost collected, cost estimated and reported.

Other related cost streams (engineering, construction, corporate charges) are then added and fused into another somewhat more detailed cost information flow (Figure 4-26).

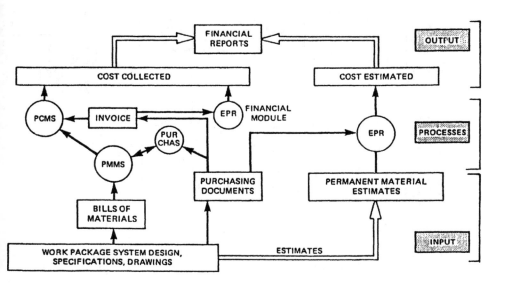

Figure 4-25
Materials Flow Diagram

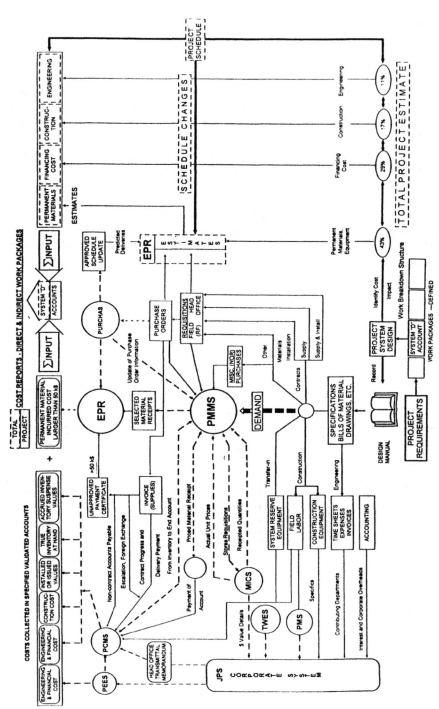

Figure 4-26

A few words should be said about the Estimating and Purchase Requisition (EPR) program. The highlights of the system are

Estimating: There is always only one current estimate on the current file. The previous estimates are stored separately on a "freeze file" and are never destroyed. The purpose of freezing an estimate is to take a snapshot of the estimate file at that particular point in time so that the estimate figures on the freeze date can be referenced later.

Purchase Requisitions: A master file is created which holds all relevant information about a purchase requisition. An adjustment file holds all the dollar and date adjustments to two types of requisitions, regular and distributed to several accounts.

Actual Expenditures: Material receipts and payment certificates are used to input incurred cost into the system.

Reporting: Over 50 types and combinations of reports are available in hard copy, on screen or graphically. All cost flow reports are in escalated dollars based on the terms of payment of the purchase order.

LEGEND

INPUT:

☐ Estimating and scheduling

☐ Other manual input

OUTPUT:

☐ Cost Reports

PROCESSES:

⟶ Manual and/or automatic

- - - ▶ Electronic transmission

☐ and ○ Computer systems

JPS	Journal Processing System
PMS	Project Manpower System
PMMS	Project Material Management System
EPR	Estimating and Purchase Requisitioning System
PURCHAS	System for Purchasing and Expediting Statistics
PEES	Project Engineering Expenditure System
TWES	Transport and Work Equipment System
MICS	Material Inventory Control System
PCMS	Project Cost Management System

Figure 4-26
Cost information flow.

4.3. TIME CONTROL

You were previously introduced to time estimating and to the development of a *time plan* (3.3 The Cost Plan - Scheduling). There we have identified the work to be done and *when* it should be done. Furthermore, we established the basis for identifying the *resources* required to perform the work.

We now must provide a measure of progress against the time plan and identify variances, analyze options and take corrective action to ensure that the objectives of the project are met.

This is time control!

The better time is managed, the more efficient a project will be. Lack of time management probably will result in failure of a project. Not meeting time objectives can be very costly in terms of financing charges or reputation in a competitive market. Time should be managed to efficiently complete specific objectives.

The PMI Body of Knowledge (3) identifies four basic processes of the time management function:

1) PLANNING, 2) ESTIMATING, 3) SCHEDULING, and 4) CONTROL

Control contains, as its components:

> "The measurement of what actually happened against what was expected to happen; what the results, or effect, will be; and if negative, the implementation of steps to prevent undesirable impacts and, if positive, the implementation of steps to insure its continuation. Control, therefore, must contain the recognition of what has been happening, and some overt action to ensure that the objectives of the project are met."

It should be emphasized that the schedule impact on the project is not only the owner's concern. Subcontractors must also pay sufficient attention to time management and adhere to established procedures in order to maintain uniformity of the process. Well documented procedures provide necessary guidance and a structured application of the system. A scheduling manual would contain:

Policy	Change Control
Basic Principles	Reporting Requirements
Definitions	Computerization
Schedule Structure	Training
Multi-Level Breakdown	Responsibilities.

When Consultants are engaged to perform major portions of design and/or construction, the basic principles outlined in a manual must be implemented. It should be the function of the project's planning and scheduling personnel to ensure that the Consultant has in place a system that embodies these principles.

It is not the intent of this chapter to teach scheduling in great depth, but to give an overview only of how schedule changes can affect project cost. There are many textbooks and training courses available that go into very specific details and are designed for the scheduling practitioner. Sophisticated computer programs are on the market that can handle tens of thousands of activities. Less sophisticated software is available for smaller projects. Even electronic spreadsheets and PC data bases can be programmed to handle basic scheduling functions.

AACE International and PMI are publishing regular software reviews in their journals.

4.3.1 Scheduling Control System

A good scheduling system

- ◆ has a standard operating procedure and a historical data base.
- ◆ uses a common format throughout the company consistent from project to project.
- ◆ ensures that important tasks are not omitted at the start of a program.
- ◆ establishes priorities; provides a yardstick against which actual performance can be measured.
- ◆ coordinates activities by contractors, suppliers, consultants, functional and project personnel.
- ◆ provides all disciplines with complete and up-to-date project coverage.
- ◆ spots potential bottlenecks in time for preventive action or for improvement of performance.
- ◆ allows trade-offs among funds, design criteria, manpower, equipment, construction methods, and time requirements for critical and non-critical areas of work.
- ◆ permits rescheduling, provides periodic evaluation of the master schedules.
- ◆ provides a historical data bank where data can be retrieved for use on future projects.
- ◆ develops model projects and plans to be used for similar projets later on.

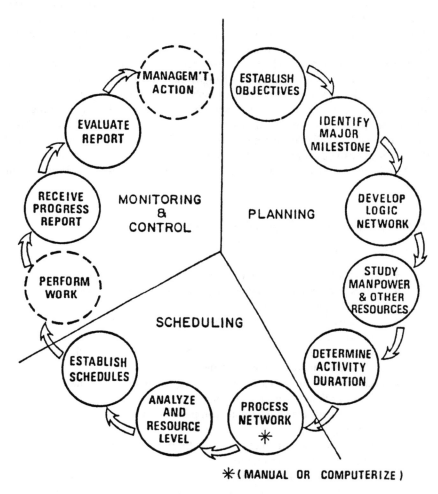

Figure 4-27
The scheduling cycle.

The Planning and Scheduling Process is similar to the cost control process. It can be depicted as shown in Figure 4-27.

Even though the scheduling cycle is generic and applies to all types of projects, we must make a distinction between large and small projects for different reasons.

Large projects do not consider resource requirements to be a major problem. It is assumed that the scheduled duration will be the actual duration. All those scheduled durations are summarized into completion dates. Duration "padding" is not allowed. This can be considered to have a 50/50 chance for the project to be completed on time (standard probability distribution).

Resource requirements are a major concern for small projects. In fact the schedule may be based on the availability of resources. In a large engineering/construction project there is a dynamic relationship between materials, human resources, configuration, time, etc. Changes, or disturbances of any kind are propagated in a manner that causes a *ripple effect* on the schedule. This is especially the case when the changes are minor.

Large changes, like the addition of major equipment are visible enough to be easily incorporated into a schedule revision. But the ripple effect due to decreased productivity because of congestion in the work area may not be as easily detectable. This also holds true for individual small changes which are numerous on large projects. They, individually, may have little impact, but their cumulative effect could lead to a schedule slippage (refer to text dealing with claims).

The examples below describe in a highly condensed way a scheduling system developed for generating stations built by a large utility(31). It is always better to "water down" or reduce the sophistication of larger systems to meet less requirements than the other way around. For large projects there are usually three major types of schedules produced:

- The Master Schedule (MS)
- The Coordination and Control Schedule (C&C)
- The Production Schedule (PS)

Schedules produced early in the life cycle of a project will only include major activities. This will be the base for the *Master Schedule*.

Later, the need for more detailed planning becomes apparent for all parties involved and is vital for progress monitoring. The WBS is used to identify those work packages for which schedules are produced. Therefore, the next lower level of the project schedule must be much more detailed in its content. Just as the project itself passes through the various phases, the project schedule must also pass through those phases. The design phase will blend into the procurement phase which in turn will blend into the construction phase. How much overlap those phases have depends on the size, duration and type of project.

The main function at the second level of scheduling is *Coordination and Control*. It provide the means for monitoring progress on a system by system basis. This is the level for resource analysis. It provides a framework for the lowest level, the Production Schedule.

Production Scheduling is used to outline the work programs to be followed by each production group in order to achieve the scheduled objectives. It deals with specific detailed activities in the near future which must be carried out within the parameters established by the resource plan. Production Schedules control the work load within each resource group and may dictate a change to the Control Schedule. This effectively integrates the production process with budgeting, scheduling and engineering.

The Master Schedule is often produced and updated manually, however, as the number of activities increases, a large data base system is typically employed.

One specific item on the Master Schedule will expand into at least one sheet of a Coordinating & Control Schedule. And one specific item on the C&C Schedule expands into at least one Production Schedule (Figure 4-29)

Changes in the Production Schedules that affect the critical path at that level will have to be reviewed by the C&C Schedule level if this affects any of their terminal events and in turn a Master Level milestone event. Figure 4-28 is a model of a typical reporting process.

As schedules are produced, actual progress is monitored, decisions for changes are made and schedules updated. Updated changes are then reported to higher levels in the organization.

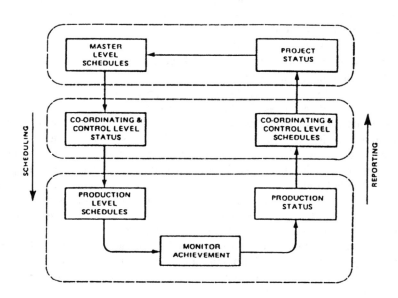

Figure 4-28
Schedules reporting model.

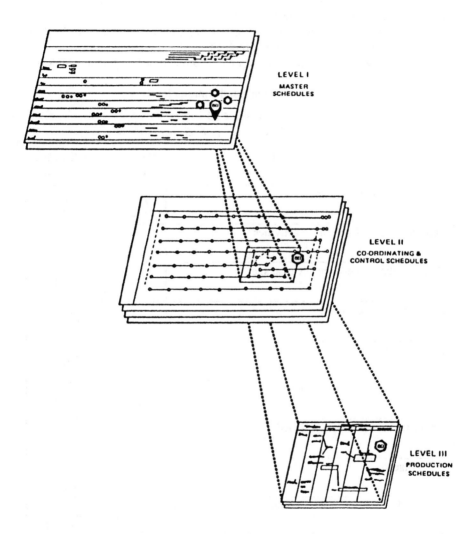

Figure 4-29
Scope of schedule details.

The method for reporting status and for controlling the flow of information throughout the management levels is designed to

♦ Provide management with precise information to make timely decisions regarding scheduling problems.

♦ Provide a uniform process for the reporting of achievement to all levels of the organization.

♦ Provide uniformity of format at the master level.

Table 4-09

Scheduling symbols

Activity or Event	Schedule Date	Actual Date Achieved
Event with event number	(50)	(50)
Terminal event of a chain of activities	[51]	
Delivery, date material received at site	Δ	(Δ)
Drawings received at site	D	(D)
Technical specifications complete	S	(S)
Tendering documents complete	T	(T)
Purchase recommendation issued for approval	R	(R)
Order placed	O	(O)
Control Event, defining a committed objective in terms of time and specific achievement for control purposes, used within a Coordination and Control Schedule.	E 17	E 17
Project Manager's Milestone Event	65	65
Key Event; turnover of completed work package to the construction organization.	KE 42	KE 42

Various scheduling symbols are used with scheduling control systems (Table 4-09). To interpret major events and milestones, the standard symbol of a circle for a node on an arrow diagram can be revised to give an easier interpretation of a flow diagram. Those symbols are very much dependent on the software used.

A general indication of progress on C&C schedules is ➠━━━━━━━➤

Activity in Progress

Following are some samples of various types of schedules and progress graphs. They are part of a scheduling manual and should be considered illustrations rather then working documents.

Project Master Schedule (Figure 4-30). This schedule provides an overview of the sequencing of major engineering and construction activities. It is a network diagram drawn on a calendar base in accordance with the content and scale of detail of the breakdown structure. Only a small excerpt is shown.

Coordination and Control Schedule Interfaces (Figure 4-31). Those schedules are network diagrams drawn for the various systems as identified by the break-down structure. Those C&C schedules are interfaced at the relevant baselines. The exhibit illustrates the method of interfacing the various C&C level schedules.

Coordination and Control Schedule (Figure 4-32). This is a cut-out of a typical coordinating and control level schedule.

The functions of the Coordinating and Control level schedules are:

1. To program the composite efforts of the individual resource groups for the entire project.
2. To help coordinate-ordinate contributing resource groups by highlighting responsibility interfaces.
3. To provide the means for monitoring progress on a system by system or structure by structure basis.
4. To generate schedule variance information required for the analysis-decision-action process on a project wide scale.
5. To provide the framework for production schedules within which the work is planned and scheduled in greater detail.
6. To provide a basis for overall project resource analysis.

Engineering and Construction Management restrain the completion dates of Control Events (CE) between higher level milestone events. This serves to establish schedule objectives for the subordinate levels of the organization. Float between Control Events can be distributed to provide flexibility of resource allocation and production planning. Restraint dates are shown in a box:

ER - 1995-08-12
FIRST STEAM

ER (early restraint) imposes a restraint on the start on all down-stream activities.
LR (late restraint) is imposed on an event restraining the completion on all upstream activities.

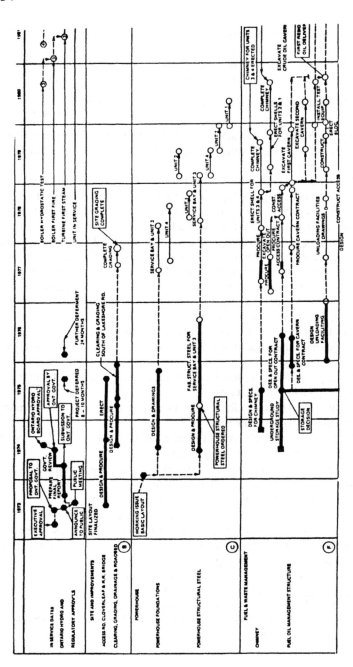

Figure 4-30

Project master schedule.

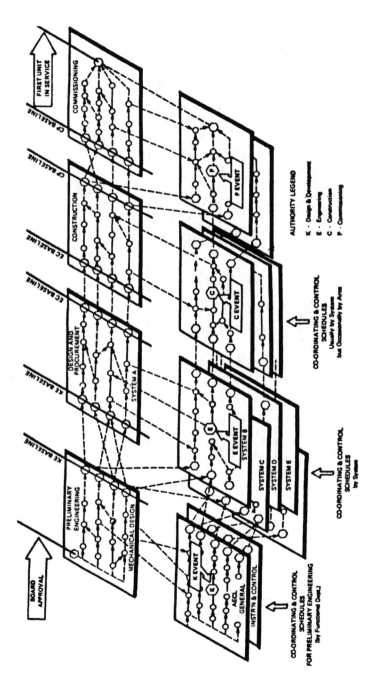

Figure 4-31

Coordination and control schedule interfaces.

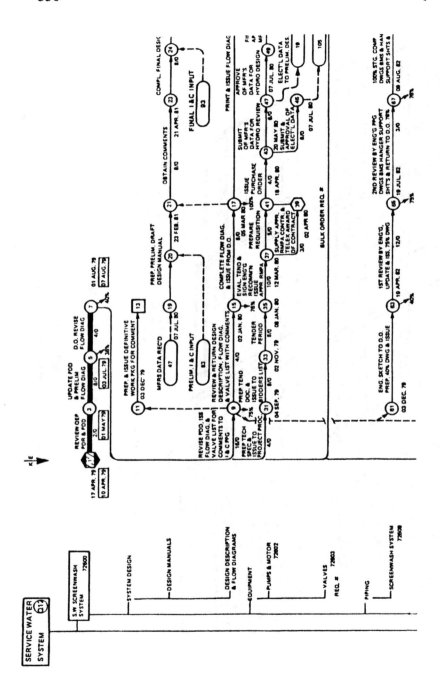

Figure 4-32

Coordination and control schedule.

Input into the C&C level is provided by the Production Level schedules. These are the schedules of the working level. The reporting process to a higher level is depicted in Figure 4-37. There is a large variety of Production Level schedules. Figures 4-38 to 4-41 show samples of those. They include

- Resource Schedule
- Profile for Production Quantities
- Profile for Manpower

The format and content of production schedules and related documents must include provision for

1. Relating actual status to planned status on the date of reporting.
2. Relating predicted start or finish dates to planned start or finish dates.
3. Stating the reason for the variances which existed at the date the report was produced and indicating intentions to correct the variances.

Production scheduling is used to identify the work programs to be followed by each production group in order to achieve the scheduled objectives. It deals with specific detailed activities which must be carried out within the parameters established by the resource plan, and may *dictate a change to the C&C schedules.*

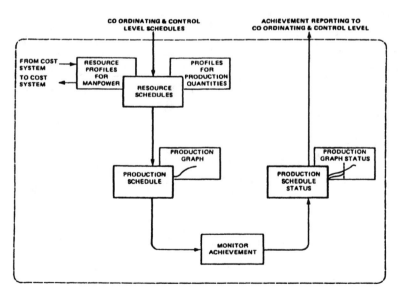

Figure 4-33
Production level reporting.

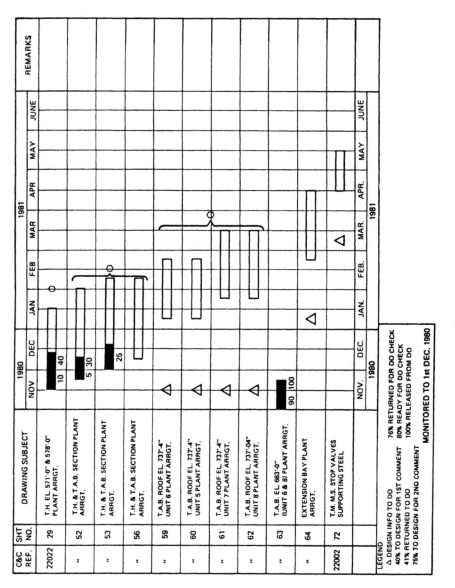

Figure 4-34
Production schedule - 1.

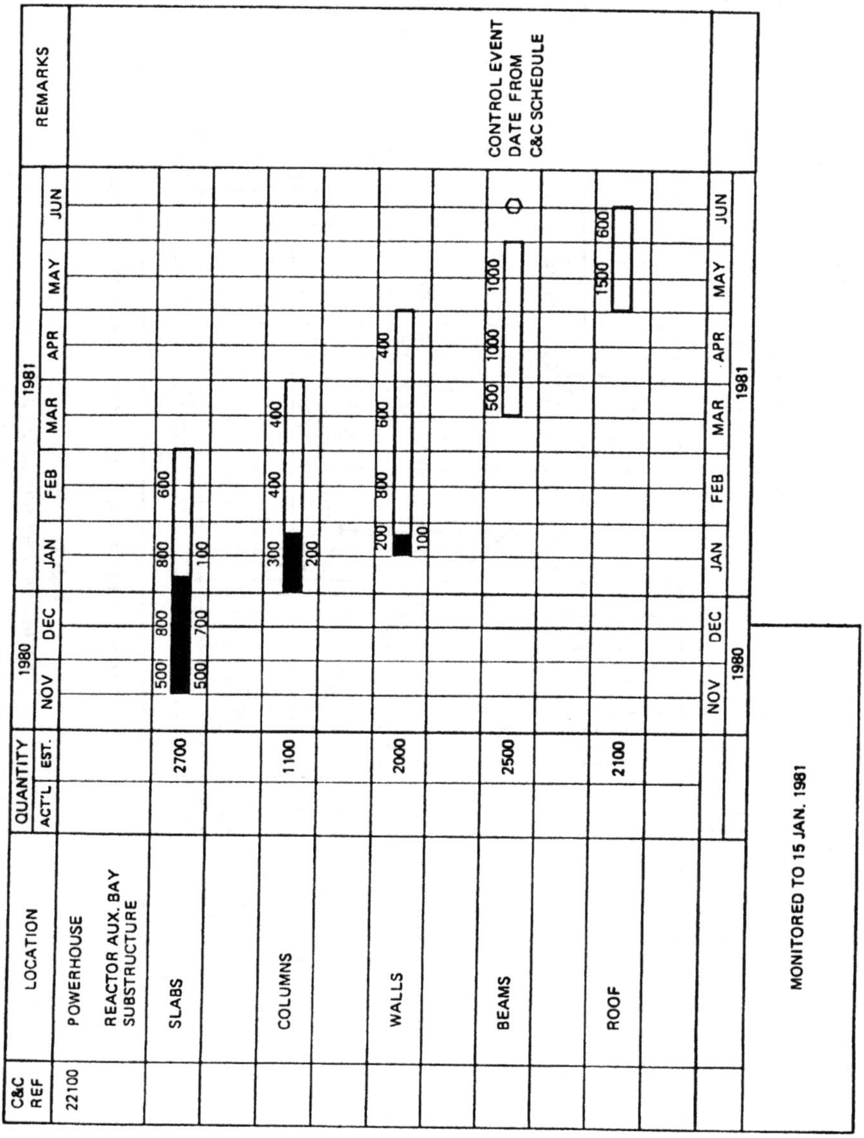

C&C REF	LOCATION	QUANTITY		1980			1981						REMARKS
		ACT'L	EST.	NOV	DEC	JAN	FEB	MAR	APR	MAY	JUN	
22100	POWERHOUSE											
	REACTOR AUX. BAY SUBSTRUCTURE											
	SLABS		2700	500 / 500	800 / 700	800 / 100	600					
	COLUMNS		1100			300 / 200	400	400				
	WALLS		2000			200 / 100	800	600	400			
	BEAMS		2500					500	1000	1000		CONTROL EVENT DATE FROM C&C SCHEDULE
	ROOF		2100							1500	600	

MONITORED TO 15 JAN. 1981

Figure 4-35

Production level schedule - 2.

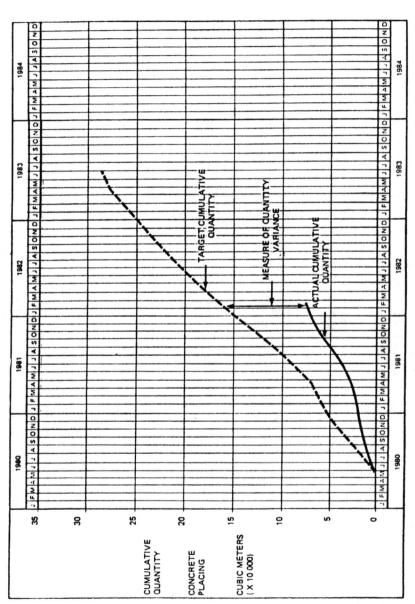

Figure 4-36

Production level schedule - 3

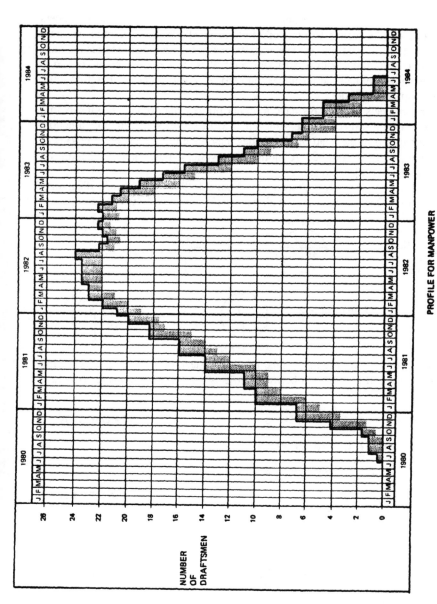

PROFILE FOR MANPOWER

Figure 4-37
Production level schedule - 4.

The examples above give some overview of a particular large project. The trend now seems to be toward smaller projects as described in the Cost Engineers' Notebook (3); "Effective Control of Multiple Small Projects." Companies must therefore develop multi-project systems. Project teams for large ventures are more "self contained" units with a sufficient pool of resources, smaller projects may have to fight to obtain the resources required. A small project is therefore much more resource oriented.

Some projects will always have a higher priority than others. Engineers may work on several projects at the same time. Most services and support functions may be handled in the head office. Trades may have to be transferred repeatedly from one project to another. A multiproject system must therefore handle the schedules of all projects in concert.

Section 3.3.4 - Schedule Constraints - Figure 3-39 shows an arrow diagram of Package P-1. This was interfaced with package P-2 (Figure 4-40):

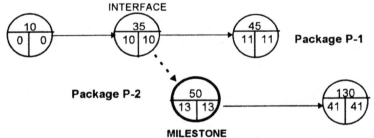

Multiple projects interface in a similar manner. Instead of interfacing between two packages, we call the restraint between projects the *Project Connector* (figure 4-38)

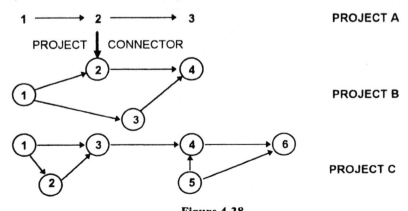

Figure 4-38
Single project connector.

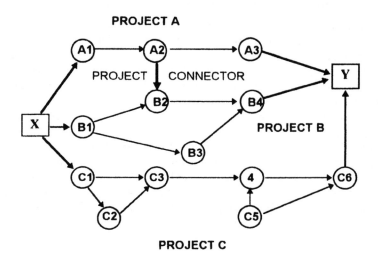

PROJECT A

PROJECT CONNECTOR

PROJECT B

PROJECT C

Figure 4-39
Integrated project connector.

Project C may not have any connector with another project. There are basically two ways to solve a multiproject network:

1. In order to combine the projects, the node numbering system must be mutually exclusive, i.e. no duplication in coding. This can be done by adding the code of the project to each number or by assigning a range of different numbers to each project (Figure 4-39).
2. If there are several projects with little interaction (very few connectors), we can treat each project independently by computing EET_j and $LETi$ with the connector being inactive.

The next computer pass checks the effect of the connector on early and late dates. The values of EET_j and LET_i may have to be replaced with the connector's early finish date (EF) or late start date (LS) respectively unless the EF of the connector is less than EET_j of the successor node *or* the LS of the connector is greater than the LET_i of the predecessor node. To inter-relate several projects needs quite a sophisticated system.

Added to this is the problem that small projects usually do not have experienced scheduling personnel. Sufficient training must be provided to those having an input into the schedule. Clearly drawn up procedures are also very important.

4.3.2 Updating Schedule Status

It is not enough to say when an activity is completed we must also be able to tell the *portion of completion* before we reach that stage. The portion completed is reported at the project status date (PSD). This date is set at predetermined intervals. Duration analysis is used to determine the status at the PSD. To better understand the calculations involved, we best look at the lowest denominator in scheduling, the work item activity. The calculations performed during the control phase are similar to those performed during the planning phase (see 3.3.4 - Basic Computations).

The values for LET_i previously calculated remain, but new values must be calculated for the EET_i. The current network is obtained by discarding all events which are predecessor events of only completed activities. Each starting event in the current network will now be calculated using its actual time of occurrence. For example, if activity 1-2 below (Figure 4-40) takes six days instead of three, the actual status after day 6 is 11 days.

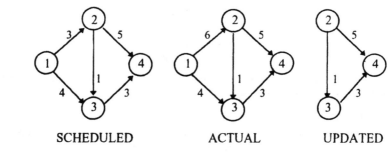

SCHEDULED   ACTUAL   UPDATED

Figure 4-40
Updated schedule status.

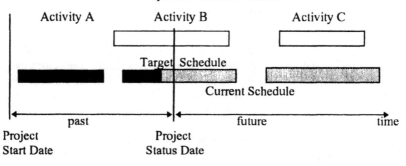

Figure 4-41
Activity status bar chart.

Activities 1-2 and 1-3 have been completed. This eliminates event 1. The network has only one starting event now, event 2. The original critical path was 1-2-4 with a duration of 8 days. The duration is now 11 days, a slippage of 3 days.

For an activity status presented in bar chart form (Figure 4-41), we need to know

1. the *location* of the activity on the project time scale in regard to
   a) the project plan (target)
   b) the current schedule
2. the state of completion of the activity
   a) activity fully completed
   b) activity partially completed
   c) activity to start in the future
3. the variance between project plan (approved schedule) and actual operation (current schedule).

Activity C has a time-scaled bar of both, the target schedule and the current schedule (forecast). Assuming a delay from the target schedule, there are three variances to be calculated (Figure 4-42):

Figure 4-42 shows the following data:

TSD = Target Start Date - Planned activity start date.
TFD = Target Finish Date - Planned activity finish date.
TD = Target Duration - Amount of time allotted as planned.
SSD = Scheduled Start Date - Currently estimated start of activity.
SFD = Scheduled Finish Date - Currently estimated finish of activity.
SD = Scheduled Duration - Estimated time needed to work on activity.
SDV = Start Date Variance - Time difference between TSD and SSD.
FDV = Finish Date Variance - Time difference between TFD and SFD.
DV = Duration Variance - Time difference between TD and SD.

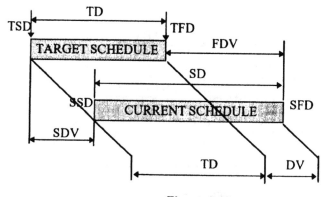

Figure 4-42
Three types of variances.

Variance Calculation:

Start Date Variance = TSD − SSD ....(1)

Finish Date Variance = TFD − SFD ....(2)

Duration Variance = TD − SD ....(3)

It should be noted that the bar chart above shows the activity's schedule duration with early start date, late start date, early finish date, late finish date and floats. The variance calculations are usually based on early dates.

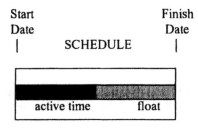

Start Date   Finish Date
SCHEDULE

active time   float

Activity B can be subjected to activity analysis. How this is handled depends on the information available. Andrew Sipos (Cost Engineering- Vol 34 /No 2) discusses a strict duration analysis method where for *a given project status* date the present activity's *actual start date* plus percent of completion or *actual start date* plus remaining duration must be available.

Target dates are necessary for variance calculations (Figure 4-43). Some new terminology is added:

PSD = Project Status Date - The date of the next workday.

ASD = Actual Start Date - The date on which activity has started.

CSD = Current Start Date - The date on which remaining work has started.

CFD = Current Finish Date - The estimated finish date.

AD = Actual Duration - Actual time worked on the activity.

RD = Remaining Duration - The estimated time needed to complete the activity.

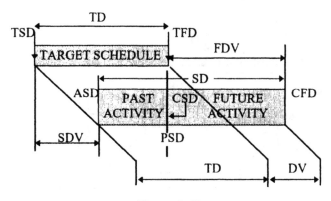

Figure 4-43
Project status date.

In addition, %C = Percent Completion is the ratio of the portion completed multiplied by 100.

We may now do some calculations based on the data that is available.

<u>Given:</u> Actual start date and remaining duration.

AD = PSD – ASD .... (4)
CFD = PSD + (RD - 1) .... (5)
%C = (SD – RD) * 100/SD .... (6)

<u>Given:</u> Actual start date and percent completion.

RD = SD – (SD * %C)/100 .... (7)
AD = PSD – ASD .... (4)
CFD = PSD + (RD – 1) .... (5)

<u>Given:</u> Actual start date only. Because no other data are available, we may assume the planned target duration TD

AD = PSD – ASD .... (4)
RD = TD – AD .... (8)
CFD = PSD + TD – AD – 1 .... (9)

If the current activity continues and the actual duration has reached or has exceeded the target duration, then the activity can not be considered completed. We add one time period to the future activity.

AD = PSD – ASD .... (4)
RD = ONE TIME UNIT (day, week)... (10)
CFD = PSD + 1 .... (11)
%C = (100 * AD)/(AD + 1) .... (12)

<u>Given:</u> Percent completion

AD = (%C * SD)/100 .... (13)
RD = SD – AD .... (14)
CFD = PSD + (RD – 1) .... (5)

<u>Given:</u> Remaining Duration

AD = SD - RD .... (15)
CFD = PSD + (RD – 1) .... (5)
%C = (SD – RD)*100/SD .... (6)

<u>Activity A</u> is history. Variance calculations have been done and are duly recorded. So are the data for activity duration and actual finish date. What we need to know now is: How do we determine the portion of work completed?

Measuring Work Progress

To measure quantities, we are dealing with a great variety of units and ratios. There are several methods to measure construction work, some of which are:

μετροv • metron > measure

Units completed	Level of effort
Start/Finish	Supervisor opinion
Cost ratio	Equivalent units

See also *Skills and Knowledge of Cost Engineering* (2).

Units Completed: This applies to repeated production of easily measured pieces of work such as m³ of concrete poured, meter of pipe installed, number of fittings done. When compared with estimated amounts, the "percent complete" can be calculated for each work item, e.g. if 30 Mg of reinforcing is placed out of a total of 40 Mg, the job is 75% complete.

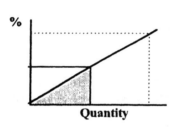

Criteria:

♦ Repeated Production
♦ Quantitative units can be *measured* easily.
♦ Units must be on the same level of effort (linear relationship).

Level of Effort: Tasks that must be done in sequence can usually be identified by their incremental milestones, measured as a percentage of the completion point. A system design could be sequenced as follows:

Basic Design completed...	30%	**(B)**
Final Design approved...	40%	**(A)**
Drawings completed...	70%	**(D)**
Material specified...	75%	**(S)**
Tenders evaluated ...	90%	**(T)**
Order placed ...	100%	**(O)**

The percentage breakdown does not necessarily mean that activities must also be incremental, they can overlap or be combined.

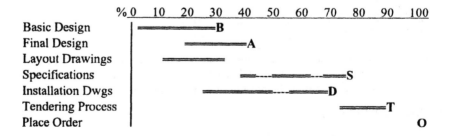

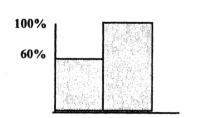

Criteria:

◆ *estimated* percentage completion
◆ cumulative progress is incremental

Task

Start/Finish: The percentage complete is quite arbitrary for activities where the start and finish of the work can be scheduled, but it may be very difficult to tell the percentage completion in between. Equipment alignment or calibration work or a planning activity are examples. An even distribution of time is usually assumed for on/off tasks (constant rate over time), i.e. start with 50%, end with 100%. For very short tasks we may want to start with 0% and end with 100%.

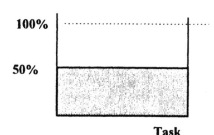

Criteria:

◆ Level of effort very difficult to monitor or estimate between beginning and end of a task.
◆ Can be modified to 0% and 100% start and finish for short durations.

Task

The above percentage completion points can be estimated by the foreman or supervisor strictly on his/her judgment.

Supervisor Opinion: With this method, the "percent complete" of the task is strictly up to the judgment of the supervisor. Examples are landscaping, cleanup, dewatering, scaffolding. The tendency here is to be overly optimistic in the beginning of a job, i.e. building up to 90% in a short time and then stay there for half the remaining time.

Cost Ratio: Tasks which involve a long period of time with various production rates or where production cannot be measured, such as project management or administration, are usually estimated in workhours or dollars. The progress is compared with the current forecast at completion, i.e.

$$\% \text{ complete } = \frac{\text{Cost or Workhours to Date}}{\text{Forecast at Completion}}$$

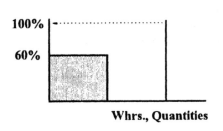

Criteria:
◆ *estimated* percentage completion
◆ cumulative progress to-date vs. forecast at completion.

Whrs., Quantities

Equivalent Units: We all know, that hybrid ratios cannot be added. The question "how many meters make a ton" seems foolish. As the saying goes, "we cannot add apples and oranges", but if we had asked:

"How many meters of #10 reinforcing steel makes a ton?", then this is a sensible question.

If we can *equate* one unit to another or several units to a *standard* one, we are able to measure progress with no distortion. Where different units of work cannot be equated readily, judgment will have to be used.

For example: An estimate could show the following breakdown for foundation work:

(1) Subtask	(2) Unit	(3) Total Quantity	(4) Workhours
Excavation	m³	5 000	800
Formwork	m²	400	1 000
Reinforcing	Mg	20	600
Concrete	m³	4 000	200
		TOTAL	**2 600**

Lets assume we are choosing m³ as the standard unit. All quantities are now converted into equivalent m³. What has been workhours previously is now equivalent m³ (column 4). The *weight* is the ratio of individual equivalent m³ and total equivalent m³. When we multiply equivalent m³ with the ratio to-date quantity over total quantity (6)×(4)/(3), we obtain earned m³ (column 7).

(1) Subtask	(2) Unit	(3) Total Quantity	(4) Equiv m³	(5) Weight	(6) To-date Quantity	(7) Earned m³
Excavation	m³	5 000	800	0.31	5 000	800
Formwork	m²	400	1 000	0.38	380	950
Reinforc'g	Mg	20	600	0.23	10	300
Concrete	m³	4 000	200	0.08	1 000	50
FOUNDATION	m³	--	2 600	1.00	--	2 100

From the above, the progress is:

$$\frac{\text{Total earned m}^3}{\text{Total equivalent m}^3} = \frac{\sum (7)}{\sum (4)} = \frac{2100}{2600} = \textbf{81\%}$$

Criteria:

♦ Level of effort is estimated for subtasks, having a variety of units.

♦ Work progress is equated to an earned value expressed in a chosen unit.

Another Example: A contractor is required to complete a 10 km section of a highway. For progress payments and statistical purposes, the Department of Highways requires a progress report by km. completed.

A status report from the superintendent shows the following to-date work progress (**WP**) in terms of quantities installed as compared to estimated Quantities (**EAC** = Estimate at Completion). For the work performed, the quantities vary so much, that a common, fairly representative unit must be chosen. For this job, there are labor and equipment, i.e. Whrs. and $/hr that would apply to each sub-task. Workhours were chosen:

(1) Subtask	(2) EAC	(3) WP	(4) EAC
Stone, sand	120 000 m^3	60 000 m^3	2 000 Whrs.
Formwork	10 000 m^2	2 000 m^2	500 Whrs.
Reinforcing	9 000 Mg	1 800 Mg	900 Whrs.
Concrete	60 000 m^3	6 000 m^3	1 200 Whrs.
Asphalt	600 loads	48 loads	600 Whrs.
		Total	5 200 Whrs.

5.2 kWhrs are now equated with a 10 km highway section or 100% completion. This can be expressed in terms of a *"weight"* ratio for each sub-task. The *earned km* can now be calculated.

(1) Subtask	(5)=10×(6) EAC,km equ.	(6)=(4)/$\sum$(4) Weight	(7)=(3)×(5)/(2) Earned km
Stone, sand	3.85	0.385	1.925
Formwork	0.96	0.096	0.192
Reinforcing	1.73	0.173	0.200
Concrete	2.31	0.231	0.231
Asphalt	1.15	0.115	0.092
	10.00	$\sum$(4)=1.000	**2.64 km**

The percent completion is (7)/(5) = 2.64/10 = **26 %**

In Engineering, including drafting, procurement, estimating etc., a similar system can be used when m² of drawings, percent of procurement activity or estimating each work package needs to have a standard work unit.

4.3.3 Resource Leveling

Realistically, no project is provided with unlimited resources. The availability of manpower, equipment, and money is limited within certain constraints for any capital project. When we produce a plan we must be certain that the resources required by each activity are available at the scheduled start time.

This is a very difficult task. To find a *truly* optimal schedule - even with the most sophisticated computer models - is hard to achieve. Much progress has been made in computerization. Most software packages today have resource scheduling capabilities, in some cases linear programming is used to allocate resources (37). Unless a project is very small, manual manipulation is not practical.

Not unlike other problems in the real world, explaining the principles of resource leveling on simple examples can not convey the complexity of computational algorithm employed in project control. Please refer to AACE Cost Engineers' Notebook "Procedures for Resource Leveling"(2) for an overview of a schedule/resource model for a large project.

In all the examples so far we have shown that all activities along the *critical path* have zero float. That means

$$EET_j = LET_j$$

If the latest allowable time for the terminal event is set by a predetermined duration T_s for the completion of the project, the slack on the critical path can also be positive or negative. This is often the case in practice. If the slack is negative, replanning is indicated. The revised definition is now

The critical path is the path with the least total slack.

We also showed *random networks* in previous examples, i.e. the network is self-contained. This is the way how scheduling basics are explained. In practice, however, the *module network* is used (Figure 4-44).

Large numbers of activities require a method of tying many schedule printouts (drawings) together while maintaining unique numbers that identify each **i** and **j** activity. The modular network, identified as ⬭ , satisfies that requirement (Figure 4-45).

Most network analyses up to now have dealt with early start times etc. Resource scheduling concerns itself with *scheduled* start times, scheduled finish times etc. through the use of *heuristic* (trial and error) programs.

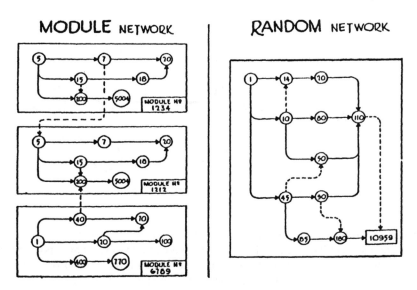

Figure 4-44
Module network.

As a simple example, assume we have to do an installation for which compressors are needed. The arrow diagram (Figure 4-51) was produced by the planning department of the company with full consideration for the crew sizes needed to satisfy the activity durations.

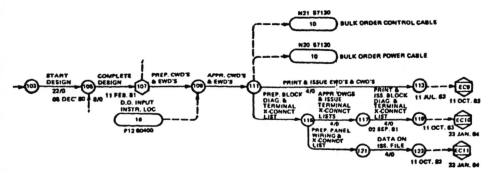

Figure 4-45
Module network example.

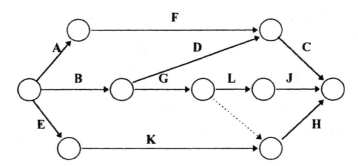

Figure 4-46
ADM resources diagram.

Utilizing the arrow diagram, Table 4-10 is a listing of the schedule.

Because activities E,F,G,J,L do not require compressors, there is no need to represent them for this resource allocation. To visualize the number of resources required, we will show the logic diagram in bar chart form (figure 4-47).

Table 4-10
Compressor installation

Acti-vity	Dura-tion	Earliest start	Earliest finish	Latest start	Latest finish	Total float	Resour-ces	Criti-cality
A	4	0	4	14	18	14	3	4
B	8	0	8	0	8	0	4	1
C	3	19	22	33	36	4	4	4
D	6	8	14	27	33	19	3	5
E	7	0	7	4	11	4	-	2
F	15	4	19	18	33	14	-	4
G	12	8	20	8	20	0	-	1
H	10	20	30	26	36	6	4	3
J	5	31	36	31	36	0	-	1
K	9	7	16	11	20	4	1	2
L	11	20	31	20	31	0	-	1

A histogram can now be drawn of those activities requiring resources in common (Figure 4-48).In this case, one must assume the resources are identical in all activities. In reality, a number of different resources are usually required. Accordingly, this function is invariably computerized.

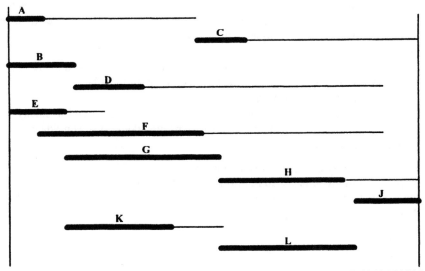

Figure 4-47
Compressor installation bar chart.

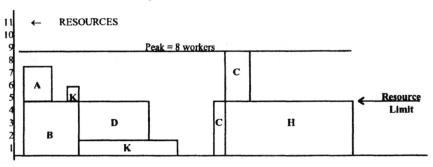

Figure 4-48
Resources histogram.

According to the superintendent at the site, only *four* compressors are available. The maximum resource allowed is therefore = 4. In order to maintain the resource constraint to four (4) resources in any one day, one must allocate the resources to the most critical paths first, continuing the allocation until the resource limit is reached.

Moving an activity downstream decreases the time available and increases the priority for that activity. If the time available becomes zero (0) and the resource limit is still exceeded, moving an activity on a more critical path might be an alternative.

When continued in this manner until all paths have been tried and the resource limits are still exceeded then either the project duration must be extended or the resource limit must be increased.

To demonstrate the heuristics used in resource leveling, we will manually go through five steps in our example. Similar steps are taken on screen when using appropriate software.

Step 1: →Activity B has top priority. Leave alone.

Step 2: →Activity K has second priority. On day 8 resources are exceeded by 1. Four days float are available to activity K. Shift K by one day, leaving 3 days float.

```
   B
4 4 4 4 4 4 4 4
                      K
              -- 1  1  1  1  1  1  1  1 ----------

.1.2.3.4.5.6.7.8.9.10.11.12.13.14.15.16.17.18.19.20.21.22.23.24.25.26.27.28.29.30.31.32.33.34.35.36
```

Step 3: →Activity H has third criticality. Resources are not exceeded. Leave alone.

Step 4: →Activity A and C have fourth priority. Looking at A first: Resources have been used by B for the first eight days. Shift start date to day 9. Now schedule C: A starts on day 9, takes 4 days. F now starts on day 13; lasts for 15 days. C cannot start until day 28.

Problem: H is using the resource limit on days 28,29,30

Solution: Shift C to start on day 31, leaving only 3 days float for this path.

```
   B
4 4 4 4 4 4 4 4            K
              -- 1  1  1  1  1  1  1  1 ----------              H
                      A                        4  4  4  4  4  4  4  4  4  4  4 ----------------------
--------------------3  3  3  3 ---------------------- F
                          0  0  0  0  0  0  0  0  0  0  0  0  0  0  0  0            C
                                                                       -------- 4  4  4 ------------

.1.2.3.4.5.6.7.8.9.10.11.12.13.14.15.16.17.18.19.20.21.22.23.24.25.26.27.28.29.30.31.32.33.34.35.36
```

Step 5: →Activity D has fifth priority, last place. Resource limit is used by K and A. Shift the start day to day 13. 15 days float left.

This is now the .Scheduled Start. resource chart. A, F and C have changed to a higher priority. The work is scheduled for 33 days with a 3-day level float:

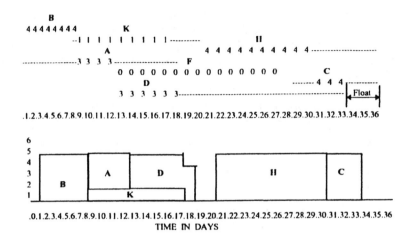

Figure 4-49
Scheduled start histogram.

There are other alternatives for consideration. Activity B may have been split. This means a change in logic and a new network to calculate. There may be several "workable" alternatives. Not all of them are optimal. Why is the solution below less optimal? This solution takes 36 days. Float is eliminated.

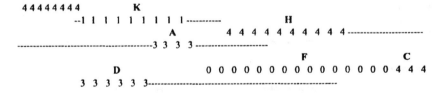

The above simple resource leveling example for equipment (EQU) becomes more complicated with human resource limitations. Looking at a production level schedule, each activity will need various disciplines. If we superimpose trade requirements on the previous example, say pipefitters (P) and welders (W), whereby compressors are needed only by pipefitters (deliberately simplified to explain the principle), we have a revised listing of the original schedule (Table 4-11).

Assuming further that *one* compressor is needed for a group of *four* pipefitters, then the workpower loading for the *maximum* equipment availability is depicted in a logic diagram (Figure 4-50) and related histograms (Figures 4-51 and 4-52).

The logic diagram shows that A, B and E are concurrent activities. The ES date for A can be moved to zero together with B. Because of the large float and low priority we can split activity D. The resulting profile is a smooth workpower loading reducing the peak from 46 to 34 workers. The equipment profile will not exceed four compressors.

Table 4-11

Compressor installation - resource limitations

Activ	Dur	ES	EF	LS	LF	TF	EQU	P	W	CRIT
A	4	0	4	14	18	14	3	12	6	4
B	8	0	8	0	8	0	4	4	2	1
C	3	19	22	33	36	4	4	16	6	4
D	6	8	14	27	33	19	3	8	4	5
E	7	0	7	4	11	4	-	-	6	2
F	15	4	19	18	33	14	-	-	12	4
G	12	8	20	8	20	0	-	-	4	1
H	10	20	30	26	36	6	4	16	8	3
J	5	31	36	31	36	0	-	-	4	1
K	9	7	16	11	20	4	1	12	6	2
L	11	20	31	20	31	0	-	-	6	1

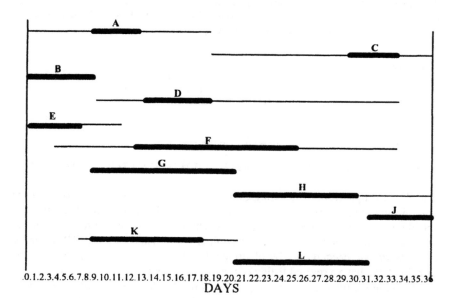

.0.1.2.3.4.5.6.7.8.9.10.11.12.13.14.15.16.17.18.19.20.21.22.23.24.25.26.27.28.29.30.31.32.33.34.35.36
DAYS

Figure 4-50
Workpower loading bar chart.

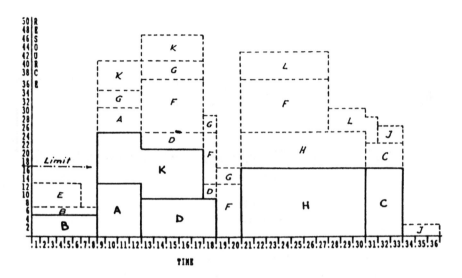

Figure 4-51
Workpower loading histogram.

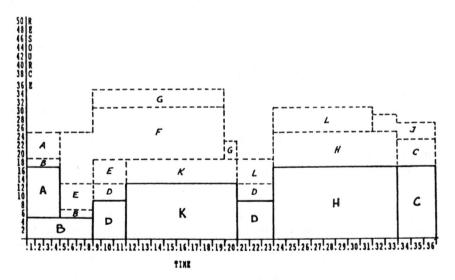

Figure 4-52
Limited resources histogram.

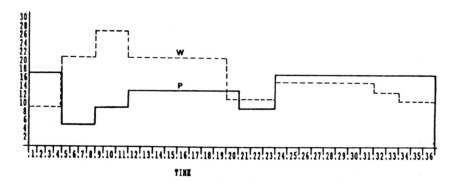

Figure 4-53
Compressor example - resource usage.

There is a general rule of thumb: *What you can draw with a setsquare and a protractor, you can calculate.* When there are hundreds of activities and several levels of schedules it would be very inefficient and costly to do resource leveling manually. We need to use a computer who "looks at" activities within a particular time interval. It then uses early start dates and total float to move scheduled dates and durations within resource limits. In our case the early start for A is moved to zero. As another example, G and K have a common late start date of 20. Their floats are 0 and +4 respectively. The early dates for both of them are within the time frame of 0 to 36 (Figure 4-52). G has no float, therefore the early start date has priority over K. Activity K needs three compressors and is within resource limits. It will be scheduled prior to its late start date. The algorithm calculates the next activity in the order of criticality and moves schedule dates to fit resource limits. If a resource level is exceeded (K plus D), the activity's early start date is pushed out. The resource usage report by trade would be entered into the next higher level of schedule (Figure 4-53). Both, the Project Management Institute and AACE International issue periodic surveys of software products (see pmnetwork for examples).

Using Linear Programming:

Resource leveling is essentially an optimization problem. We want to *minimize* the resources used at any point in time. The LP model discussed in Section 1.4.1 would have a constraint of > 4 compressors. Variable activities (durations and float) would be the input (x) into the LP model. Based on the SIMPLEX method, special software is used to calculate the allocation of resources (EXXON applies LP to the scheduling of drilling operations). A drawback seems to be the high setup cost and the need for a high speed computer (37).

Q #4.04: A building has to be erected. You are asked to draw up a logic diagram with "milestone" events. The designers will give you a block diagram with the following building components:

♦ Blocks A, B, D, E and G take one unit of time.
♦ Blocks F and J take three units of time.
♦ Blocks C, H and K take two units of time.

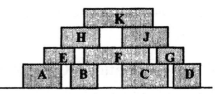

Number of blocks can be laid concurrently, e.g. A, B, C, D can start together. The supporting blocks must be in place before a higher block can be laid.

A) Draw the arrow diagram with a minimum number of nodes.
B) Show the forward and backward time calculation at each node with a heavy line for the critical path.
C) What is the total float for block E?
D) If block E took 3 time units, what would be the effect on the duration?
E) Describe two features which critical path computer reports should have to make them effective for management control.

A #4.04: A) and B)

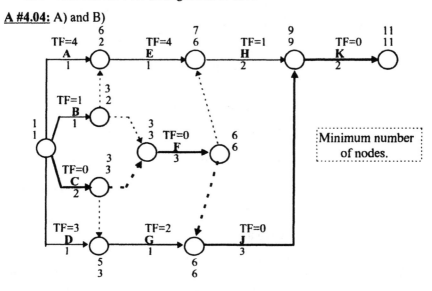

C): Total float for E is four time units.   D): There is no effect on duration.
E): 1. Suitable level of detail for each level of management.
    2. Managers receive reports on only the work for which they are responsible.

4.4. COST CONTROL PROCESSES

Some Control Processes have already been introduced in Sections 4.2.2 and 4.2.3.
We will now expand on that and elaborate in more detail on the various aspects of
those processes.

4.4.1 Conventional Cost Control

Projects are under control only if four basic elements are under control:

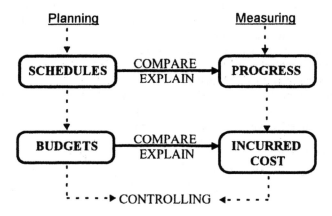

Figure 4-54
Basic cost elements.

SCHEDULES   are time scaled plans for the execution of a project,
                                while
PROGRESS   is the measure of headway made when carrying out
                                these plans
BUDGETS   are a quantity of funds allocated for the performance
                                of a specified amount of work, and
INCURRED COST   deal with the measurement of the consumption of
                                these funds.

The simple diagram above compares the actual timing of activities with the predictions made in the approved schedule. It also compares the dollars spent with
the budgeted amount if a cost flow is available. In some cases, progress is measured as percentage completion. The remaining work is then estimated and a new
forecast produced.

This indicates, that we can have two different variances:
- ♦ variance between actual cost and estimated cost at a specific point in time,
- ♦ variance between the approved total cost estimate or budget and the new forecast.

The above is often referred to as *conventional cost control*.

Estimates are usually prepared on the basis of a schedule bar chart. The bar chart shows functional elements in accordance with an established work breakdown structure. (Figure 4-55).

A dollar value is then attached to each bar. This "pricing" of the schedule is the cost plan. The approval of this plan will constitute the budget.

Progress is measured against the schedule, and actual expenditures against the budget. Several reviews of the estimate will take place during construction phases such as

- ♦ *Concept phase:* Unit cost method, capacity curves, equipment ratios.
- ♦ *Early design:* Layouts, flow sheets, equipment lists, gross floor areas.
- ♦ *Schematic design:* Unit rate by element, elevations, electrical diagrams, set of specifications.
- ♦ *Pre-bid phase:* Unit price by trade, installation drawings.

The frequency and production of estimate reviews depends on the type of project and management policy.

When the cost run below the estimate, we have a favorable variance. This may be good news to the manager who has to explain budget variances to the owner. He should be advised, however, to check the physical progress on the job. The project may be behind schedule and expenditures are delayed.

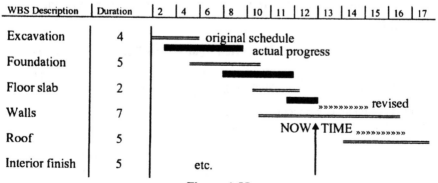

Figure 4-55
Schedule base for pricing.

Another frustrating experience a manager can have is wrong cost reporting. If the actual reported cost are based on cash flows, major payments due may have been delayed, and have not been reported to him on an *accrual* basis. Similarly, if actual cost exceed the budget, thereby indicating an unfavorable variance, the project may run ahead of schedule and quality may have suffered.

If the project is run in a functional manner, the scheduling and costing functions are organizationally separated. It is therefore difficult to relate progress in time with the consumption of funds. This is a disadvantage of the conventional cost control.

It is often quite difficult to communicate, when different terms are used for the same idea or vice versa. Some of the words one frequently encounters are:

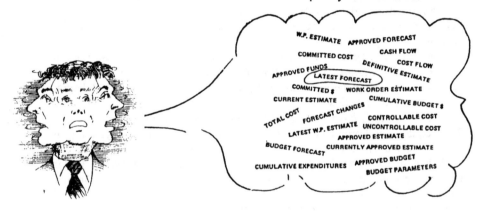

There are probably many more. For the purpose of better understanding, we will look at the difference between the three terms: Estimate, Forecast and Budget:

Estimates

Estimates are the result of a review of total costs likely to be incurred.

For example, based on preliminary data, a vacuum cleaning system for a job was estimated to cost $ 100 000 with the following cost flow:

Delivery of major components	June 15	$ 40 000
Installation	August	$ 20 000
	September	$ 10 000
	October	$ 10 000
	November	$ 10 000
Delivery of minor components	Sept. 15	$ 10 000

The date of the estimate was February 15.

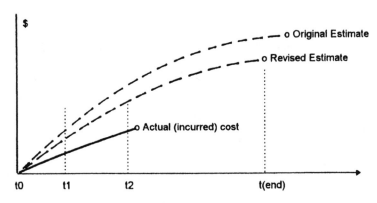

Figure 4-56
Definition - estimate.

A quarterly review of the estimate (May) showed that the total cost are only $ 80 000 and with a shorter duration:

Delivery of major components	June 15	$ 30 000
Installation	August	$ 20 000
	September	$ 10 000
	October	$ 10 000
Delivery of minor components	Sept. 1	$ 10 000

The effect this cost element has on the total project is reflected in the revised *estimate cost flow*. The next quarterly review is scheduled during the month of August when all cost components of the project are updated. Any delays in delivery and installation between May and August will lower the incurred cost flow, causing a favorable variance during this period.

That means even though actual expenditures have a strong influence on estimate reviews, some elements of the estimate are not necessarily hinged on actuals. (Figure 4-56).

Budgets

The budget figure is based on the estimate and is frozen over a specific time period, usually one year.

Budgeting for large projects is an iterative process starting from zero *at the beginning of each time period.* (Figure 4-57).

Different corporations have different budget procedures. This is only natural because their mandates, their mode of operation and funding sources are different.

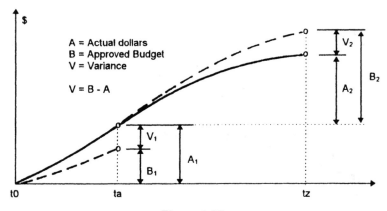

Figure 4-57
Definition - budget.

Even among similar projects, budgets have diverse functions. What we want to discuss here is a typical example of a conventional budget for a long duration construction work program.

AACE and PMI define "BUDGET" as a planned allocation of resources. It is also referred to as an estimate of funds planned to cover a fiscal period. It may take six months to establish a budget for the following year, a forecast for the next two or three years and a final cost estimate for the end of the project.

Below is a typical example of a budget review process in which estimates are reviewed annually, based on a monthly review of certain packages (scheduled for regular annual review or spontaneous review when drastic changes occur). The variance is based on the difference between the latest estimate and "actuals" obtained from cost accounting, which is the recorded consumption of funds (Figure 4-58).

Many groups and organizations within the corporation are involved to prepare the budget. In order to provide an effective basis for planning and control, annual spending and resource parameters are developed. These parameters are based on corporate constraints, and any expected major changes in the work program. Within those parameters, project management's major responsibility is the control over the funds (figure 4-59). Control means:

◆   To know how much must be spent within an acceptable variance and when.
◆   To obtain appropriate approval for additional funds to be spent when changes are identified.

It is, therefore, important to establish a viable time-cost relationship within the framework of work package estimating and cost collection.

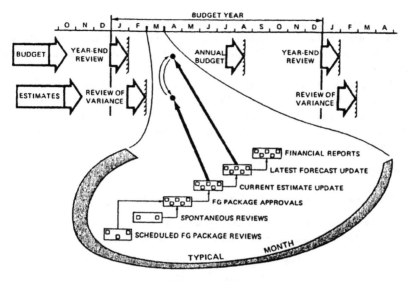

Figure 4-58
Budget review process.

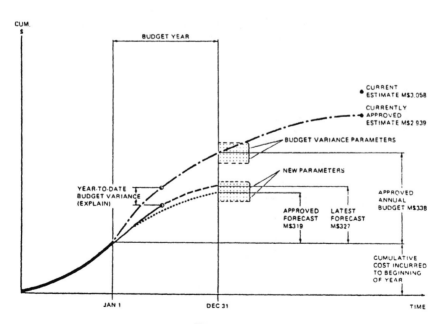

Figure 4-59
Conventional cost flow.

The project budget submission must indicate the work to be carried out, when it will be carried out and its estimated cost. Here again, the prime responsibility for evaluating the work program for the budget year rests with the holder of the work package.

Figure 4-59 shows an example of a conventional cost flow and budget for a three billion dollar megaproject lasting several years. The forecast to completion in this case is called the current estimate. Here, the manager of the project has been given discretionary "budget variance parameters" above and below the approved budget.

Forecast

The forecast is a projection into the future and is hinged on actual cost.

It is also defined as a method for extrapolating trend curves (Figure 4-60).
The word *forecasting* describes exactly what forecasting is:

> When a point in time has been reached along a path, we are "casting forward" or projecting the future path, based on the experience of the past.

A farmer who predicts a mild winter uses observations of the past, i.e. dew worms still active in November, to forecast the weather for the next season.

A gun barrel will "forecast" the bullet to reach a target. If we measure the bullet moving along the gun barrel as predetermined data, we can trace its path mathematically and predict where it will hit the target.

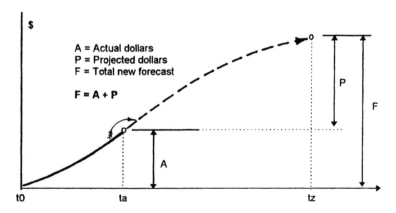

Figure 4-60
Definition - forecast.

How well it hits the target is *qualitative* forecasting. Where it hits the target, is classified as *geographical* data. If we shoot that gun a hundred times and establish a statistical frequency of "nearness" to the established target, this is classified as *quantitative* data.

Statistical forecasting deals mostly with the collection of data based on differences in time which apply to the various elements in a specific category. For example, the value of sales for the XYZ company over a five-year period may be classified by month and by aggregate value of sales for each monthly period. This is a *chronological* type of classification with *periodic* data. Periodic means that data is building up on a non-duplicative or *mutually exclusive* total over the five years. When duplicative data is used, such as the number of items in inventory where the same items are included in every count, it is called *point* data. All data for chronological forecasting can be plotted on a *time-series* scale.

A forecasting model has the following components:
- The trend component
- Seasonal adjusted component
- Cyclical pattern
- Random fluctuation

Because a project deals with the consumption of resources over a specific period of time, the component we are mainly interested in is the *trend* component. Trends should have a relatively long-term time-series also called the *secular* trend with data collected over many years (econometric model). Project data are collected over a short period of time with a short *forecasting horizon*. The forecasting horizon is defined as the number of periods projected into the future (Figure 4-61).

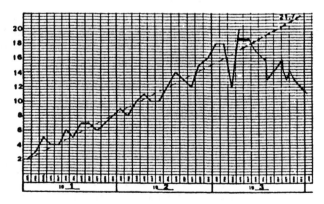

Figure 4-61
Time series forecasting.

Time-series data is first collected in tabular form, e.g.:

Period	Computer Cost in k$		
Month	Year 1	Year 2	Year 3
Jan	2	9	18
Feb	3	8	18
Mar	5	10	12
Apr	4	11	20
May	4	10	
Jun	6	10	
Jul	5	12	
Aug	7	14	
Sep	7	13	
Oct	6	12	
Nov	7	15	
Dec	8	16	

It is more difficult to analyze data in table form than graphically.

A time-series plot will give us a better "picture" of the data's behavior.

The first two years are obviously showing an *upward* trend, whereby year 3 has an undetermined trend. Since the data are not scattered widely, a view of the graph will clearly give us an overall upward trend. Using regression analysis (Section 1.2.0 The Estimating Line), the statistical trend to the end of year 3 is shown as a broken line with approximately 22 k$ by the end of the third year.

Judgmental Forecasting

Looking at the example above, we have assumed that historical past data will form the conditions for future projections (quantitative forecasting). This is not true in cases where the time horizon is large and future conditions are independent on the past and must be studied on their own (qualitative forecasting).

If in the estimating example above (see Figure 4-56), major components for the vacuum cleaning system are delivered on July 15, and the cost incurred are reported on July 30, the cumulative cost flow would be

ACTUALS	FORECAST	
July $ 30 000	August	$ 50 000
	September	$ 70 000
	October	$ 80 000

To do this, the costing system must have the capability to constantly update incurred cost in relation to the project cost flow. The total project cost flow can be made up of thousands of items. It would not be cost effective to monitor each and everyone of those. The concept of *materiality* should apply (see 3.4.2).

4.4.2 Pareto's Law

A project manager needs to be selective in evaluating cost information. Costs incurred, that are beyond his control, such as interest payments, escalation, government regulations, etc. are *monitored* by him, but not *controlled*. Furthermore, many costs can be summarized, especially for those items, where final commitments have been made.

In establishing a cost control system, the idea is to isolate and control in detail those elements with the greatest potential impact on final project cost, with only summary level controls on the remaining elements.(20) That means, that cost control should be approached as an application of *Pareto's Law*. Pareto's Law states that

> **in any series of elements to be controlled, a selected *small* fraction of items in terms of number of elements will always account for a *large* fraction in terms of effect.**

Conversely, the majority of the items will be of relatively minor significance in terms of effect.

Vilfredo Pareto, economist, sociologist and mathematician, was born in France in 1848 and died in Switzerland in 1923. He developed methods of mathematical analysis of social and economic data and devised a cyclical theory of society. His chief work was "Mind and Society" in which he expressed contempt for democratic methods.

Pareto's Law implies that effort, time, money and other resources to be spent in estimating and controlling a project should be allocated among areas or items in proportion to their relative importance or value.

For example, out of approximately 2 000 purchase orders issued for a large construction project, only 430 were in excess of $ 50 000 at the time of monitoring. But those 430 accounted for 98% of the total dollar value. It would therefore be more expensive than it is worth to manually input all 2 000 records into a costing system.

Maintaining only one quarter of the total number of purchase orders greatly reduces the amount of work and computer storage with little variance in the total value covered.

Pareto's Law is often used in estimating when the cost of producing the estimate is a consideration. It is called

"The 80/20 Rule",

whereby 20% of items to be worked on are the vital few that account for 80% of the project cost (Figure 4-62).

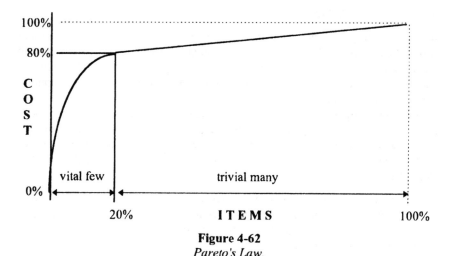

Figure 4-62
Pareto's Law.

4.4.3 Budgeting and Trending

Conventional cost control is based on what costs ought to be and what they are
actually predicted at any point in time. Deviations from the standard (budget) are
examined periodically. On engineering and construction projects the project cost
status is usually shown under the following headings:

Cost Code: This is the cost account or WBS code under which the costs are
categorized.

Description: Not all cost reports will have this category.

Budget: It is the standard against which cost performance is measured. It is
sometimes called the "Approved Estimate."

Revised Budget: This shows the monetary impact of approved changes on the
original budget.

Actual Cost To Date: These are the incurred cost at the status date. Those cumulative costs are taken from the job cost ledger. They are supposed to be collected on an accrual basis, which includes accounts payable.

Commitment To Date: Costs are committed when a purchase order is issued or
a contract is signed because the owner is required to pay the vendor or contractor
in accordance with their agreement. Commitments should also reflect cancellation
charges.

Percent Complete: This is an estimate of the portion of the work completed
(see Section 4.3.2 - Measuring work progress).

Forecast Final Cost: The latest prediction what the total cost will be at completion is entered under this heading.

Variance: To measure project cost performance, the variance is the difference between the forecast at completion and the revised budget. To include the changes in project scope, the forecast is compared with the original budget. Variance can also be expressed as the difference between actual cost to date and the value of work performed.

Some Thoughts on Budgets

Budgets have essentially two purposes:

1. They are a limiting device with a punitive undertone. This makes them in some way both behavioral and technical. Foremen and supervisors are judged severely by their performance relative to their established budgets (bottom line). This type of budget is rigid. It is often used by Governments or organizations with stable environments to control spending. The application of "Zero Base" budgeting is a good example.
2. The other type of budget is more flexible. It is used for planning and resource allocation. It is based on economic principles and open to alternatives. It contains within its structure the means for reasonable corrections. This flexibility is desired because conditions may change and patterns emerge that have a different effect on the life cycle of the project than originally expected. This budget is open for further evaluation, not the end of evaluations. It is more applicable to entities with unstable environments such as capital construction projects.

When the cost status shows the heading "Revised Budget", it means that management have applied a system of control that correct exceptions or variances *before* they occur. A detailed review of the impact on the total project life must be made by the corporate decision makers before budget changes take place. That does not mean the initiation for budget changes flows only from top to bottom, it is usually the other way around. Functional management sometimes must be "motivated" to respond positively to problems or negative trends identified by the project organization (38).

Because a project passes through several phases during its lifetime the early "order of magnitude" estimate is too unreliable to be used as a measure of performance. At a later phase when the scope is determined in more detail and the budget is finalized can we use the budget as a yardstick for comparison with the forecast and actual cost.

Trending

Assuming we obtain the following simple cost status report:

WBS Code	Budget	Revised Budget	Actual to Date	% Compl.	Final Forecast	Variance (3-6)
1	2	3	4	5	6	7
A-10	1100	1100	10	2%	1100	0
B-15	4800	5000	950	13%	6000	(1000)
B-20	3000	3200	3100	100%	3100	100

A-10:

Work on the package A-10 has just started. It is too early to establish a trend at 2% completion. The budget has not been revised, neither has the forecast at completion.

B-15:

The budget was revised, most likely due to a scope change. The job cost ledger shows $950 incurred. The 20% completion indicates that a review of the work package was done for the first time which resulted in a negative variance of $1000 (over revised budget). Any variance outside an established parameter must be analyzed and explained what caused it. Budget parameters for work packages are set by the project manager and are usually about 5% of the estimate. Explanation of variances can be very difficult.

 The causes must be quantified and described in such a way that corrective action can be taken. It would be wrong to say: "The variance is caused by an increase in cost" or any other such generalization.

Variance explanations must not be like steer horns: A point here, a point there and a lot of bull in between.
(Toastmasters' expression for some speeches)

The variance review must be specific, such as

Variance		Explanation
$ 400	Scope Change:	Construction drawings revised to show five additional pipe hangers caused by rerouting a pipe run.
($ 200)	Estimate Change:	The valves ordered were less costly than estimated.
$ 700	Unforeseeable Changes:	Government imposed new safety rules for scaffolding. Additional material required.
$ 100	Minor Miscellaneous Changes:	None
$1000		

It is very likely that the new safety rules for scaffolding will affect several other work packages. The project manager may take action and initiate a method study to investigate different, less costly methods of work area access (e.g. movable platform).

The "value of work" can be estimated by multiplying actual cost with the ratio of budget over forecast (12)

$$\text{value of work} = \frac{5000}{6000} \times 950 = 792$$

The present rate of performance shows that the value of work done is only $ 792 or $ 158 less than the actuals indicate at this point in time. Value of work is also called the *estimated* cost of work done as compared to the *actual* cost of work done. The physical work performed was measured to be 13%. This is the portion of the forecast that was "earned", i.e.

$$6000 \times 0.13 = \$ 780$$

B-20:

This work package is 100% complete, therefore the forecast is equal to the incurred cost and $ 100 below the revised budget. The next status report will show a revised budget of $ 3100 and no variance.

More on Trending

Variances are reported over time to indicate a "trend" for control purposes. In case of large work packages with long durations it is easier to interpret trends graphically. Seldom will past performance on a job continue at the same rate of progress into the future (non-linear). Remedial action taken to improve performance will have an effect on trending. *It is always better to improve the aim instead of moving the target.*

Plotting a *status index* is another way of trending. Assuming an installation contract is summarized by four functional groups. The relative effort is estimated as follows (Table 4-12):

Table 4-12
Relative effort

Function	Relative Effort	Work Packages
Civil	10%	Clearing, excavating, foundations, structure, architectural, etc.
Mechanical	40%	Equipment, piping, heating, etc.
Electrical	30%	Distribution, lighting, etc.
Instrumentation	20%	Controls, communication, etc.

Table 4-13
Status index

FUNCTION	RELATIVE EFFORT	PERCENT COMPLETE	RELATIVE EFFORT
Civil	10%	90%	9
Mechanical	40%	50%	20
Electrical	30%	20%	6
Instrument.	20%	15%	3
		TOTAL	38%

The physical progress of the work is now measured as percent completion and multiplied with the relative effort to obtain the percent complete for the total contract (Table 4-13). If the cost expended were 14 M$ with a total contract value of 50 M$, then

$$\frac{\text{Actuals}}{\text{Budget}} = \frac{14}{50} = 28\%$$

That means roughly that the work is 38% complete with 32% of the budget consumed. The ratio of the two will give us a status index

$$I = 38/28 = 1.36$$

What does the index mean?

Up to 0.5	=	Resources are inadequate; bad estimate; review project performance !
0.5 to 0.8	=	Activity needs corrective action.
0.9 to 1.0	=	Performance in line with plan.
1.0 to 1.3	=	Performance better than planned.
1.3 and up	=	Activity has too many resources, ahead of schedule.

This, done over a period of time, will give us a trend indicator:

Reporting Period	Jan	Feb	Mar	Apr	May	Jun	Jul	Aug
Activity A	1.4	1.3	1.1	1.0	0.8	0.8	1.0	1.0
" B							take action	
" C				too many resources				
etc.								

4.4.4 Cost/Schedule Integration

It would be an ideal world if we could estimate cost on exactly the same basis as we produce schedules.

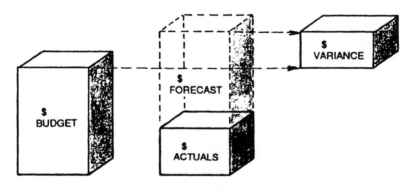

Figure 4-63
Conventional cost variance.

This may be possible on smaller building projects or in the manufacturing industry,. where detailed networks are available *before* the work starts and where costs have been estimated by activities or pay items.

However, the larger the project is, the more difficult it is to match expenditures with scheduled activities. During the earlier phases of a project the costs are estimated and monitored mainly by functional elements. The budget estimate is compared with the latest forecast. Under conventional cost control, budgets and schedules are usually not integrated (Figure 4-63).

A combination of optimistic forecasting and improper accounting will usually produce an overly favorable view of final cost. Relating "planned" cost only to actual expenditures does not indicate how well a project is performing.

We said previously that *schedules* are time scaled plans and *progress* is the measure of headway made when carrying out these plans (see Figure 4-54).

Network-tied Cost

Progress measures physical quantities or workhours against the project schedule. The actual work completed can be plotted against the schedule time frame. The Y-axis depicts quantitative measures such as quantities installed and/or delivered, workdays, percent completion. Workdays can be actual calendar days or calendar/earned work days based on the critical path. If the critical path shows a schedule duration of seven months, the planned calendar work takes 213 days (365×7/12). The "earned" work days can be based on the critical path progress percentage. The variance can then be used as a trend to forecast the project-at-completion status (39).

Table 4-14
Network-tied trend 1

(1) Mid-Month	(2) Scheduled Quantity	(3) Installed Quantity	(4) Earned Value	(5) Variance (2 - 4)	(6) Schedule Trend
Dec.	900	850	850	50	10450
Jan.	1850	1600	1600	250	10650
Feb.	3500				
Mar.	4900				
Apr.	6600				
May	8550				
Jun	10400	(schedule	base)	Forecast	= 11125

Table 4-15
Network-tied trend 2

(1) Mid-Month	(2) Scheduled Quantity	(3) Installed Quantity	(4) Earned Value	(5) Variance (2 - 4)	(6) Schedule Trend
Dec.	900	850	850	50	10450
Jan.	1850	1600	1600	250	10650
Feb.	3500	3000	3200	550	10950
Mar.	4900				
Apr.	6600				
May	8550				
Jun	10400	(schedule	base)		
Jul				Forecast	= 11445

This type of schedule forecasting by means of earned value lends itself to graphic plotting when using a schedule spreadsheet. It will make the trend curve easier to view in relation to the base line.

Using the data from Figure 4-64 we can tabulate the *cumulative* data as shown in table 4-14. The schedule trend is the variance added to the base. The forecast at the monitor date is calculated automatically by means of linear regression.

29% of scheduled time has been consumed. Installed quantities increased at the end of the following month to 3000 m^3 with 31% of the work along the critical path completed. The earned value is now 3200 m^3. The schedule trend is now 10950 m^3 and the new forecast 11445 m^3 equivalent (Table 4-15).

With the trend still increasing and 3/7=43% of scheduled time consumed, the forecast equivalent is converted to eight months (rounded). This is reported on a graph similar to the one shown below (Figure 4-64). The extension of the schedule by one month will affect the critical path and thereby the higher level control schedule

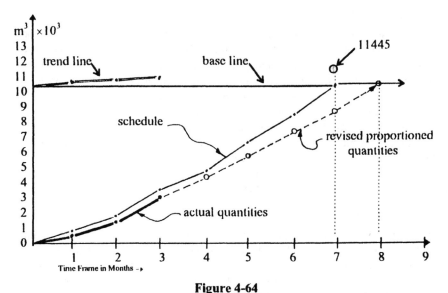

Figure 4-64
Network-tied cost graph.

A decision will have to be made what (if any) action to take. The control schedule may have enough float for this work package that the extended completion date has no impact. (see schedule crashing later in this Section).

In order to control cost on a *network* basis, the estimate needs to be prepared by the activities indicated on the network plan. It is therefore essential, that the work breakdown structure subdivides functional and system elements into work items and activities.

Hira N. Ahuja (12) divides network based cost into four main categories:

1. Rigidly tied cost
2. Semirigidly tied cost
3. Loosely tied cost
4. Partially tied cost

The functional element "substructure" for example, needs to be broken down into slabs, columns, walls, beams etc. in order to support activities of the CPM scheduling program. This is sometimes hard to do and may in some cases not receive the support by team members in the organization. Quite often estimators are not familiar with network planning and may not have enough time available for this application. Furthermore, some packages may not lend themselves at all to network costing.

Rigidly Tied Cost

If we are able to distribute all our cost over the various activity durations in a network, we have an ideal cost control situation. On larger jobs with many activities, this detailed breakdown may not be practical. To simplify, several work items can be joined for a normal activity such as installing a platform. We therefore have

  ♦   Cost based on a single work item, tied to an activity, or
  ♦   Cost based on several work items for a single activity, or
  ♦   Cost based on a pay item for many activities.

The third possibility above may be more practical for accounting or cost collection purposes, because costs are assigned to a pay item. Several activities may be included in that pay item. The costs are then distributed over the activities between the start event and the end event of the pay item.

Semirigidly Tied Cost

Relating cost to a broader time span, whereby a number of activities are combined between selected events makes the application of cost control much easier. The breakdown of cost is less rigid. Those selected events are the cost milestones. Cost milestones are not necessarily the same as time milestones. The disadvantage here is, that cost/schedule integration is somewhat more difficult. Even though cost are not assigned to individual activities, a realistic cost distribution over the time span between two cost milestones could be attempted. For example, consider an event

52 = Start of foundation
68 = End of foundation work

as cost milestones. The cost distribution between those events is not linear (figure 4-64). It results in a cumulative cost distribution as depicted in figure 4-65.

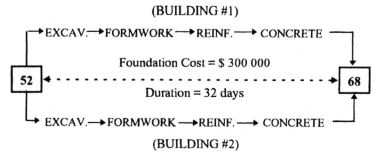

Figure 4-65
Semirigidly tied network.

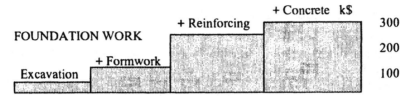

Figure 4-66
Cumulative cost distribution.

Loosely Tied Cost

Here, only the start event is linked to a time span of a determined duration. Therefore, the time span is fixed with an estimated dollar value identified. The network can be updated independently of changes in the CPM network, i.e. when a starting date shifts, so does the cost profile. The advantage is, that costs for projects that need contingent plans can be updated easily. Even though we are still linked to the schedule, we are getting further away from a true schedule/cost integration.

The amount of $90 000 is tied to event 68 which is the start event for the 90 days duration of the downstream activity. The cost are estimated independently without tying into the successor event. Only the start event is monitored for cost control (Figure 4-67).

Partially Tied Cost

When only selective cost are tied down to the CPM schedule, the costing becomes even more flexible. For example, the owner may furnish materials and a contractor does the installation. In this case, the contractor's network would show physical progress in order to process progress payment certification, while the owner may only need a bar chart to show material delivery dates. Very few changes in the network affect the cost distribution.

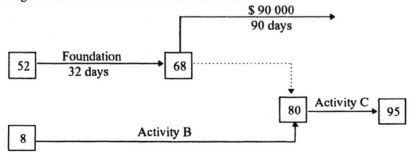

Figure 4-67
Loosely tied network.

4.4.5 The Earned Value Concept

Progress measures physical quantities or workhours against the Project Schedule. If we measure this progress in relation to the dollar budget, we obtain the Budgeted Cost of Work Performed (BCWP).

If we measure the consumption of funds (cost incurred) in relation to the schedule, we obtain the Actual Cost of Work Scheduled (ACWS).

Therefore, at any point in time we should obtain a cost picture that is conducive to control (Figure 4-68).

The *earned value* concept integrates cost and schedule for measuring *overall* project performance. This system is somewhat more complicated than the systems used for a conventional budget process. Computerization, however, helps to deal with the additional data elements that are required. Those who are exposed to the "earned value" concept for the first time must not be intimidated by the many new expressions and acronyms. The basic fundamentals are relatively simple.

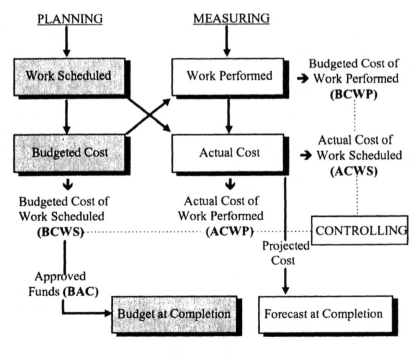

Figure 4-68
The earned value model.

Major system acquisitions by governments and large corporations have resulted in the need for very detailed and formal instructions and documentation that are difficult to understand by the uninitiated. Those cost/schedule control systems criteria (C/SCSC) are working well for incentive type government contracts with a high risk factor. But for most project managers a simple, broad based application of the "earned value" concept, comparing the *planned* value of the work against the *earned* value of the physical work accomplished and the *actual* cost incurred will summarize project cost performance quite well.

Earned Value Data Elements

Budgeted Cost of Work Scheduled **(BCWS)**
This is the cost of the approved plan, both, budget and schedule. It is the base line against which project performance will be measured. It can be reported in tabular form or plotted on a cumulative dollar/time scale.
Budgeted Cost of Work Performed **(BCWP)**
It is the budget applicable to the work actually accomplished. It is the value that can be attributed to the completed work. The value is based on quantitative measures or percent complete, i.e. earned value.
Actual Cost of Work Performed **(ACWP)**
These are the cost monitored as actual expenditures by the accounting office for the work done (incurred cost).
Budget at Completion **(BAC)**
This is the total management approved cost for the completed project. It includes all direct and indirect costs, whether they are controllable or uncontrollable plus contingency reserve, profit or fee if a contract. BAC is the cumulative BCWS.
Estimate (or Forecast) at Completion **(EAC)**
It is the sum of actual cost of work performed (ACWP) to-date plus the forecasted cost to complete the project.
> *Unfortunately, this expression (EAC) is widely used. The term (FAC) Forecast at Completion would be more appropriate. For many projects, especially smaller ones, "Estimate" and "Budget" have the same meaning. (Please compare figure 4-56 with figure 4-60).*

Note:

All COST above do not need to be expressed in dollars, but also in work-hours when applicable

To report the *performance status* of a project, indices can be calculated for easy interpretation by management:

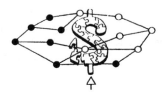

Earned Value Indices

Actual Percent Complete **(APC)**

This index is the relationship between the actual cost of work performed and the current forecast at completion.

$$(APC) = (ACWP) / (EAC)$$

Planned Percent Complete **(PPC)**

It is the relationship between the budget value for what was scheduled and planned to-date and the total budget.

$$(PPC) = [(BCWS) / (BAC)] \times 100$$

Cost Performance Index **(CPI)**

Compares the budget with cost incurred. Greater than 1.00 is good performance.

$$(CPI) = (BCWP) / (ACWP)$$

Schedule Performance Index **(SPI)**

It compares the budget for those items of work that were scheduled to be accomplished as of the status date with the budget for the work that was actually accomplished as of this same date. Values greater than 1.00 indicate more volume of work is being done than was originally planned.

$$(SPI) = (BCWP) / (BCWS)$$

Schedule Variance **(SV)**

This is a measure of slippage from schedule. A positive (SV) indicates good schedule performance i.e. ahead of schedule (often measured in workhours).

$$(SV) = (BCWP) - (BCWS)$$

Cost Variance **(CV)**

It is a measure of cost overrun or underrun for the work accomplished to-date. A positive (CV) is good cost performance.

$$(CV) = (BCWP) - (ACWP)$$

Example 1:

To demonstrate Earned Value calculations, here is a simple fictitious exercise:

Nine stone blocks are assembled into a three-level pyramid. The blocks cost $10.00/kg to install. Each block at the low level is estimated to weigh twice as much as each block installed at the middle level, which weighs twice as much as the top level block, which weighs 10 kg. The schedule calls for a progress of 20 kg/day.

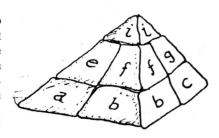

1) Show the scheduled cost flow BCWS:

a,b,c,d = 40 kg × 4 = 160 kg lower level (2-day installation)
e,f,g,h = 20 kg × 4 =   80 kg middle level (1-day installation)
    i = 10 kg × 1 = 10 kg top   ($^1/_2$-day installation)
Scheduled duration = 250 kg/20 = 12.5 days

Block	a	a	b	b	c	c	d	d	e	f	g	h	i		
Days	1	2	3	4	5	6	7	8	9	10	11	12	13	14	15
BCWS	2	4	6	8	10	12	14	16	18	20	22	24	25		

2) The pyramid builder had some problems after the second day. Delivery of the second block was delayed by two days. It was also planned that there will be no work performed on day 11 (slave appreciation day). All other blocks can be installed as planned. The actual cost per unit for the pyramid turned out to be 10% less than budgeted for (discount). Show BCWP and ACWP.

Block	a	a	-	-	b	b	c	c	de	d	-	e	f	g	h	i
Days	1	2	3	4	5	6	7	8	9	10	11	12	13	14	15	16
BCWS	2	4	4	4	6	8	10	12	14	16	16	18	20	22	24	25
ACWP	18	36	36	36	54	72	90	108	126	144	144	162	180	198	216	225

3) After the eighth day, show the variance report:

SV	SPI	CV	CPI	BAC	EAC	PPC	APC
- 400	0.75	+ 120	1.11	2500	2250	64%	48%

Example 2:

Below is another example which carries us through the phases of a civil contract in regard to earned value reporting:

Define Scope, Design, Specify:
    The work package B2, (Main Building Foundation) for a hydraulic plant needs to be finished by the fall of 1992. Maximum pouring capacity is considered to be 200 m^3/ day.

Develop a Plan, Schedule, Estimate:
    Tenders will have to be invited for the installation of concrete foundations including supply of ready-mix. Because of pending price increases for aggregates, it will have to be a cost-plus contract (actual material and labor cost plus profit).

January 2,1992:

Owner's Schedule:	Start 1992-06-01, Finish 1992-08-31
Quantity:	Double quantity in July.
Cost Estimate:	Quantity: 10 000 m^3, unit cost: $ 100 per cubic meter.

Table 4-16
Earned value - civil contract example -1

	June	July	August	Total
Scheduled percent completion	25%	50%	25%	100%
Estimated installed quantities	2500	5000	2500	10 000 m³
Monthly cost in k$	250	500	250	1000
Budgeted Cost of Work Scheduled	250	750	1000	(cum)

Table 4-17
Earned value - civil contract example -2

1992-06-30 Report:

		June	July	Aug.	Sept.	Total
Revised Quantities:		1500	5500	2500	500	10000 m³
BCWS	k$:	250	750	1000	1000	1000 (BAC)
BCWP	k$:	150				
ACWP	k$:	150				
Forecast (cum.) k$:			700	950	1000	1000 (EAC)

Report on Status:

SV	SPI	CV	CPI	BAC	EAC	PPC	APC
- 100	0.60	0	1.00	1000	1000	25%	15%

February 1992:

Owner awards contract to lowest bidder showing present cost:
Material and Labor = $ 83.33 + 20% profit = $ 100/m³.

1) What is the expected cost performance? See Table 4-16.

2) What is the Budget at completion?   → 1 000 k$

June 30, 1992:

As measured and reported by the cost engineer in the field, the contractor has installed 1 500 m³ of concrete in June. In preparing the June cost report to the project manager, the contractor told the cost engineer that he can increase quantities
by 10% in July, but any change in August would increase labor cost considerably
(overtime etc.).

Because this work package has enough float, the manager decides not to increase quantities, but rather have the schedule slip into September.

3) Measure Performance, Review and Revise (Table 4-17).

July 15, 1992:

At this time, the contractor submits an invoice, showing material and labor costs
to be $100/m³. Back-up for the increase was included.

Table 4-18

Earned value - civil contract example -3

1992-07-31 Report:

		June	July	Aug.	Sept.	Total
Revised Quantities:		1500	5500	2500	1500	10000 m³
BCWS	k$:	250	750	1000	1000	1000 (BAC)
BCWP	k$:	150	700			
ACWP	k$:	150	840			
Forecast (cum.) k$:				1140	1320	1320 (EAC)

Report on Status:

SV	SPI	CV	CPI	BAC	EAC	PPC	APC
- 50	0.93	-140	0.83	1000	1320	75%	64%

July 31, 1992:

Excavation inaccuracies (owner revised drawings) caused an overall increase in quantities. The new estimate is 11000 m³. Additional work will have to be performed in September.

4) Update Cost Performance and Forecast (Table 4-18).

5) Variance Analysis:

Cost Variance:

Costs are higher than budgeted for because of an increase of unit price by $20/m³ and a revision of quantities due to excavation inaccuracies.

Schedule Variance:

Still falling behind schedule, but improving. Not on critical path for one month. Installing remaining 1500 m³ in September should not be a problem for contractor.

4.4.6 Cost Effect on Schedule Crashing

How do we assert control over the cost for a network that is based on the approved schedule?

Saving construction dollars is a commendable goal for a construction manager. There are situations, however, when additional expenditures are used in some areas of work in order to benefit the overall objective of the project.

Schedule/cost optimization is a way to systematically reduce durations while incurring the least amount of cost increases. The reduction in duration is accomplished by buying time along the critical path where it can be obtained at the least additional cost.

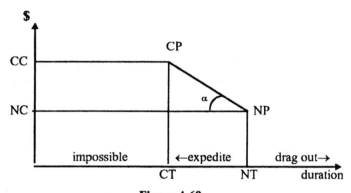

Figure 4-69
Crash point graph.

A reduction in duration is accomplished by the addition of resources, which increases project cost. Extra effort must not be wasted on non-critical work.

As additional resources are placed on an activity to reduce its duration, inefficiencies and waste are driving costs upward until a point is reached where no further reduction in time is possible. This is the CRASH POINT. It represents the minimum duration between events (figure 4-69).

A cost/time ratio can be calculated as follows:

$$\text{slope } \tan\alpha = \frac{\text{crash cost (CC) - normal cost (NC)}}{\text{normal time (NT) - crash time (CT)}}$$

In deciding which critical activities to reduce, those with the smallest cost/time slopes should be chosen first.

When multiple critical paths are present in the CPM network, the one with the least additional cost combination is the one that should be chosen. Once the selection is made, the activity is reduced by one time unit, and a network update is prepared. Several iterations of the procedures for establishing cost/time slopes by selecting activities to reduce and update networks may be needed until the desired duration is achieved or until all the activities on any one critical path are set to their crash times.

As shown previously, costs go up when a schedule is speeded up or *"crashed".* To determine the most cost effective action that can be taken when a schedule has to be crashed, the minimum durations for all activities and their costs have to be tabulated first. This is best demonstrated by an example (Table 4-19). It can be represented by a "normal duration" network (Figure 4-70).

Table 4-19
Schedule crashing example 1

Activity Letter	Normal Duration	Min. Crash Duration	Crashing Cost/Day	Normal Cost
A	8	6	$ 50	$ 800
B	6	3	60	600
C	4	2	70	600
D	5	3	30	300
E	10	6	40	800
F	6	4	80	900
			TOTAL	= 4 000

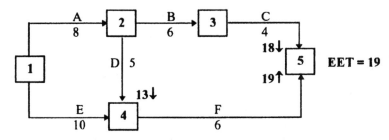

Figure 4-70
Crashing diagram 1.

The network, used as an example for crashing, is very simple. It has only six activities. The most cost effective combination is immediately obvious to the experienced observer.

The reason we go through various steps in this example is, that most networks have hundreds of activities and the result can not be obvious. The optimum cost must be calculated by the computer, which is not capable of judgment. A computer goes through thousands of calculations, considering *all* possibilities, even those, the human mind would not consider. Douglas R. Hofstadter in his book *"Gödel, Escher, Bach"*, a metaphorical fugue on minds and machines, discusses the subject of programmed computers as compared with the human mind. The development of very sophisticated so-called "artificial intelligence" (AI) type of future computers, he argues, could reach the level of human reasoning and creativity by also giving up the speed and accuracy of calculation present computers have. In other words, this AI computer would be very slow in adding figures and may even make adding mistakes like humans because it must resort to higher-level rules that often override logic.

If the project end date (day 19) is *advanced by three days*, what are the combinations in crashing various activities?

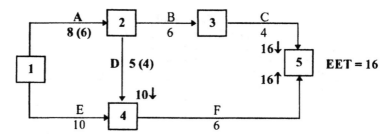

Figure 4-71
Crashing diagram 2.

Table 4-20
Crashing schemes

A	B	C	A	D	F	E	F	Schemes	Proj. Cost
8	6	(2)	8	(4)	(4)	10	(4)	C D F	$4 330
8	6	(2)	8	(3)	(5)	10	(5)	C D F	4 280
8	(5)	(3)	8	(4)	(4)	10	(4)	B C D F	4 320
8	(5)	(3)	8	(3)	(5)	10	(5)	B C D F	4 270
8	(4)	4	8	(4)	(4)	10	(4)	B D F	4 310
8	(4)	4	8	(3)	(5)	10	(5)	B D F	4 260
(7)	6	(3)	(7)	5	(4)	10	(4)	A C F	4 280
(7)	6	(3)	(7)	(4)	(5)	10	(5)	A C D F	4 230
(7)	6	(3)	(7)	(3)	6	10	6	A C D	4 180
(7)	(5)	4	(7)	5	(4)	10	(4)	A B F	4 270
(7)	(5)	4	(7)	(4)	(5)	10	(5)	A B D F	4 220
(7)	(5)	4	(7)	(3)	6	10	6	A B D	4 170
(6)	6	4	(6)	5	(5)	10	(5)	A F	4 180
(6)	6	4	(6)	(4)	6	10	6	A D	4 130

To program the combinations for crashing various activities, we may use the following criteria:

1.   No path must exceed the sum of 16.
2.   Each path shall generate a combination of durations from normal to minimum crash.
3.   Where activities are duplicated in different schemes, durations for those activities will be copied for each trial run.

Accordingly, the crashing schemes would then be as shown in Table 4-20. The computer now picks the path that results in minimum cost, which is path AD.

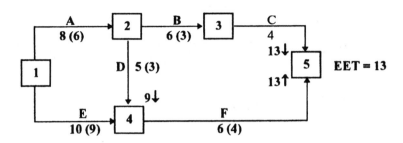

Figure 4-72
Crashing diagram 3.

This results in the following additional cost:

```
A @ 2 x 50 = 100
D @ 1 x 30 =  30
    Total      130   therefore program cost = $ 4 130
```

If we want to reduce the schedule to the practical limit, we must first find the critical path for minimum crash durations. Another scheme (Figure 4-72) will have an EET of 13 days. We then look at the other activities for minimum cost similar to the above procedure.

This results in the following additional cost:

```
A @ 2 x 50   = 100
B @ 3 x 60   = 180
D @ 2 x 30   =  60
E @ 1 x 40   =  40
F @ 2 x 80   = 160
    Total       540   therefore program cost = $ 4 540
```

Schedule Crashing - Example 2

Consider *three* systems to be designed and specified in parallel, then *one* contract for the three systems subsequently to be awarded and installed. Durations are in days (Figure 4-73 and related Table 4-21).

1) Assuming a two-day reduction in schedule, what is the increase in cost? (see Table 4-22).

What does it cost for a 5-day reduction? (see Table 4-22 and Table 4-23).

(*) = Asterisk in tables indicates a critical path.

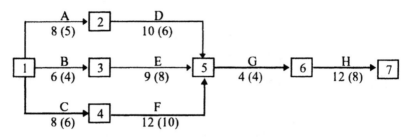

Figure 4-73
Crash Diagram - Example 2

Table 4-21
Crashing example 2 - durations and cost

Activity	NT	EET	LET	CT	NC(k$)	CC(k$)	CC/day
A	8	0	2	5	16	20	1333
B	6	0	5	4	18	20	1000
C*	8	0	0	6	32	33	500
D	10	8	10	6	50	56	1500
E	9	6	11	8	13.5	16	2500
F*	12	8	8	10	12	20	4000
G*	4	20	20	4	8	8	0
H*	12	24	24	8	72	86	3500
					221.5		

Table 4-22
Crashing - 2 and 5 day reductions

Activity	NT	CT	CC/day	2 days D	2 days CC	5 days D	5 days CC	5 days D	5 days CC
A	8	5	1333	8*	0	6*	2666	8*	0
B	6	4	1000	6	0	6	0	6	0
C*	8	6	500	6*	1000	6*	1000	7*	500
D	10	6	1500	10*	0	10*	0	10*	0
E	9	8	2500	9	0	9	0	9	0
F*	12	10	4000	12*	0	10*	8000	12*	0
G*	4	4	0	-	-	-	-	-	-
H*	12	8	3500	12*	0	11*	3500	8*	14000
			Total		1000		15166		**14500**

Table 4-23
Crashing - 5 day reductions

Activ-ity	NT	CT	CC/day	5 days D	5 days CC	5 days D	5 days CC
A	8	5	1333	8*	0	7*	1333
B	6	4	1000	6	0	6	0
C*	8	6	500	6*	1000	6*	1000
D	10	6	1500	10*	0	10*	0
E	9	8	2500	9	0	9	0
F*	12	10	4000	12*	0	11*	4000
G*	4	4	0	-	-	-	-
H*	12	8	3500	9*	10500	10*	7000
			Total		11500		**13333**

Reducing activity H by the available four days will be more cost effective than reducing it by only one day. There are further possibilities to crash the schedule by five days (table 4-23). The minimum cost of crashing for a schedule reduction of five days is therefore **$ 11 500**.

The *minimum reduction time possible* to reach the crash point is 8 days. This would reduce the durations to

C(6); F(10); A(6), H(8) or
$1000 + $8000 + $2666 + $14000 = **$ 25 666**

The larger the network is, the more difficult it becomes to recognize the best combination by just viewing the diagram. Computer application is essential for this type of analysis. After minimum penalty costs have been determined for every possible daily deduction, crash costs can now be listed, i.e.:

Cost/time
36 days = $ 221 500 ratio/10^3
35 days = $ 222 000   0.5
34 days = $ 222 500   0.5
33 days = $ 226 000   1.5
32 days = $ 229 500   2.0
31 days = $ 233 000   2.3
30 days = $ 236 500   2.5
29 days = $ 241 833   2.9
28 days = $ 247 166   3.2

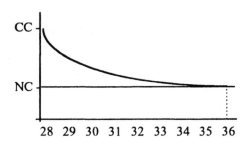

Cost Flow

Assuming a *5-day week,* the NC cost flow can be calculated on the early event time and late event time (or in between if required).

$$\text{The conversion factor is} \quad \text{cost} = \frac{\text{NC}}{\text{duration}} \times \text{days / week}$$

Even though it is automatically computed for the bottom line cost flow, it may be beneficial to the student to show the procedure schematically (Figures 4-80 to Figure 4-82). For example, activity A is calculated for the EET as $ 16 000/8 × 5 days = $ 10 000 for the first week and for the second week as $ 16 000/8 × 3 days = $ 6 000.

This is shown schematically on a weekly Gantt Chart with daily durations from which the cost flow can be calculated as a .bottom line..

Normal cost (NC) are depicted in Figure 4-74 for the early event time (EET) The duration is 36 days.

Normal cost (NC) are depicted in Figure 4-75 for the late event time (LET) As expected, the cost flow is not as heavily loaded in the early stages.

The *minimum cost* of crashing has a duration of 28 days. This compressed cost flow for the crash point is shown in Figure 4-76.

Weeks	1		2		3		4		5		6		7	
activity	A				D				G		H			
EET														
D/week	5	3	2		5		3	2	4	1	5		5	1
k$	10	6	10		25		15	0	8	6	30		30	6
activity	B			E										
EET														
D/week	5	1	4		5		5							
k$	15	3	6		7.5		0							
activity	C				F									
EET														
D/week	5	3	2		5		5							
k$	20	12	2		5		5							
cost flow	45		39		37.5		20		14		30		30	6

=221.5 k$

Figure 4-74
Normal cost - EET.

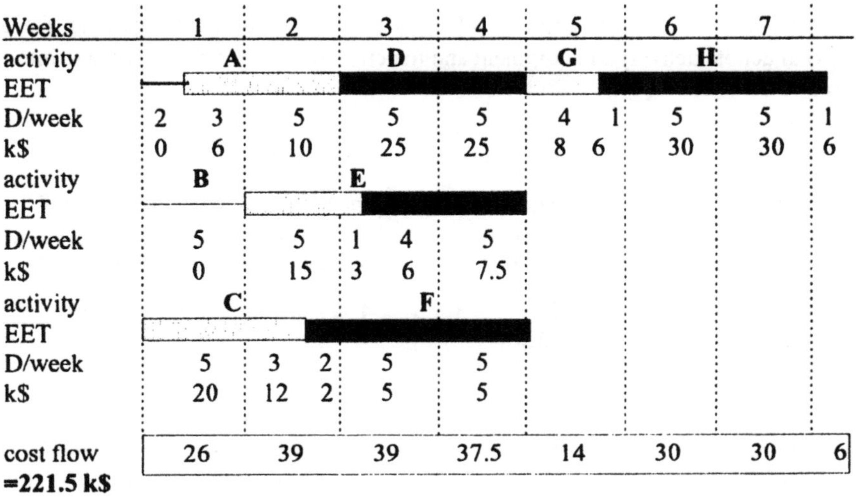

Weeks	1	2	3	4	5	6	7
activity	A		D		G		H	
EET								
D/week	2 3	5	5	5	4 1	5	5 1
k$	0 6	10	25	25	8 6	30	30 6
activity	B		E					
EET								
D/week	5	5 1 4	5				
k$	0	15 3 6	7.5				
activity	C		F					
EET								
D/week	5 3 2	5	5				
k$	20 12 2	5	5				
cost flow	26	39	39	37.5	14	30	30 6
=221.5 k$								

Figure 4-75

Normal cost - LET.

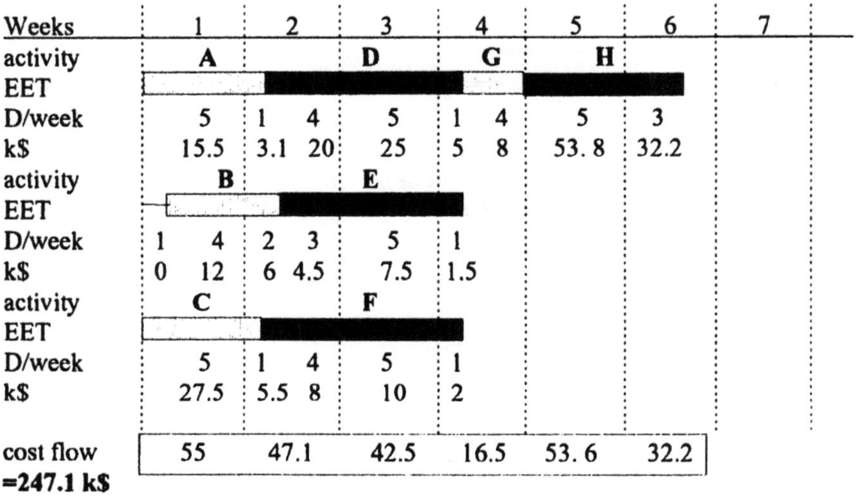

Weeks	1	2	3	4	5	6	7
activity	A		D	G	H		
EET							
D/week	5 1 4	5	1 4	5	3	
k$	15.5 3.1 20	25	5 8	53.8	32.2	
activity	B		E				
EET							
D/week	1 4 2 3	5 1				
k$	0 12 6 4.5	7.5 1.5				
activity	C		F				
EET							
D/week	5 1 4	5 1				
k$	27.5 5.5 8	10 2				
cost flow	55	47.1	42.5	16.5	53.6	32.2	
=247.1 k$							

Figure 4-76

Compressed cost flow - LET.

Activities between events 1 and 7 of Figure 4-74 (normal cost - EET) could be considered the labor content of a WBS package. If we use the Greek letter Delta (Δ) to denote delivery of equipment and materials we can enter this package into a higher level bar chart:

Assuming the costs of the materials and equipment are k$ 30. This amount will be added to the cost flow at the point of scheduled delivery during the fifth week.. The durations in weeks will automatically be converted into dates (Figure 4-77).

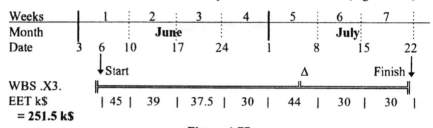

Figure 4-77
Cost flow - equipment added.

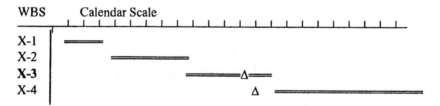

Figure 4-78
Scheduled cost flow - WBS.

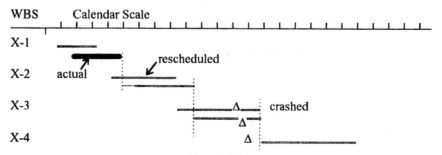

Figure 4-79
Rescheduled cost flow - WBS.

The same has to be done for all other packages of work. Part of the estimated project bar chart may look similar to Figure 4-78.

The cost flows of all packages will be vertically added to obtain the total project cost flow. The project had started two weeks late. Since our package X-3 is on a critical path, it was decided to crash the duration of the package. The project in-service date must not be delayed. We show the actual performance as a solid line on the bar chart (Figure 4-79).

4.4.7 Fast Tracking

Toronto desperately needed an additional airport. It took the politicians many years to finally sign a Development Agreement by the end of 1987, followed swiftly by ground breaking five months later. Standard construction projects start with design work that lead to detailed drawings, which are then submitted to a prime contractor for costing.

The contractor lets the sub-contracts for building trades, civil and utility construction; and then the work begins. Building an air terminal usually requires between three and five years of design work, six months of contract tendering, and three years of actual construction.

DESIGN, SPECIFY   TENDER   CONSTRUCT

The strong desire to produce revenue quickly, let to the decision to get the terminal built in record time. This leads to a process that trades money for time. It is called *fast track construction.*

Under fast track, preliminary drawings are submitted to the contractor and work begins almost immediately on the broad features of the building. No details are available yet. The more detailed designs are drafted, while the preliminary ones are actually being built, folding two processes into almost one time period.

DESIGN, SPECIFY, TENDER

CONSTRUCT

Occasionally, conflicts between preliminary designs that have been built and the detailed ones coming on stream will appear and designers that are working on different parts of the project concurrently will interfere with each other. The role of coordination must be emphasized and very strong management is essential.

The best organizational set-up for fast tracking is a task force whereby the project manager is in charge of design and construction:

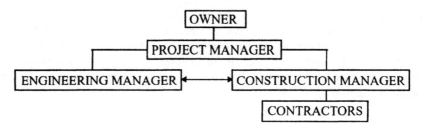

The demand for flexibility is paramount. For example, the use of steel rather than concrete wherever possible will become a design requirement. Invitation to tender is issued on incomplete drawings, which could result in costly change orders.

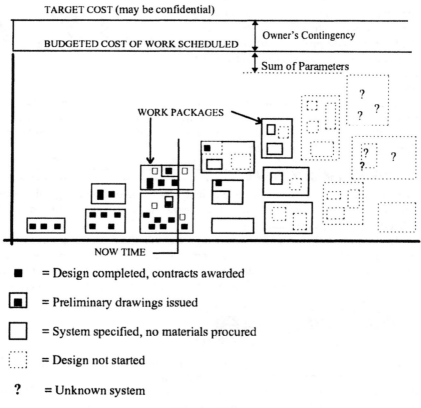

Figure 4-80
Fast track cost flow.

Slower progress on one package of work will not delay another. In fast-tracking, the project schedule and estimate exist from the beginning to the end, but remains under constant review. It is important to be conscious of the fact, that work packages are in various states of completion as the project unfolds (Figure 4-86).

This type of fast-track project management carries with it a greater than normal risk. Cost monitoring and reporting are very important and crucial to the success of the project. Immediate decisions must be made on the basis of incoming reports issued on a weekly or even daily basis.

Because of the many unknown factors in the beginning, the "Order of Magnitude" estimate (also called "Concept Estimate") carries a high margin of error. This is the type of estimate, which should be produced as a range estimate, being prepared by means of the Monte Carlo method. The accuracy of the estimate normally ranges from plus 50% to minus 30%(8). The owner's reserve or contingency should therefore match the risk factor.

The work breakdown structure includes responsibility assignment and accountability of an individual engineer for each package. This includes the responsibility for estimating and trending. The approved estimate for each package becomes the budget within which the individual has to operate.

A variance parameter is given to him which he must not exceed without the approval of the project manager. This parameter could be looked at as a "mini contingency", which will be reviewed as the job progresses. The more detailed the estimate, the smaller the contingency. The sum of all package parameters will be a portion of the total project contingency. As more information becomes available for each package, the estimates are updated:

Design Stage	Estimate	Parameter		Parameter
System Requirements	Concept	+ 30%		
Preliminary	Budget	+ 15%		**BUDGET**
Detailed	Definitive	+ 5%		
Complete	Final	0%		

As mentioned previously, some packages may already have been constructed or are in the construction process while others have not even been designed. Updated estimates, that are not within the budget will have to be reviewed. The question should be asked if cost could be cut by modifying the design. If the design is at a stage when bids are evaluated, and their quotes are above the budget of the contract package, the budget can be revised, if within the variance parameter. If not, it may be possible that more acceptable bids can be negotiated.

If this is not successful, the package holder should investigate the possibility of a change in design, its cost impact and, if favorable, prepare new bid documents. It is an iterative process for each package and the project.

As the project progresses, the latest estimates for each package are summarized and compared with the approved funds.

At this level, the decision to continue with the project or to cancel it has to be made when no more funds are available. Project managers must always be conscious of the requirement to meet a tight schedule. Any schedule delay will alternately delay the return on investment.

Fast tracking can also be applied when a starting date is delayed, but the end date must be fixed.

Start delays can have many reasons. There may be lack of adequate funding, environmental approval delays and others.

On March 13, 1993 the Darlington Generating Station unit 4 was started up two months earlier than the latest schedule indicated. Every month saved to startup meant a saving of about 20 M$ per month in interest.

A time/cost/benefit trade-off analysis must be done to justify fast tracking (see 2.2.4 Corporate Decision Making). We must achieve an overall reduction along the critical path of the schedule. This can be done by overlapping activities which will result in higher labor cost (learning curve), earlier equipment deliveries and changes to existing contracts Equipment purchased too early with preliminary design specifications could result in the purchase of inadequate equipment or in paying extra for capacity that is not needed. Those added cost must now be compared with the benefits of an earlier in-service date. In addition to the reduction in interest payments and inflation cost which can easily be calculated, there is also a reduction in the other indirect costs, including

♦   Administrative Cost
♦   Taxes and Insurance
♦   Depreciation
♦   Cost of Supplies and Rentals

Benefits are also gained from earlier realized income. Full production can begin sooner, benefiting from the early use of funds.

As an example, assume a product is to earn $ 50 000 per week in profit. With a four weeks reduced in-service date and a 10% investment rate, the interest earned is $ 908.00.

Configuration Management

Project management must install a control technique that can reduce excessive engineering changes. Especially on fast track projects, some companies may bid on proposals at 50% below their cost, betting they can make up the difference later with engineering changes.

Configuration management is a structured process for a formal review and approval of configuration changes. This is usually a group of experts representing the customer, contractor and owner working as a committee. Effective configuration control will benefit both, the customer and the contractor.

4.5 PROCUREMENT AND CONTRACTS

Procurement is the acquisition of external goods and services. A *contract* is the legal agreement between buyer and seller to provide and to pay for those goods and services. The owner must decide what goods and services should be purchased and which should be produced internally. This *make* or *buy* decision is a very important part of project cost control. It is not the purpose of this book to discuss in detail the complex procurement functions of a project and related purchases and contracts. For this, specialized text is available on the market.

However, an *introduction* to project procurement management in general may be necessary to understand the cost control aspect of the subject.

4.5.1 Contracting

A contract can be described as a promise or a set of promises that the law will enforce. When obligation only rests on one party, the contract is said to be *unilateral* (an offer one can not refuse). When obligation rests on each of two parties, a *bilateral* contract exists. It is desirable to have contracts in writing. Documented contract terms override verbal agreements. A simple contract involves five essentials:

1. *Offer and acceptance:*
   Once an offer has been accepted, it is irrevocable. An offer must be accepted within a reasonable time period or it lapses.
2. *Legal capacity of the parties:*
   Classes of persons that are legally incapable or legally excluded from binding themselves by contracts lack capacity.
3. *Mutuality of mind:*
   Both parties must understand the arrangements in the same way. Misrepresentation, non-disclosure, fraud, duress, or undue influence can void a contract.
4. *Consideration:*
   The promise must be of some value in the eye of the law. It must be either present or future, it cannot be past.
5. *Lawful object:*
   The purpose of the contract must be legal. Illegality may invalidate the whole or the illegal parts of the contract.

The legal relationship of a contract may come to an end in different ways such as

- ♦ *Lack of Performance* (as stipulated in the contract)
- ♦ *Repudiation* (refusing to continue)
- ♦ *Agreement* (freedom to end a contract mutually)
- ♦ *Frustration* (unexpected event for which neither party is responsible. Accrued rights and liabilities remain in effect by future obligations are terminated).
- ♦ *Breach of Contract* (refusing or making it impossible to fulfill obligations).

Where a contract has been breached by one party, the other party can claim *damages*. If an actual loss has not occurred, but there has been an actual breach of contract, nominal damages only can be recovered. When actual losses are the result of circumstances that could not have been reasonably foreseen at the time of the contracting, they cannot be recovered. Settlement for damages caused by nonperformance or breach of contract constitute compensation to the injured party, not punishment.

Tendering

An owner usually is not qualified to plan, manage and construct an entire project. He therefore may hire an engineer to design and direct the work for him and engage contractors to build the facility. Contract documents must be prepared, consisting of drawings, specifications, general conditions and form of agreement. On that basis, tenders are invited for the construction of the work.

A tender is an offer to carry out specific work upon stated terms. An owner who invites tenders is not obligated to accept any of them. A tender may be withdrawn at any time before acceptance. However, once a tender is accepted, the owner cannot make changes in the scope of the work before a formal contract has been entered into.

Payment (Consideration)

The usual provisions in contracts with respect to time and method of payment are

- ♦ in advance
- ♦ upon completion
- ♦ by periodic installments (work progress)
- ♦ at a specified time

When applying for *final payment*, the contractor states that he has no more claims against the owner.

Owners are required by law to withhold part of the value of the work to provide a fund for possible lien claimants (this requirement may differ under different jurisdiction).

A contractor is therefore only paid the amounts stipulated in a progress or final certificate less the statutory holdback, which is payable when the holdback period expires. If any extra work is to be done over and above what has been stipulated in the original contract, it requires a new contract. This can be circumvented by adding a contingency which provides the owner with a right to order extra work. The terms of payment for this extra work must be specified in the original contract.

Change orders can be initiated in addition to and outside the scope of the work called for in the contract. The contractor is not compelled to perform additional work without obtaining the owner's new agreement for payment. An *injunction* may be granted in certain circumstances to prevent a contracting party from doing what was contracted not to do.

Extra work can be initiated within the scope of the work when conflicts arise between the plan and the execution of the work such as material shortages, interference, schedule delays, crew assignments, supervisory conflicts, organization disputes.

The owner often requires the contractor to provide a *performance bond* to reduce any loss sustained as a result of defaults by the contractor. To avoid litigation, a *bank letter of credit* is usually preferred by the owner. The onus to recover lost money is then left to the bank, not the owner.

The Contracting Process

There are basically eight steps in the contracting process (1):
1. Execution strategy
2. Contracting strategy
3. Validation of contractor capability
4. Analysis and selection of incentives
5. Analysis of cost liabilities or impact of risk
6. Contract language
7. Contractor selection
8. Contract administration

Contractual arrangements can have a significant impact on cost. Owners can achieve cost savings through improved contracting techniques. The owner should match his objectives and resources with those of the contractor. Some owners use incentive clauses in an effort to achieve better contractor performance. Owners must be aware that most successful contracts have one fundamental characteristic in common:

**Thoughtful and meticulous preparation by the owner *before*
the contract is let.**

Both parties should recognize each other's goals and capabilities. Good communication and mutual trust benefit both, owner and contractor.

To reduce risk, an owner should only select contractors who can demonstrate that their procedures, systems and personnel capabilities are adequate to control schedules, costs, materials and that they have a good quality assurance program.

The owner's concern about cost control is quite different from that of a contractor. It depends entirely on the type of contract what each ones involvement is. To satisfy the objectives of both, *owner* and *contractor,* compromise is sometimes necessary.

Types of Contracts

Once an owner decides on a contract, s/he can pick from a wide variety of single or multiple contracts. The responsibility for cost control ranges from the *stipulated* or *fixed price* contract where the contractor has primary cost responsibility to the *reimbursable* or *cost plus* in which case the owner shares cost-responsibility with the contractor. A fixed price contract usually keeps cost within the budget, but the completion date is not taken very seriously by the contractor. A cost plus contract will cut project time because construction starts while engineering is still incomplete. To be successful, the owner must apply modern management systems to control the cost.

Owners generally award contracts separately for design or construction. If a contract is let for design, construction and commissioning it is called a *turn-key* contract.

The breakdown below (figure 4-81) is an attempt made by Ahuja (12) to categorize various types of contracts that may have different names and expressions such as "fixed price", "target price", "fixed fee indirects", "cost sharing", "cost plus award", etc. in literature.

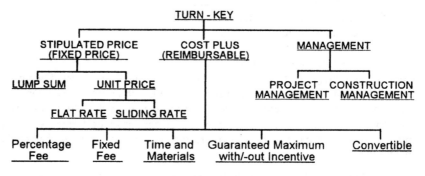

Figure 4-81
Types of contracts.

Owners and Contractors

Below is a summary outlining cost implications on various types of contracts (1):

Fixed Price Contract: (stipulated price)

This type of contract is considered to be the best incentive for the contractor to control costs and thereby enhance productivity. However, the owner must accept the responsibility for providing a complete contract (scope, schedule, quality, site conditions). Because the contractor wants to cut cost and maximize profit, quality of work may not be his priority. It is in the owner's best interest to maintain some degree of influence through contract administration, inspection, and monitoring contract performance. Even though the owner wants to place as much cost responsibility on the contractor, there is a danger, that an owner may also place inordinate liabilities on the contractor for certain risks over which the contractor has little control. Lump Sum contracts often encourage claims for extras by the contractor. There are variations to the fixed price contract:

Unit Price: The contract is bid on the basis of estimated quantities. The owner assumes the risk of quantity variations by providing additional compensation if actual quantities vary significantly from those in the bid schedule.

Fixed-Price by Package (Series of Fixed-Price): The work is subdivided into a series of fixed-price contracts, awarded when the design for that package is complete. This allows for "fast-track" construction. Part of the project can be built while design continues on other portions, normally reducing total construction time.

There are some disadvantages to the owner:

♦ The owner must define the scope of the work carefully for each contract and become significantly involved in both, coordination and administration. This usually requires a larger owner's staff or the hiring of a construction management firm.

♦ Contractors may interfere with each other, resulting in expensive delays. It is also difficult to assign priorities or fix responsibility for problems, including labor disputes because more than one contractor is involved. This may give rise to claims against the owner.

♦ The overall cost of the project may increase because of the added overhead of the multiple contracts.

Fixed-Price with Escalation: The owner assumes the risk for future changes in prices of certain specified materials or for overall inflation. It provides an alternative to the cost-reimbursable contract during unstable economic conditions or for long term projects, because it still gives the contractor an incentive to use his resources efficiently.

Cost-Reimbursable Contracts: (Cost Plus)

The owner runs the largest economic risk with a cost plus contract because he pays all of the contractor's allowable costs including tools, temporary facilities, home-office expenses, and profit or fee. Where the contractor is charged with engineering and materials management responsibilities, these can also be fully reimbursed by the owner, depending on the type of contract.

For an expenditure to be made, the questions of need must be satisfied and fully auditable. The authority to determine need depends on the contract conditions. When an owner grants this authority and the risks that go with it, the contractor has more freedom to act, but the owner must make certain that the contractor adheres to his commitments. Performance measures and reporting procedures must be effectively implemented. This type of contract can offer a negative encouragement to contractors to be wasteful and inefficient if tight cost control is not maintained.

Four basic types of reimbursable contracts exist:

Cost-plus-percentage-fee: The owner assumes the full risk for cost and schedule. Constant audits are required by the owner. The contractor has no incentive to control cost, in fact, he may be rewarded for overruns. The only advantage is, that it may permit an earlier start of construction and gives the owner a greater degree of control over project execution. Because of bad experience with some defense contracts, the Armed Services Procurement Act of 1947 prohibits the use of cost-plus-percentage-of-cost contracts. The owner may establish a ceiling which the contractor may not exceed except at his own risk or by specific approval by the owner.

Cost-plus-fixed-fee: The fixed fee, once established, does not vary with actual cost. Renegotiation of the fixed fee can lead to controversy. There is still little incentive for the contractor to control schedule performance and productivity.

Cost-plus-time-and-materials (Direct-cost reimbursable plus fixed-fee for indirects): The risk to the owner is reduced by approximately the amount of the indirects (usually 15%). Direct labor, permanent plant equipment and construction materials are compensated by the owner at cost.

Guaranteed-maximum: (with or without incentive). The contractor makes a commitment not to exceed a set cost limit for the project. Even though there is some incentive for the contractor to stay below the limit, the owner must still be very much involved in cost control. He may give the contractor some incentive in splitting the benefit of the "below-the-ceiling" dollars saved. This usually depends on the height of the ceiling.

Convertible: Owner and contractor share the risk in proportion to the uncertainties of the specifications. Under great uncertainties, contractors are unwilling to quote a reasonable lump sum. In this case, the contract starts as cost-reimbursable and is converted to fixed-price as engineering proceeds and the project's scope becomes fully established. The owner must be vigilant and monitor productivity until the fixed-price conversion takes place. Stipulation of the fixed-price may be a matter of bargaining rather than competition.

Management Contracts:

The owner hires a *project manager* who develops a task force to manage the project through all stages, from concept to the in-service date. This includes policies and procedures, budgeting, design, procurement, contract administration and project control.

The project manager hires architects, engineers and services staff. He is directly responsible to the owner. Cost control is delegated to the project manager who will be monitored by means of regular progress reports to the owner.

When *construction* management is used on a project, a construction manager, hired by the owner, will be in charge of the physical installation of permanent facilities. He will be responsible for construction planning, estimating and cost control. He will give general direction to contractors and report progress regularly to the owner.

A very important part of any long duration contract must be the inclusion of a *cancellation clause.* There was a case familiar to the author where a cancellation clause was not provided for contracts supplying expensive, prototype equipment. The owner paid a percentage of manufacturing cost based on scheduled milestones of completion (performance incentive). A situation developed whereby low demand and very high operating cost made the project unprofitable. Studies were initiated to look at two alternatives: Delay or cancel. Studies of this magnitude take time and cannot be kept a secret. When the project was finally canceled, the manufacturing progress of the supply contractors had been extremely accelerated. The owner paid a severe penalty for not having a cancellation clause in the contract.

Controlling Contract Cost

When an owner decides to contract s/he will already have identified the type of facility, the required operational date, project life, reliability and supporting facilities. The scope and sequence of work will then form the base for the preparation of the bid package.

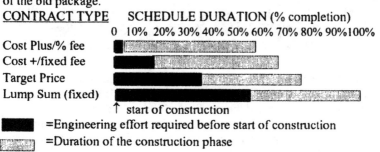

Figure 4-82
Type of contract - effect on schedule.

The type of contract will have an effect on schedule duration for the start of construction and completion of the project. This is crucial in understanding the control of total project progress and cost (Figure 4-82).

What is the cost impact of controllable risks to the owner for two major contract types, the fixed price and the cost- plus (reimbursable)?

Controllable Risks	Fixed Price	Cost-Plus
Labor productivity	low	high
Scope	high	low
Indirect cost	low	high
Construction quality	medium	medium
Safety	medium	medium
Schedule	high	high
Labor relations	low	low
Project Management	low	low

The owner, opting for a Cost Plus (reimbursable) contract will need Cost Control staff in his organization to monitor the contractor's performance. The owner must make certain, that the contractor's project cost engineer is experienced and qualified and has enough influence within the contractor's organization to affect any changes that are necessary to keep the cost in line.

A contractor dedicated to the principles of control will have well documented standard procedures. Those procedures must include not only cost collection and reporting, but also analysis, rectification, and projected action. A prospective client must look at the contractor's methods of cost collection and reporting. The work breakdown and cost code should be well defined and described in detail. It is often desirable to integrate the contractor's and the owner's computerized costing systems. Electronic transfer of data is definitely preferable over manual transfer.

The contractor must be capable to perform the required estimating function. It is the owner's responsibility to outline those requirements. This includes a comprehensive cost trending program and the handling of change orders. In general,

"owners can improve cost effectiveness by taking a more discerning approach to the contractual arrangements for their projects. The terms of a contract should embody the interest of both parties, recognizing the goals of each and the ability of each to control and reduce specific risks and costs." (1)

Tax Treatment of Contract Income

Cost conscious contractors are well advised to know the applicable tax rules of *the country where income payments apply* before they enter into a contract. They can negotiate the terms of payment such that it will give them a tax advantage. As an example, the following applies to *Canada* (readers from other countries may be interested to compare):

For many years Revenue Canada permitted contractors to compute income from contracts on either

♦   the completed contract basis, or
♦   the percentage of completion basis.

Contractors who have always used that method are permitted to continue on that basis for "lump-sum" contracts. New rules apply now:

All amounts which have been or *could have been* billed under the contract terms during the taxation year must be included in income, such as

a) progress billings for work performed
b) unbilled amounts with respect to work performed during the year which could have been billed
c) advance payments received for work yet to be performed.

The old method of income reporting can be maintained but must consistently be used from year to year. Whence a contractor changes to the new basis of accounting, Revenue Canada will not permit the contractor to change back again.

Under the new method holdbacks become legally receivable when an engineer's certificate has been issued or the expiration of the Mechanics Lien period, whatever is last. This could mean that payment from holdbacks must be reported as income even though the money has not been collected.

All outlays made or expenses incurred during the year are deductible in the year, including the unbilled cost of work in progress. However, a contractor may claim a reserve for work not done at year's end but for which the customer has been billed (i.e. downpayment).

Example:

A cost plus contract on which a plumbing contractor had started work in November 1989, half-completed by Dec. 31, 1989, and completed May 15, 1990. Cost totaled $45 000 of which $22 500 were incurred in 1989. The contract provided that he would render bills monthly equal to the cost of the work performed that month plus 20%. Each bill was subject to a holdback of 15%, payable after the work was approved by the customer's architect. The architect's approval was issued in October 1990, and the amount of the holdback was paid to the contractor in January 1991.

The income from this contract would be partially taxable in each of the 1989 and 1990 taxation years.

To summarize:

Cost Plus Contract:

Total cost to contractor:	$ 45 000
Start work:	Nov. 18, 1989
50% completion:	Dec. 31, 1089
100% completion:	May 15, 1990
Contract Provisions:	Monthly cost plus 20% on work performed.
Holdback =	15% payable at certification.
Architect's approval:	Oct. 10, 1990
Bills sent	15 days after completion of work each month.
Payment received	30 days after billing.

Income tax to be paid on the following net income:

1989:	Cost incurred	$ 22 500 = 50% of total cost
	Profit billable	4 500 = 20% of cost incurred
	Total billable	$ 27 000
	Holdback	(4 050)= 15% of total billable
	Cost incurred:	(22 500)= 50% of total cost
	Net Income	**$ 450** Taxable amount in 1989
1990:	Cost incurred	$ 22 500 = 50% of total cost
	20% profit	4 500
	billable holdback	4 050 = 1989 holdback approved
	Total billable	$ 31 050
	Actual cost inc.	(22 500)
	Net income	**$ 8 550** = Taxable amount in 1990

No income on this contract is reportable in 1991, although the holdbacks were not collected until that year. The hold-backs became legally receivable when the architect issued his approval in 1990. The date of collection is not affecting tax reporting.

4.5.2 Change Orders

Any changes or additions to a contract or changes within the scope of a contract will result in additional expenditure. In order to keep those cost to a minimum, it is incumbent upon the owner to *optimize contract language and specifications.* This way, many changes are avoidable. However, we are dealing with predictions of future performance when a contract is initiated. In spite of all the efforts we may put into the awarding of a contract there will always be changes from the original assumptions made in the beginning.

Changes affect the total cost of a project. They impede the orderly flow of work. The avoidance and the management of changes is therefore a very important part of cost control.

Project management should plan at the very beginning of a project to carefully design and implement a support system of documentation that will minimize later disputes, disagreements, conflicts and other problems.

In regard to contracts, the inclusion of a change order clause will allow for modifications to the scope of work after a contract has been signed. *Change clauses allow the owner to make changes* in return for an adjustment to the contract. The contractor will make the changes before pay ments are made by the owner.

In practice however *a change is often identified and priced by the contractor* for approval by the owner. Because the contractor is usually not allowed to continue with the work until the approval date, change orders must be processed expeditiously. This forces the owner on the one hand to pay the contractor promptly for any work satisfactorily completed. On the other hand, any errors, omissions, disputes can cause delays, bad will and decreased productivity.

The Environmental Protection Agency(40) found that

- ♦ For 40 to 50 percent of the change orders reviewed, the reviewing agencies had to request additional information from the grantee in order to process Change Orders because the initial documentation was inadequate.
- ♦ The reviewing agencies had difficulty reviewing claims in a timely manner because of their complexity and the lack of national guidance on claims.

For changes to be orderly resolved between owner and contractor, a methodically designed change order procedure should be included in the contract's general conditions section. It should be agreed on during the preconstruction phase by both, the contractor and the owner.

This includes how changes should be priced, how a change order should be processed, how it is routed, reviewed and approved, and the time allowance for processing. Beside the more obvious direct cost incurred for changes (labor, material, subcontracts), how distributed costs and overheads (utilities, supervision, tools, office and temporary facilities and insurance) should be treated must also be documented.

A "systems" approach to deal with change order management, using flow chart and responsibility matrix, is the most effective approach.

Change Initiation

A change may first become apparent to a contractor who happens to notice a conflict between drawings or with local codes or laws. An owner may find that the specified system will not function as intended without making changes. When an owner initiates a change, a *written change authorization* must be produced in a systematic and orderly manner.

The control of changes in both, lump sum and reimbursable projects, is the owner's responsibility. An owner who permits members of his project team to verbally authorize changes by means of conversation with the contractor is violating a fundamental rule of cost control.

Verbal authorization of changes at any management level must be immediately backed up by a formal written CA (Change Authorization). In some situations a contractor may refuse to proceed with his work without a change initiation; but he should not spend any time or money on a change without prior owner approval, except in emergencies.

Change Order Processing within the Project Scope

A clear and precise procedure must be established in order to avoid changes which cannot be justified by stated criteria. An example of stated criteria may be:

- Laws or regulatory requirements have changed
- Financial project evaluation make modifications necessary
- Site conditions dictate unexpected changes
- Specified equipment is not available
- Substantial improvement in safety can be realized
- A clear economic or performance advantage over the life cycle of the project can be substantiated without causing a cascade of other changes.

All additional work must be evaluated on the basis of their impact on schedule and cost. In case of a unit-price contract, it is the owner's duty to outline to the contractor the resulting changes in the work. The contractor will then estimate the cost/schedule impact and submit a quotation to the owner's representative (general superintendent) showing quantities, unit prices, or lump sum in sufficient detail for the owner to evaluate and approve. The owner will keep a running account of all changes made to specific contracts so that costs and savings can be applied to the contract price. Since the owner's budget is the base line for controlling project cost, a budget review will become necessary if the overall changes exceed the contingency allowance.

Changes Affecting the Project Scope

In addition to changes within the scope of the contract, additional work that affects the scope can be requested by the owner. For major scope changes that may affect the final cost or in-service date, detailed economic evaluation may be necessary to obtain head office approval of the change. In severe cases this could result in project refinancing, contract cancellation, or project deferral.

The contractor may refuse to perform the work without a supplemental agreement. Negotiations for extra work usually result in considerable cost to the owner. It is advisable that provisions for extra work be included in the original contract.

Claims

Project conditions may give rise to claims by the contractor against the owner. Claims occur when a contractor feels that a breach of contract situation has arisen because of the owner's failure or refusal to perform a duty owed by him under the terms of the contract (12). The owner may also claim for damages against contractors, A/Es, or consultants for noncompliance with agreements incl. drawings, schedules and specifications. When claims arise, the work flow on the project will be interrupted. Preventing claim situations is in the interest of both, the owner and the contractor. Basic causes for claims are

- ◆ Schedule variations
- ◆ Field conflicts
- ◆ Access delays
- ◆ Change Orders
- ◆ Accelerations
- ◆ Work out of sequence
- ◆ Owner furnished materiel delay
- ◆ Superior information withheld

Many claims, although relatively minor by themselves, can have an expensive "ripple effect" on other components of budgetary and schedule commitments (impact cost).

There have been many papers presented by organizations such as the American Association of Cost Engineers (AACE) and the Project Management Institute (PMI) dealing with the subject of claims. They all seem to express the opinion that losses to the contractor for unexpected interferences with their work can be obvious but must be substantiated by keeping a record of each occurrence that may result in a claim, such as history sheets, minutes of meeting, memos.

History Sheets:

The contractor must have a procedure whereby the foreman keeps a record of unusual job situations which may give rise to claims.

Those history sheets are assembled by the contractor, summarized and attached to a formal claim request. This claim request may or may not be accepted for compensation. Negotiations with the usual accusations and denials will take place and a following final agreements will hopefully result.

```
ABC CONSTRUCTION COMPANY                         Date 1995-12-12
                                                 Job #  EH 136
                        History Sheet # 184
Project Name:  Cable Pan Run                     Foreman: Joe Doe
Location: Maple Creek
Equipment on Site: Sky Jack, Job Boxes
─────────────────────────────────────────────────────────────────
                 Number of Trades      Hours    Equipment Hours
Floor: 4th       4 Laborers            15        2 material lift
Area: B5 - D7    1 Journeyman          4         ...............
                 ...........           ....      ...............
─────────────────────────────────────────────────────────────────
TIME LOSS and why:
           Arrived at working area for the night shift. Job boxes were delivered
           by A/E crew during the day and stored where we were supposed to work.
           Superintendent not able to find crew with lifts to remove boxes. ........

PROBLEMS ENCOUNTERED AND CORRECTIVE MEASURE TAKEN:
           It took us 3 hours to get lifting equipment from another site to move boxes.

REMARKS:   This is part of an ongoing problem /
              Joe Doe                          ~~~~~
           Signed (Foreman)         Verified (Superintendent)
```

Figure 4-83
Claims - history sheet.

WITHIN THE CONTRACT (NON-COMPENSATORY)	NORMALLY EXPECTED INTERFERENCE
NEGOTIABLE	INTERFERENCE DUE TO UNEXPECTED ADVERSE GENERAL SITE CONDITIONS
CLAIM CONSIDERED JUSTIFIABLE (COMPENSATORY)	EXCESSIVE UNFORESEEABLE INTERFERENCE CAUSED BY OTHERS.

Figure 4-84
Interference claims.

The history sheet (Figure 4-83) would include such items as

- ♦ Project identification, job number and location
- ♦ Name and signature of foreman and date of event
- ♦ The equipment, material or tools involved
- ♦ The time loss in workhours, hours of idle equipment
- ♦ A description for the reason of time loss
- ♦ The problems that were encountered and who was responsible.
- ♦ The corrective measures taken and notification to the owner's representative and overall remarks.

To process claims, we could categorize three basic types of claims (figure 4-90):

The contractor is expected to include normal construction interferences in its bid price. He can visit the site to look at the conditions and also consider the reputation of the owner in regard to similar projects. This would be categorized under

Normally expected interference.

For this, claims would usually not be accepted for compensation by the owner even though the history sheet may show those interferences.

There are occasions when the contractor feels strongly that adverse site conditions are beyond his control and cause him considerable losses.

This may include haphazardly placed material by others in busy work areas (material interference) or owner supplied materials not available for scheduled work; pilferage, vandalism, sabotage, which the owner should have been able to prevent and other interferences that could be debatable as to the amount of losses.

This would be categorized under

Interference due to unexpected adverse general site condition

Those claims are usually negotiable.

The third category can be classified as

Excessive unforeseeable interference caused by others.

These are predominantly cases of owner's active interference or passive neglect. This includes imposed priority of work activity causing demobilization, information such as schedules or drawings not passed on (superior knowledge undisclosed) or revised without a change order or extra work recognition.

Legitimate claims should be settled expeditiously by the owner, because it affects the owner' s cash flow, which is a very important consideration. If it comes to *litigation*, the courts will usually award compensation for loss of productivity (cost impact) in addition to specific losses claimed.

Reviewing a Claim

A very important part of the claims process is the review of the claim by the owner's representative, usually the architect or engineer. A fair owner will endeavor to evaluate all claims objectively; but the engineer, as the designer and supervisor of the job and acting in behalf of the owner, may feel a bias toward the owner in settling a claim. He may even be the subject of blame in the claim. If the engineer is paid a percentage of the job cost, merely approving claims will increase his income. Another problem is that the contractor relies on the owner for his livelihood (loss of future work) and may be very reluctant to antagonize the owner or his agent. Those conflicts can be avoided if the owner hires an independent consultant for a fixed fee.

In summary, if changes, modifications, additions, and extras are taken lightly, if they are only verbally authorized and not properly evaluated, no matter what else is done right, the job is running out of control and will result in a proliferation of claims.

4.5.3 Equipment and Materials

Procurement and Materials Management

At any major project, material and equipment costs are a large component of construction projects (about 60% in many cases). In *fact they are the largest single element of project cost*. The breakdown of capital cost for a particular nuclear generating station was (14)

Permanent Material and Equipment	43%
Financing Cost	29%
Construction (Field) Cost	17%
Engineering	11%

Major equipment cost are relatively easy to estimate, categorize, and control. Their delivery dates or progress payments are known quite early in the planning stage.

Equipment sizes are initially based on project requirements and capacities (see Sections 3.1.1 & 3.4.4). This is the planning stage, when we determine equipment needs, research the availability of critical materials, develop a bid list for major equipment, and prepare a procurement program and delivery schedule.

Modern technology allows us to integrate information into a total materials management network. This means that the materials management system is meshed with the overall project management system. This requires the function of a coordinator who will treat suppliers and vendors as partners.

The *procurement function* itself can be summarized as

PURCHASING: SUBCONTRACTS:
Pricing of Proposals   Pricing of Proposals
Suppliers' Prequalification   Prequalification Survey
Inquiries to Vendors   Inquiries to Subcontractors
Expediting   Expedite Quotations
Review Submitted Bids   Analyze Quotations
Coordinate with Stakeholder   Accept Bid or Negotiate
Negotiate if Necessary   Select Subcontractor
Select Supplier   Change Orders
Issue Instruction Notices   Contractual Matters
Contract Administration

VENDOR DATA:

Vendor Inspection: Engineer/Vendor Liaison:
Quality Control   Access to Information
Welder Qualifications   Personal Contacts
Inspection Reports to Owner   Written Confirmations

Traffic: Field Procurement:
Movement of Goods   Construction Equipment
Special Transport Equipment   Quality Control, Inspection
Auditing   Construction Subcontracts
Loss, Damage, Shortages
In-transit Status
Bill of Lading
Freight Rates

Traditional material cost monitoring systems are usually integrated with other purchasing and expediting programs. They compare estimated vs. actual cost often only after order completion. Furthermore, many materials are ordered in bulk, which makes distribution to cost packages extremely difficult. Adequate cost accounting procedures are a major consideration in controlling material cost.

Below is a sample of a procurement activity flow of materials and equipment ordered by the home office (Figures 4-85 and 4-86):

System Requirements

A good system must
♦   Integrate material management into project planning (overall coordination)
♦   Train personnel in using computerized systems to its greatest advantage.
♦   Timely control information to respond to impact of design changes and changes in specification.
♦   Expedite suppliers' delivery promises.
♦   Maintain effective performance standards.

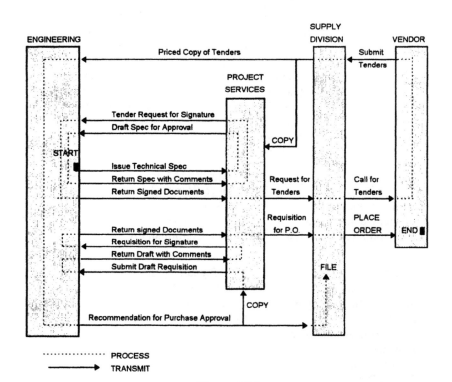

PROCESS
TRANSMIT

Figure 4-85
Procurement activity flow.

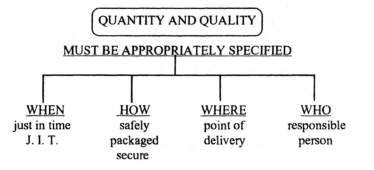

Figure 4-86
Materials management criteria.

Organization:
1. Define responsibilities for all activities
    clearly understood
    consistently assigned
2. Write procedures
    consistent with project procedures
    not in conflict with others
3. Establish standards
    to keep track of staff levels (transfers)
    to compare performance against plan.

Planning:
1. Prepare schedules for material management activities
    must reflect interdependence of related project activities (in time, but not too early)
    test activities for cost effectiveness
    develop budgets for performance measures
    develop procedures for incorporating changes (assess impact)
    assure facilities can handle required craft materials on job site.

Execution:
- ◆ Make effective use of computer-aided systems
- ◆ Expediting and inspection personnel should participate in the selection of suppliers.
- ◆ Use bidders' past performance (performance list)
- ◆ Prepare expediting plans for every order based on updated schedules
- ◆ On critical item, obtain suppliers' total production schedule and make shop visits.
- ◆ Document quality assurance (QA) and quality control (QC) requirements for inspectors to monitor compliance with specifications.
- ◆ Execute an expeditious and economical transport of materials and equipment to the job site and protection against damage.
- ◆ Observe procedures for handling and storage on job site (minimize handling).

Materials management for a *field organization* should have the following objectives:

- ◆ Avoid material shortages
- ◆ Make the most economical purchase
- ◆ Maintain optimum (low) inventory levels
- ◆ Minimize surplus in stock
- ◆ Know dollar value of inventory at any point in time
- ◆ Maintain optimum material dollar distribution
- ◆ Keep historical records

Computerization, even for small projects, should be designed to include
• demand • supply • schedules • costs • receipts • usage
Starting with the design function, materials are specified, bills of materials, equipment lists, instrumentation schedules are given the necessary codes and divided into items that are ordered by the construction forces or by the home office (engineered items). Computerized systems are processing the information for use by various organizational units. The computer will interface with other computerized system not only internally, but also with some major suppliers, if compatible.

4.6 CONSTRUCTION COST CONTROL

When compared with overall project cost the labor content of *on-site field cost* is high. It stands to reason that optimization of human resources will result in considerable cost savings. Controlling the cost of labor is essentially a productivity issue which will be dealt with in chapter 6. Other areas where cost can be optimized include

♦ Location of temporary construction facilities
♦ Storage and movement of materials
♦ Equipment selection and replacement
♦ Control of subcontract work
♦ Construction indirects (field cost)

4.6.1 Facility Layout

We can raise productivity on-site through an effective layout of construction support facilities and efficient material handling. To achieve a high level of productivity, it is important to do a very thorough study on the layout of construction facilities (27).

Depending on the size of the project, the basic design group should make allowance for space needed to build the capital project. Designers must make use of the knowledge and experience of construction personnel to look at both, the permanent *and* the temporary facilities of the plant. A rough evaluation of the size of property needed may even be necessary *before* the land is bought.

The Systems Approach

In general, a systematic approach requires gathering and analyzing data, assessing alternative solutions for particular circumstances and specifying preferred solutions for decision making. This approach is the basic requirement for the layout study of temporary facilities. Chapter 7 will deal with this approach in more detail including the application of process charting which is based on a matrix showing a set of activities occurring between defined INPUT and OUTPUT boundaries, and a hierarchy of various levels of activities.

Layout Relationships: An old but effective method to layout construction support facilities is the use of color coded templates placed over base drawings (28). The size of the templates that represent true scale facilities and their spatial interdependencies is easily obtained by means of flow diagrams and relationship charts. Instead of the old-fashioned templates, computer aided design (CAD) or specialized software is effective in the production of layout drawings.

Below is a simple example of a movement pattern that occurs between two floors of a building. This sort of display can emphasize poor movement patterns, e.g. excessive length and back-tracking. Changes in activity sequence or locations may be suggested as a result (Figure 4-87).

Movement density diagrams give a good indication of expected heavy traffic frequencies on an individual or aggregated basis (Figure 4-88).

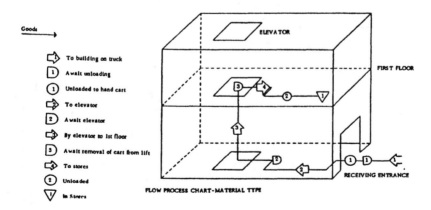

Figure 4-87
Layout - movement patterns.

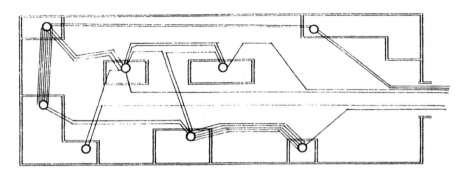

Figure 4-88
Layout - density diagram.

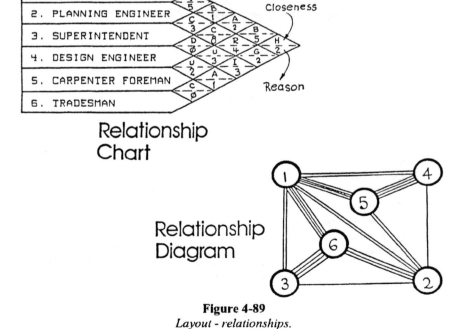

Figure 4-89
Layout - relationships.

Relationship Charts: They are produced to show how important it is to have areas close together or apart. This helps to place work areas and storage in the best pattern. From this, the relationship diagram is derived. It gives the best arrangement of functions and their spatial proximity according to strength of relationship ranging from "closeness essential" to "closeness undesirable". The number of linkages is proportional to the closeness rating (Figure 4-89).

To summarize, there are two main purposes for process diagrams and flow charting:

1. It gives contributors to the layout of construction support facilities a structured and logical way of planning productively.
2. Used as a discussion paper, it entices input and cooperation from all disciplines affected.

The *systems* approach will make it possible to come up with a better layout resulting in better support services and thereby increased construction productivity.

Figure 4-96 shows a skeleton of a chart that can be used as a template for the design of an efficient layout.

The diagram shows four layers (multiple input) with increasing details and dependencies. It is similar to the concept of the work breakdown structure. If B20 deals with layout relationships, relative closeness could be one of the "C" items, which in turn may even require a further breakdown.

Similar to the practice in computer programming, there are usually spaces left when numbering the codes. This allows for inserts during the review. To pick an example, let us follow

$$A1 \rightarrow B10 \rightarrow C40 \rightarrow D160$$

Description:

A-1 = Facility location.

B-10 = Size of facilities.

In some cases the craft maximum or project peak work force requirements determine the size of the various facilities. (Savings can be realized however, if peaks are taken care of by portable or easily removable facilities).

C-40 = Equipment Requirements.

Requirements depend on project type, size, schedule, amount of site fabrication, environmental conditions, contract requirements, distance to supply centers etc.

D-160 = Warehouses.

Material availability and the risk of not having supplies ready when needed will determine the size of warehouses.

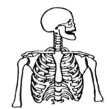

This *skeleton* is a good thought facilitator which entices input and cooperation from all disciplines affected.

INPUT (list of required facilities):

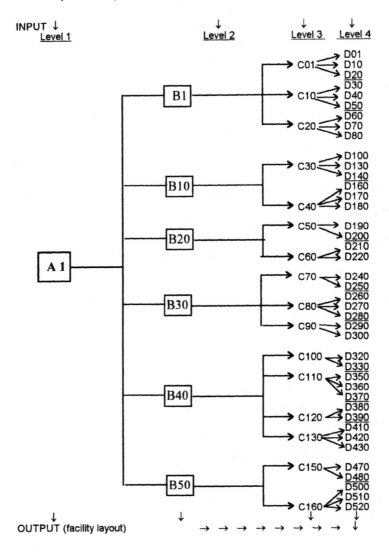

Figure 4-90
Facility layout chart.

Maintain the System

After having reviewed all possibilities in such a structured way, we have now installed the best arrangement of facilities, effective storage and efficient movement of materials, equipment and labor. Can we now rest and let construction run its course? No! Even though we cannot cost effectively change already built warehouses, offices and shops, we still can improve on methods.

We must constantly keep an eye on how the system performs
by measuring the level of production and productivity.

Following is an example of finding the best location for temporary facilities by using the cost of transportation.

Example: Reinforcing steel of various sizes arrive at a warehouse near the construction site. The custom designed cutting and bending takes place at a steel fabrication shop which can be located either at

$$B = 100 \text{ m}, \quad C = 80 \text{ m, or } D = 60 \text{ m}$$

from the warehouse location A. The prepared steel is then send to a temporary storage location to be retrieved for installation at the job site Z (Figure 4-91).

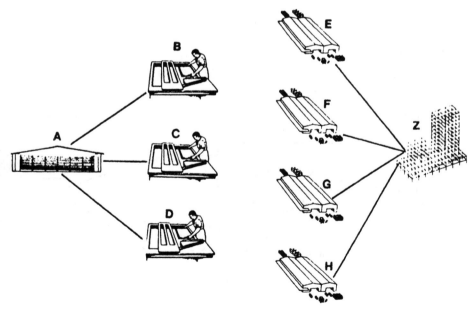

Figure 4-91
Transportation example.

There are four possible temporary storage locations available, E, F, G, and H. Those are the distances in meter from the temporary storage:

From temp. storage	B	C	D	Z
E	50	20	60	40
F	40	60	80	100
G	60	10	50	80
H	100	80	30	60

To locations

Based on once per week shipping, the average monthly cost of traveling is $ 2.00/m. Find the optimum layout of facilities, using both, a tabular (Table 4-24) and a diagrammatic (node-arc configuration) method (Figure 4-92).

Table 4-24
Transportation example - spreadsheet

	B↓	C↓	D↓		Paths	Additions	
A →	10	8	6	Z →	B	C	D
E	5	2	6	4	19	14	16
F	4	6	8	10	24	24	24
G	6	1	5	8	24	17	19
H	10	8	3	6	26	22	15

The smallest path addition is $C_A + C_E + Z_E = 14$ or $ 280

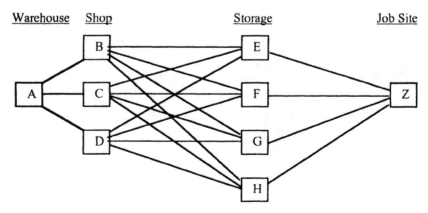

Figure 4-92
Transportation path diagram.

Table 4-25
Transportation example - minimum cost

Storage to Job Site		Shop to Job Site		Warehouse to Job Site	
Route	Cost	Route	Cost	Route	Cost
E-Z	80 →	B-E-Z	100+ 80 =180	→ A-B-E-Z	200+180=380
F-Z	200	B-F-Z	80+200 =280		=480
G-Z	160	B-G-Z	120+160 =280		=480
H-Z	120	B-H-G	200+120 =320		=520
		C-E-Z	**40+ 80 =120**	→ **A-C-E-Z**	**160+120=280**
		C-F-Z	120+200 =320		=480
		C-G-Z	20+160 =180		=340
		C-H-Z	160+120 =280		=440
		D-E-Z	120+ 80 =200		=320
		D-F-Z	160+200 =360		=480
		D-G-Z	100+160 =260		=380
		D-H-Z	60+120 =180	→ A-D-H-Z	120+180=300

Tabular Method

To find the optimum layout, the cost/meter are immaterial unless they are variable. The unit cost of travel is identical in this example. We only need to be concerned with distances. To simplify the calculation further, all distances can be divided by ten, if the input is done manually. The formulas for tabulation are:

$$B_A + B_{E \to H} + Z_{E \to H} \; ; \; C_A + C_{E \to H} + Z_{E \to H} \; ; \; D_A + D_{E \to H} + Z_{E \to H}$$

Table 4-25 shows that Route A-C-E-Z is the least costly route to transport the steel.

4.6.2 Storage and Transport of Materials

The problem above established the best facility location by optimizing transportation cost. Most problems are not that simple in practice. The movement of materials depends on the property of materials to be handled (weight, size, shape etc.) and the type of transportation equipment needed. In addition to location, the capacity of the shop and the size of the storage facility together with supply and demand constraints make it desirable to use linear or dynamic programming (see Section 1.4.1).

Let us somewhat modify the previous transportation problem by saying that we have three fabrication shops that are able to cut and bend a maximum of 420 pieces of steel bars per month. Those shops are portable units (prefabricated), each having different output capacities.

Similarly, the storage facilities are also of different size, but all together they must be able to store those 420 bent bars to satisfy the installation schedule.

Shop Output			Storage Capacity			
B	C	D	E	F	G	H
144	120	156 = 420	96	108	132	84 = 420

Quantities enter the picture now. We should therefore do the costing by quantity. The monthly cost of travel in the previous example was based on four trips per month at an estimated cost of $ 2.00/m. For example, shop B can receive material for 144 bars. They are delivered once every month, costing $ 2.00/4 = $ 0.50 per month per meter. Transportation cost from warehouse A to shop B are

$$100 \times 0.5 = \$ 50.00 \text{ for 144 bars or } 50/144 = \$ 0.35 \text{ per bar}$$

The shipping from shop to storage is restricted by both, what the shop can supply and what can be stored. Here again, deliveries can be made once per month, i.e. $50m \times \$ 50.00 = \$ 25.00$ from B to E or $ 0.26 per bar. See Table 4-26 for a summary of those calculations in form of a spreadsheet (cost per bar):

The installation schedule requires a weekly pickup from the storage for delivery to the job site. The cost per bar for transportation from E to Z is $40 \times 2/96 = \$ 0.83$. This will make the total travel cost per bar along route A-B-E-Z

$$0.35 + 0.26 + 0.83 = \$ 1.44$$

What we want is the shipping schedule that optimizes the travel cost.

Table 4-26
Transportation - cost per bar

	E	F	G	H
Units	**96**	**108**	**132**	**84**
Cost from B to	0.26	0.19	0.23	0.60
Travel to Z from	0.83	1.85	1.21	1.43
Cost from B to Z	1.09	2.04	1.44	2.03
A to B cost	0.35	0.35	0.35	0.35
Cost A-B-Z over	**1.44**	2.39	1.79	2.38
A to C cost	0.33	0.33	0.33	0.33
Cost A-C-Z over	1.42	2.37	1.77	2.36
A to D cost	0.19	0.19	0.19	0.19
Cost A-D-Z over	1.28	2.20	1.60	2.19

By using the unit cost above we can define the decision variables. Let us call X_{BE} the variable for the quantity shipped from B to E. If Z is the *objective function*, then

$$Z_{min} = + 1.44X_{BE} + 2.39X_{BF} + 1.79X_{BG} + 2.38X_{BH}$$
$$+ 1.42X_{CE} + 2.37X_{CF} + 1.77X_{CG} + 2.36X_{CH}$$
$$+ 1.28X_{DE} + 2.20X_{DF} + 1.60X_{DG} + 2.19X_{DH}$$

There are two *constraints,* one of which is the manufacturing capacity for each shop:

B: $+ X_{BE} + X_{BF} + X_{BG} + X_{BH} \leq 144$
C: $+ X_{CE} + X_{CF} + X_{CG} + X_{CH} \leq 120$
D: $+ X_{DE} + X_{DF} + X_{DG} + X_{DH} \leq 156$

the other is the capability of the storage facility to meet the demand of the job requirement:

E: $+ X_{BE} + X_{CE} + X_{DE} = 96$
F: $+ X_{BF} + X_{CF} + X_{DF} = 108$
G: $+ X_{BG} + X_{CG} + X_{DG} = 132$
H: $+ X_{BH} + X_{CH} + X_{DH} = 84$

There are only three "supply" constraints and four "demand" constraints. This gives us $3 \times 4 = 12$ decision variables and $3 + 4 = 7$ constraints. It is generally not practical to calculate the optimum solution manually. There are, however, short-cuts possible to the simplex method of linear programming under certain conditions, one of which is the minimum cost method.

The fact that all *constraint coefficients* are either zero or one and that 1 appears only once for the supply constraints and once for the demand constraints gives us that special condition when a manual solution can be obtained without too much effort (table 4-27). The slack was left off.

It is advantageous to understand the basics of the processes before one uses the computer's black box.

It allows us now to present objective function coefficients in a matrix form combined with a summarization of supply and demand constraints (Table 4-28).This is the initial transportation tableau.

Table 4-27
Transportation -constraint coefficients

	BE	BF	BG	BH	CE	CF	CG	CH	DE	DF	DG	DH
B	1	1	1	1								
C					1	1	1	1				
D									1	1	1	1
E	1				1				1			
F		1				1				1		
G			1				1				1	
H				1				1				1

Table 4-28
Transportation Tableau #1

	E	F	G	H	Supply
B	1.44	2.39	1.79	2.38	144
C	1.42	2.37	1.77	2.36	120
D	1.28	2.20	1.60	2.19	156
Demand	96	108	132	84	420

Table 4-29
Transportation Tableau #2

	E (X)	F	G	H	Supply
B	1.44 (X)	2.39	1.79	2.38	144
C	1.42 (X)	2.37	1.77	2.36	120
D(*) →	1.28 (X)	2.20 (*)	1.60 (*)	2.19 (*)	156
	96 ↓		60		(156)
Demand	96	108	132	84	420
	(96)		(60)		

Table 4-30
Transportation Tableau #3

	E (X)	F	G (#)	H (@)	Supply
B	1.44 (X)	2.39	1.79 (#)	2.38 (@)	144
		108		36	→(144)
C	1.42 (X)	2.37	1.77 (#)	2.36 (@)	120
			72	48	→(120)
D(*) →	1.28 (X)	2.20 (*)	1.60 (*)	2.19 (*)	156
	96 ↓		60 ↓	↓	→(156)
Demand	96	108	132	84	420
	(96)	(108)	(132)	(84)	→(420)

To start off, we look for the lowest cost cell, which is $X_{DE} = \$ 1.28$. We now allocate as many demand units as possible to this cell. With a supply of 156, we can assign the total 96 demand units to this cell. This takes care of the total demand for the E column which we can now cross out. This will be shown as (X) on the matrix (Table 4-29).

The next lowest unit still assignable is in cell $XDG = \$ 1.60$. The supply from D is 156 units , 96 of which were used, leaving us 60 units to use up the D supply of 156. We can enter those into the DG cell which meets part of the demand for 132 units. Since the supply from the D shop is exhausted, we can cross out this line, indicated as (*) in the tableau (Table 4-29).

The G demand is reduced to $132 - 60 = 72$. The next lowest amount in a cell still available is $X_{CG} = 1.77$ (Table 4-30). Entering 72 units in CG will satisfy the demand for G completely (shown as (#)) and reduces the C supply to 48 units which we will enter into CH. The next lowest cost cell is BH ($2.38), into which we enter 36 units, satisfying the H demand (@). Only the F demand is left now into which we put 108 units.

Please note, that we have only $3 + 4 - 1 = 6$ cells occupied. If it should happen that a row and a column have the same allocation, i.e. supply equals demand, less than six cells will be occupied. In this case, we arbitrarily assign a zero to any unoccupied cell.

The present solution is only the first step in finding the best routing schedule. Of the 12 decision variables only 6 are in the initial tableau. We must test the other six cells if their inclusion will give an improved solution.

We need only one unit to observe the effect on the outcome (Table 4-31). Assuming we test the empty cell CE. Adding one unit (+1) will increase the demand to 97 and the supply to 121. Since this is not acceptable, we have to "balance" the matrix by adjusting occupied cells, i.e. reducing by 1 cells DE=(−1) and CG=(-1). This will affect DG=(+1) with a net cost effect of

$$CE = +1.42 - 1.28 - 1.77 + 1.60 = - \$ 0.03$$

Table 4-31

Transportation Tableau #4

	E		F	G		H	Supply
B	1.44		2.39	1.79		2.38	144
			108			36	
C	1.42	+1	2.37	1.77	-1	2.36	120
				72		48	
D	1.28	-1	2.20	1.60	+1	2.19	156
	96			60			
Demand	96		108	132		84	420

The net effect for all the other open routes is:

BE: $+1.44 - 1.28 - 2.38 + 1.60 - 1.77 + 2.36 = -0.03$

CF: $+2.37 - 2.39 - 2.36 + 2.38 =$ zero

DF: $+2.20 - 2.39 - 1.60 + 2.38 - 2.36 + 1.77 =$ zero

BG: $+1.79 - 2.38 - 1.77 + 2.36 =$ zero

DH: $+2.19 - 2.36 - 1.60 + 1.77 =$ zero

The total net effect on the system after the first iteration is -0.06 which is very close to an optimal solution. This results in total shipping cost of

Route	# Of Bars		Unit Cost	Total Cost
BF	108	×	2.39	= 258.12
BH	36	×	2.38	= 85.68
CG	72	×	1.77	= 127.44
CH	48	×	2.36	= 113.28
DE	96	×	1.28	= 122.88
DG	60	×	1.60	= 96.00
	420			= 803.40

Routes BE and CE can be subject to improvement. In order to save 3 cents per unit shipping cost we will consider adding units to the lower transportation cost cell CE = \$1.42. Looking at Table 4-31 above, we can see that the least number of units for negative transfer (-1) are in cell CG. We will now transfer those units to cell CE (Table 4-32).

To balance the tableau, DE becomes 24 and DG 132. Going through the evaluation again:

CG: $+1.77 - 1.60 - 1.42 + 1.28 = +0.03$

BE: $+1.44 - 1.42 - 2.38 + 2.36 =$ zero

CF: $+2.37 - 2.39 - 2.36 + 2.38 =$ zero

DF: $+2.20 - 1.28 - 2.39 - 2.36 + 1.42 + 2.38 = -0.03$

BG: $+1.79 - 2.38 - 1.42 - 1.60 + 2.36 + 1.28 = +0.03$

DH: $+2.19 - 2.36 - 1.28 + 1.42 = -0.03$ for a total net effect of zero.

Table 4-32

Transportation tableau #5

	E	F	G	H	Supply
B	1.44	2.39 108	1.79	2.38 36	144
C	1.42 + 72	2.37	1.77 - 0	2.36 48	120
D	1.28 - 24	2.20	1.60 + 132	2.19	156
Demand	96	108	132	84	420

For total shipping cost:

Route	# Of Bars		Unit Cost		Total Cost
BF	108	×	2.39	=	258.12
BH	36	×	2.38	=	85.68
CE	72	×	1.42	=	102.24
CH	48	×	2.36	=	113.28
DE	24	×	1.28	=	30.72
DG	132	×	1.60	=	211.20
	420				801.24

It appears that this may be the best solution, however, there are still two negative values in DF and DH. That means there could still be room for improvement.

The next iteration will then look like Table 4-33. And the total cost:

$$108×2.39 + 36×2.38 + 96×1.42 + 24×2.36 + 132×1.60 + 24×2.19 = \textbf{\$ 800.52}.$$

This is the best improvement we can make because we have been moving from a total negative net cost effect to zero. The next iteration would result in all cells equal to zero except cell DE = + 0.03 for a total cost of $ 801.24.

This result can be shown in the form of a node-arc diagram (Figure 4-93). It can be linked to the program that does the calculations for the transportation problem. Generally, Material handling cost are difficult to determine. By observing the activities of material handling we will notice that those activities contain a large portion of the labor cost.

In addition to an optimal job layout based on transportation cost there are many areas of concern that are not as easily measurable. To solve those problems it is advisable to use a systems analysis procedure (14). This is a larger scale results oriented type of search for solutions often called work study or method study (see chapter 7). Material handling is not an isolated system.

Table 4-33

Transportation tableau #6

	E		F		G		H		Supply
B	1.44		2.39		1.79		2.38		144
			108				36		
C	1.42	+	2.37		1.77	-	2.36		120
	72						24		
D	1.28	-	2.20		1.60	+	2.19		156
	0				132		24		
Demand	96		108		132		84		420

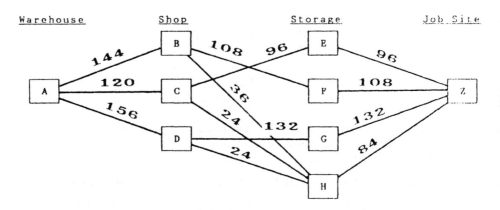

Figure 4-93
Transportation problem - optimum solution.

Materials handling must be fully integrated into the overall materials management concept.

Storage

It would be an ideal situation when materials and equipment reach the right place at the right time and in the right condition for installation. There are many reasons why this is hard to achieve. The cost of interrupting the flow of materials in the daily production must be weighed against the cost of temporary storage. Savings realized by buying materials in bulk must be weighed against storage cost. There is a distinction to be made when purchasing "off-the-shelf" material easily obtainable and specially fabricated material which cannot be replaced easily.

Materials can be stored indoor or outdoors. Materials stored outdoors must be protected not only from the weather, but also from theft or damage. All material must be labeled, the location recorded and placed such that it is as close to the work location as possible. All material must be easily accessible, kept off the ground and made ready for retrieval. This will avoid undue expediting and searching. This all needs good work planning and housekeeping.

Materials delivered to the job too early and in larger than needed quantities will clutter the work area and cause interference. To avoid this, *a good inventory control system should be established.* A project management system usually recognizes two basic processes for the accrual of material supplies:
   1.   Priced material receipts
   2.   Vendors' invoices (supply progress)

Dollars for priced material receipts are debited directly to an inventory suspense account. The system must be set up such that we take full advantage of accrual accounting and the identification of received material cost to the proper end account. To do this, the material identification in terms of the work breakdown structure must already be done early in the process when the drawings are prepared. Assuming the estimate for a work package is based on a mechanical system which requires various sizes and description of pipe.

The pipe account may be identified as

71519 Domestic Water Piping: →300m copper @ 25 mm diameter

→500m copper @ 19 mm diameter.

The Bill of Materials (BM) for this system and other systems that need copper pipe will specify the requirements as needed. That may amount to a total of

→2000m copper @ 25 mm

→3000m copper @ 19 mm

The installation schedule will show the need for materials including copper pipe overlapping during time intervals:

Deliveries	▽		▽		▽
Account #	71519	71520			71548
Quantity	300 | 500	200 | 300			200 | 500
Account #		71539		71532	
Quantity		800 | 1000		500 | 700	

The purchasing department decides to issue two bulk orders of 1300m @ 25 mm and 1800m @ 19 mm to take care of the first year's supply and 700m @ 25 mm and 1200m @ 19 mm for the second year. Unless we can find a way to identify incurred cost by work package, we cannot control the cost as estimated.

The Bill of Materials should therefore identify the portion allocated to the various system accounts e.g.

Deliver January 1st

(1)   1300m @ 25 mm copper tubing ......$ 9800.00

Distributed to accounts

71519   = 23%

71539   = 62%

71520   = 15%

(2)   1800m @ 19 mm copper tubing ......$ 8100.00
     Distributed to accounts
               71519   = 28%
               71539   = 55%
               71520   = 17%

The cost for bulk orders go into an inventory suspense account at the time of delivery. When the material has been installed, it is issued from inventory to site and to the direct end account. Collecting the cost to the end account will allow the individual who is responsible for the work package a valid comparison of cost incurred vs. the estimate.

Inventory

At the heart of a computerized material management system is the inventory control. As mentioned previously, the better we control the flow of materials into and out of inventory, the less storage space is needed, which in turn improves cash flow and enhances productivity. Below is an outline of a material management system describing briefly how inventory controls can be set up.

Let us follow the data flow through the system for a hypothetical piece of material. The need for the material is normally noticed first by the design group. This could happen during preliminary design, or when a flow diagram is produced, or when a detailed drawing is complete and quantities are taken off. The need for materials is input to an integrated project material management system which we may call PMMS (see 4.2.4 Cost System Flow Diagram).

This material is input under its stock code number and required delivery date as an item on a Bill of Materials or Equipment list or Instrument Schedule etc. Usually the need for other items in the same part of the plant will be input at the same time. The system notes the demand for all those items and produces a Bill of Materials for that part of the plant. A required date for each unit on each Bill, based on the construction schedule, is also input to the system.

     If the material is a common item, it will probably be called up on several Bills of Materials. PMMS will add up the demand from each Bill and report demand on a time scale based on the scheduled dates for each unit. If the material is issued for a purpose which is not called up on a BM, the issue appears as a NON-BM. Therefore, the demand is the sum of the quantity called up on Bills of Materials plus the quantity consumed at the site which is not specified on a BM. A foreman may request extra items (field added) or extra quantities on Bills (Extra to BM).

PMMS has files of current inventory quantities and so can satisfy material control requirements by providing Supply - Demand status reports and predicting status at any given future date (Table 4-34). Suppose for example, that various Bills of Materials have called up 600 pieces of material with an extra demand of 50 The warehouse has received 700 (a) and now has 50 on hand (b). On September 30, 50 more are supplied, giving a projected inventory of 100 (c). A BM demand scheduled for October 1 depletes inventory to 65 (d) and another for April 30 projects a negative inventory of 35 (e). A delivery of 100 scheduled in May returns the quantity on hand to 65 (f). PMMS will flag the situation and point out the need to advance the May delivery date. A July demand for 165 leaves the inventory at minus 100 (g).

If all of the expected demands and deliveries in the example had been input as BMs and outstanding procurement records, PMMS would flag the negative inventory situations at (e) and (g) calling attention to the need to advance the May 15 delivery date and to begin some procurement action before July 15 to forestall the 100 shortage at (g).

When the material is received at the project it is checked against the requisition and the receipt is input to PMMS under its stock code. The system will increase inventory and decrease outstanding procurement. It will also calculate a new average unit price for the current inventory now in the warehouse. Next, a priced material receipt is produced for the Accounting Department, giving quantity and type of material received, pricing data, and the *inventory* account number to which the material is charged. A foreman who wishes to draw material from the warehouse uses a terminal near his work location.

Table 4-34

Inventory status

Situation	Date	B of M Issue	Demand Quty	Demand Cumul	Supplied Quty	Supplied Cumul	Projected Inventory
(a) Existing	now	various	600	600		700	100
(b) Extra BM Demands	to date	Extra	50	650		700	50
(c) 50 more delivered	30 Sept			650	50	750	100
(d) Demand for 35	01 Oct	at hand	35	685		750	65
(e) Demand for 100	30 Apr	at hand	100	785		750	-35
(f) 100 more delivered	15 May			785	100	850	65
(g) Demand for 165	15 July	at hand	165	950		850	-100

This will show the demand for each item and also how much has been issued against each project unit (if the project is broken down into units). If the foreman wants to confirm that his materials are in stock, he can obtain the current inventory status for any stock code on his bill.

THE PROJECT MATERIAL MANAGEMENT PLAN

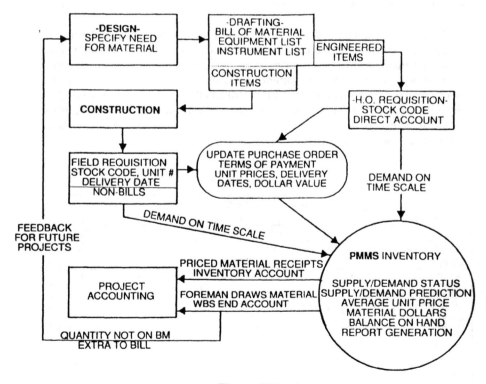

Figure 4-94
Material flow diagram.

A clerk then inputs the quantities needed plus the required delivery time and location. If material is needed that is not on the Bill, this can also be input. It registers against the Bill and will be displayed as "field added" items. If he has asked for more material that was on the original Bill, this will show later as "Extra to BM". These features give feedback to Engineering about actual quantities used, so that design records and estimates can be updated. The issuance transaction distributes the cost of the material to the end account, thereby updating the cost incurred for the benefit of the work breakdown structure.

Furthermore, the issue transaction generates the data needed to produce a "picking-packing" slip at the warehouse. PMMS has all storage locations for material on file, and so the picking part of the slip has the items arranged in order of storage in the warehouse. A packing slip will be delivered to the foreman with the material. The system produces material receipts, deals with sales taxes and trade discounts, distributes the cost of receipts to end accounts and computes total inventory value or the value of a particular group of inventory items. Input of receipts and issues to PMMS enables management to achieve the inventory control objectives to control the material cost of the project.

The diagram (Figure 4-94) depicts the major information flow as it pertains to a typical inventory control system:

4.6.3 Work Equipment

The work equipment used for heavy construction forms a major part of a job. Equipment used for highways, pipe lines, dams etc. can cost more than one third of the total construction cost. For buildings, a contractor may expend approximately 10 to 15% on work equipment.

When selecting the needed equipment, a contractor must make the decision to buy or to lease the equipment.

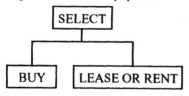

Buying the equipment gives the owner the security that it is available when needed. Leasing is merely another method of financing. Renting is used for short periods and intermittently usually for seldom needed equipment. There may be scheduling problems with renting.

This must be taken into consideration as a risk factor. The decision to buy or rent is very much dependent on the frequency of usage. Compound interest calculations are used to determine the cost when purchasing or renting equipment (see 1.3.1). When money is borrowed to purchase equipment, the interest rate is usually known. If the company pays cash, the "opportunity" interest rate must be used in calculations. That means evaluation is based on the interest obtainable when investing the cash instead of using it.

Jelen (5) suggests to use continuous compound interest to estimate rates. When a present value (P) is converted to a uniform flow of money (F), then the value of the flow is

$$F = \frac{1 - e^{-in}}{in}$$

and for the capital recovery amount (R'), which Jelen calls "unaflow"

$$R' = P/nF$$

Q #4.03: A scraper costs $ 100 000 to be paid for in cash. Interest of 10% is assumed over a 15 year service life.

A #4.03:

$$in = 1.5 \; ; \; e^{-in} = 0.22313 \; ; \; F = 0.5179 \; ; \; R' = 100\,000/15 \times 0.5179 = \$\,12\,872$$

or a monthly cost of $12\,872/12 = \$\,1\,073$

Q #4.04: A rental company charges $ 3000/month for a similar scraper. The contractor considers renting the equipment for eight months instead of buying it. He calculates that the average maintenance cost over a service life of 15 years will have a present value of $ 30 000 if he buys the scraper.

A #4.04:

$$1 - e^{-in} = 0.77687 \; ; \; F = 0.5179 \; ; \; R' = 130\,000/15 \times 0.5179 = \$\,16\,734$$

compared with $ 24 000 for the rented equipment.

Q #4.05: Assuming the contractor had a choice to buy the $ 100 000 scraper with a service life of 15 years or another scraper costing $ 120 000 with a service life of 20 years, both at an estimated 10% annual interest. For comparison, we should use capitalized cost, whereby

$$K = P/(1 - e^{-in}) \; \text{ for n years duration}$$

A #4.05: Calculation for K_{15}, when in = 1.5

$$1 - e^{-in} = 0.77687 \; ; \; K_{15} = \$\,128\,722$$

Calculation for K_{20}: when in = 2.0

$$1 - e^{-in} = 0.86466 \; ; \; K_{20} = \$\,138\,782$$

The 15 year purchase is more economical.

Many other factors have to be taken into consideration. Different operating cost are incurred for rental vs. leased equipment. Salvage value has to be estimated on purchased equipment, aging or obsolescence due to technical advancemet, depreciation etc. need to be considered.

Replacements

The maintenance and repair costs increase with age and the cost for new modern equipment may also increase. The contractor must devise a strategy to forecast a replacement of the existing equipment. This strategy plays an important part in cost control. Contractors that look at replacement problems at the time they occur are taking a great risk. Evaluating the old existing equipment is done periodically, usually before the end of each year. Past performance will assist in estimating future cost which will determine when a replacement should take place (5).

What is involved is a procedure that accepts the conditions of the present situation and allows the contractor to determine the correct sequence of decisions from this point on. That means *decisions are made in stages* (end of the year). Let

$X_{i,j}$ = Decision associated with stage i,j
$Y_{i,j}$ = Variables associated with stage i,j
$f_{i,j}$ = Value of decision criteria at stage i,j
$C_{i,j}$ = Cost associated with $f X_j Y_i$

$$.....Y_{i-1} \longrightarrow \boxed{i} \longrightarrow Y_i \to .... \quad Y_{j-1} \longrightarrow \boxed{j} \longrightarrow Y_j \longrightarrow$$

with X_i, Z_i on stage i and X_j, Z_j on stage j.

The output Y_j at the final stage j is related to the input Y_{j-1} and the decision variable X_j. The cost function Z_j is also dependent on Y_{j-1} and X_j.

The decision problem is set up to establish a recursion relationship, i.e. we are linking the optimal decision for the last stage to the optimal decisions for the prior stages. Dynamic recursion for Z is

$$Z_i \text{(minimum)} = C_{ij} - f_j$$

Example: A machine has a service life of 15 years. For the sake of simplification let us assume that the total cost, including purchasing, repairs and maintenance, and salvage is estimated over the last four years:

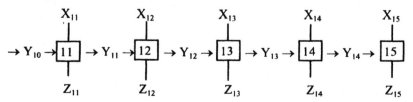

The relevant total cost were calculated and are shown in k$ below.

$$C_{11,12} = 16 \quad C_{12,13} = 16 \quad C_{13,14} = 10$$
$$C_{11,13} = 30 \quad C_{12,14} = 30$$
$$C_{11,14} = 44$$

We now solve for Z_j and search for the minimum amount. We set all the cost for Z_{15} = zero, then, at the end of the year 13

$$Z_{14} = C_{13,14} + C_{15,15} = 10 + 0 = 10$$

At the end of year 12 we have two alternatives, either buy a new machine and replace at the end of year 13, costing

$$C_{12,13} + C_{13,14} = 16 + 10 = 26$$

or we could keep the machine to the end of the 13th year, costing $C_{12,14} = 30$.

At the end of the 11th year we have the following alternatives:

$$C_{11,14} = 44$$
$$C_{11,13} + C_{13,14} = 30 + 10 = 40$$
$$C_{11,12} + C_{12,14} = 16 + 30 = 46$$
$$C_{11,12} + C_{12,13} + C_{13,14} = 16 + 16 + 10 = 42$$

Replacement at the end of the 13th year is the best alternative.

For larger problems with many stages, we should note, that the matrix for C_{ij} grows in a triangular shape:

	1	2	3	4	5	6
6	1,6	2,6	3,6	4,6	5,6	0
5	1,5	2,5	3,5	4,5		
4	1,4	2,4	3,4			
3	1,3	2,3				
2	1,2					
1	0					

For a six-year duration i goes from 1 to 5. Therefore, the number of elements are

$$(1+i)(i/2) = 6 \times 2.5 = 15$$

Instead of dynamic programming, there are other methods to optimize equipment replacement cost. The decision tree can be used in conjunction with capitalized and "unacost" calculations.

Q #4.06: A contractor's truck starts to have rust problems. Because of this, the salvage value declines sharply; $ 10 000 now, $ 2500 next year and $ 1000 the year after. Operating and repair cost are not expected to increase very much, for the next two years an estimated $ 3500 and $ 4000. A new truck will cost $ 38 000, expected to be in operation for 10 years with little salvage value.

Operating expenses are $ 2000/year. Assuming 10% interest, should the contractor replace the truck now?

A #4.06: After *one* year, the capital cost of the truck is

$$10\ 000 - \left(\frac{3500}{1.10} \times \frac{1.10}{1.10 - 1} \right) = \$\ 75\ 000$$

Operating cost are $\$\ 3500/1.10 \times 11 = 35\ 000$; for a total of **$\ 110\ 000**
For *two* years

$$10\ 000 - \left(\frac{3500}{1.10^2} \times \frac{1.10^2}{1.10^2 - 1} \right) = 7107 \times 5.76 = 40\ 950$$

$$3500/1.1 \times 5.76 = 18\ 327$$
$$4000/1.1^2 \times 5.76 = \underline{19\ 041}$$
$$\text{Total} \qquad \$\ 78\ 318$$

Capital cost of the *new* truck:

$$38\ 000 \times \frac{1.10^{10}}{1.10^{10} - 1} = 38\ 000 \times \frac{2.59}{1.59} = 61\ 845 \text{ plus}$$

$$38\ 000 \times \frac{1.10^{10} - 1}{1.10} = 38\ 000 \times 15.93 = 31\ 875$$

$$\text{Total} \qquad \$\ 93\ 720$$

Should the truck be replaced?
It would appear that the purchase of a new truck is more economical after the first year; but keeping the old truck for another two years will be more economical than buying a new one (78 k$ vs. 94 k$).

Leasing

A few words should be said about leasing. No general rule can be given as to advantages or disadvantages of leasing vs. debt financing. When capital is "freed" it must be effectively used for more cost effective purposes in order to make a long term lease economical. Similar to short term renting, the leased equipment must earn its keep. Long term leases usually draw a heavy penalty if canceled. Another important consideration is the tax treatment of lease vs. ownership (depreciation).

When evaluating whether to lease or buy equipment, we must also be aware that investment is a function of time (cash flow). For example, if a computer has a technological life of four years and cost $ 10 000 with a 15% discounted cash flow rate of return, we may estimate that the share of net income will differ depending on reinvestment funding.

If we assume the following cash flow

Year	1	2	3	4	
Estimate A	7000	3000	2000	576	(accelerated reduction)
Estimate B	2000	3000	5000	4731	(retarded reduction),

then the discounted cash flow (DCF) is (see Section 2.1.3 Project Evaluation):

Year	0	1	2	3	4
Estimate A	10 000	−4500	−2175	−501	±0
Estimate B	10 000	−9500	−7925	−4114	±0

The differing book value obviously has an effect on the depreciation and thereby the tax treatment of the purchased computer. For ease of calculation, the example below assumes an equal annual amount.

Example: The $ 10 000 computer can be disposed of for $ 2000 at the end of the fourth year. It can be maintained for $ 1000 per year and the benefit of net-working with others is estimated to be $ 2000 per year. The company uses "sum-of -digits depreciation. Its present-worth factor is 8.3013 E-01 (5). If the interest rate is 10% p.a. after a 40% tax, what is the equivalent uniform end-of-year pur-chasing cost (unacost) of the computer? For "unacost" see Section 1.2.1 - Simple Interest Calculations.

$$10\ 000(1 - 0.40 \times 0.83013) \times 0.31547 = \$ 2107$$

The disposal income is

$$- 2000 \frac{1 - 0.40}{(1.10)^4} \times 0.31547 = - \$ 258$$

Maintenance and benefit:

$$(1000 - 2000)(1 - 0.40) = \underline{- \$ 600}$$
$$\text{Total cost} = \$ 1249$$

To *rent* an equivalent computer would cost $3000 per year payable at the begin-ning of the year. The lease cab be renewed each year at the option of the lessee.

This option has the advantage of having the choice to buy an up-to-date new advanced computer. This is estimated to be presently worth $ 6000. Unacost for the rented free service computer is

$$300 \left(1 - \frac{0.40}{1.10} \right) \times 1.10 = \$ 2100$$

The advantage of delaying the purchase together with the tax benefit is

$$- 6000 (1 - 0.40)(0.31547) = - \underline{\$ 1136}$$
$$\text{Total cost} = \$ 964$$

It is therefore more economical to rent.

CHAPTER 5
COST REPORTING

The objectives of this chapter are:

1. To demonstrate the importance of having an expert Cost Management System installed as a requirement for timely reporting that is conducive to action.
2. To establish a reporting structure that avoids overlap and duplication through the use of the work breakdown structure.
3. To identify misleading information that may unintentionally enter the system and give warning.

It is the purpose of reporting to communicate. This communication must be true and unbiased. The collected facts must be summarized properly in a vertical and lateral network to inform the decision maker.

Introduction

It ain't what you don't know that makes you look like a fool,
it's what you do know that ain't so.

Old Appalachian Proverb

The report is the knowledge module on which decisions are based. Cost reporting is the output of a Project Costing System. The reports reflect the adequacy of the system. They are the "messengers" of information generated by the data that was input into the system. It is the *quality* of the processes that convert the data into usable reports.

The chapter shows how reports relate to a control sequence. The cybernetic aspect of graphic presentation and the type of charts and graphs are shown in this chapter including the rules that help to identify graphic distortions.

It is the purpose of reporting to communicate. This communication must be true and unbiased. The collected facts must be summarized properly in a vertical *and* lateral net work to inform the decision maker.

According to the dictionary, to report is "...to give formal account or statement of..., to inform...". This has the connotation that reports are written to inform a superior authority. While this is true for a major part of any reporting system, information is also reported (transmitted) laterally.

The choice of the reporting structure should be based on an optimal information flow within the type of organization and mode of operation. A network type reporting and even free linkages can be of great advantage, especially in a construction environment.

Network Structure   **Tree Structure**

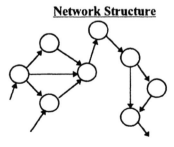

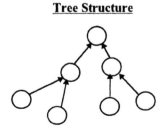

PROJECT TEAM   FUNCTIONAL HIERARCHY

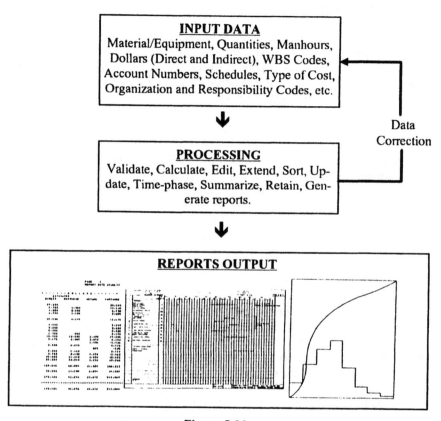

Figure 5-01
Report processing.

The success of any cost control system depends on accurate monitoring of the various cost related information, its reporting to the individuals responsible for those areas within the organization and for a critical analysis of performance. A proper reporting stream allows the tracing of reasons for variances by management to the source of the variance from the plan and its cause.

The type and frequency of reports depends very much on the nature of project, the organization (owner or contractor) and style of management. Weekly or even daily reports may be needed for the construction manager and his/her staff installing a large pipeline, whereby monthly reports are sufficient for the owner contracting out work for a power development.

The project report is the result of processed input data (Figure 5-01).

5.1 DOCUMENTATION

Accurate information must be used in a timely manner to produce reports that are easily comprehended and are conducive to action (see Section 4.2.2 - Cost Information Flow). Project Cost Reports must contain incurred cost and should not be confused with financial reports (see Section 4.1.1 - Cost vs. Cash).

There are three basic types of reports:

1. TABULAR,   2. DESCRIPTIVE,   3. GRAPHIC.

The *tabular* form is the most popular because it is the easiest to produce. Most information are generated by computer processing. By means of selecting, summarizing and formatting data, any desired combination of reports can be tabulated. Because it is easy to produce, the amount of information presented must not be overwhelming and should be should be quite selective.

The most important aspect of a tabular report (or any report for that matter) is its validity. Software packages often "standardize" rows and columns. It is possible for a user with little understanding of cost engineering to produce impressive professional looking project reports that result in more questions than answers. Low-cost software can do more damage than good because some software designers lack basic project management expertise (41).

Descriptive cost reporting is preferred when various explanations have to be given. Here again, the content of descriptive reports must be to the point and objective. Descriptive reporting can also be verbal. Narratives often accompany tabular reports to explain cost variances. They are also used to explain trends.

The *graphic* form is the most effective but the least understood format. Charts can often be misleading. This subject will be dealt with in more detail later.

A good management reporting system would make use of all three types of reports. The mix depends on the level of management within the organization. The higher the level, the more graphic the presentation can be.

The work breakdown structure lends itself to a hierarchy of reports. Thus performance information is available to *all* holders of a work package at a lower level adding responsible participation and commitment to all levels of project management during the stages of a project life cycle. It provides for a concept of cost summarization without sacrificing lateral information exchange. This needs a sophisticated information integration where reports do not necessarily look alike but are designed to meet the need of the user.

The software that is required must allow users to customize their own reports, which means selecting, sorting and presenting the information needed at the next level (Figure 5-02). This does not mean that management receives reports with different *formats* from the various groups. On the contrary. Own formats are used for *internal* reports where flexibility is desirable. *(Major electric utilities in Canada are using OPEN PLAN to meet those requirements).*

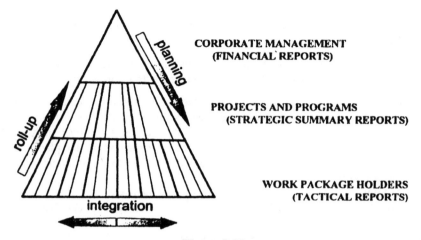

CORPORATE MANAGEMENT
(FINANCIAL REPORTS)

PROJECTS AND PROGRAMS
(STRATEGIC SUMMARY REPORTS)

WORK PACKAGE HOLDERS
(TACTICAL REPORTS)

Figure 5-02
Integrated reporting.

Integration means that engineering, scheduling, procurement and construction are not separate functional entities operating in isolation. They all are working within an integrated cost management and reporting system.

Control Sequence Documentation

In reference to the elements of Cost Control (see 4.2.3 - The Control Cycle), this is how reporting fits into the control sequence:

Control Sequence	Documentation
State Objective	Project Requirements
Define Scope	Project Specification
Develop a Plan	Logic Diagram, Charts
	Cost Flow Estimate
Measure Performance	Progress Report
	Budget Comparison
Analyze Variance	Narrative Trend Report
	Recommendations for Improvement
Assess Impact	Cost Analysis Report
Forecast Trend	Earned Value Report
Issue Status Report	Management Reports
Take Action	Instruction Notices,
	Production studies

Because of the accelerated capabilities of computers, reports can be so diverse, that it is not prudent to show specific examples of all the reports above. Project reporting is usually achieved through menu-driven report generator systems. This allows for standard and modified formats, which depend on the user's requirement for information.

Because of extended data base capabilities, information can now be accessed and reports generated with greater variety.

It means for example, that at any level of the organization the demand / supply situation at any point in time could be accessed, or a report on outstanding purchase orders be issued. Reports must contain the answers for critical analysis of performance in time to take action. Any significant variance from the plan is analyzed by those responsible and the reason given in a *Cost Analysis Report*.

For example:
$50 000 -   Design Change: Pump capacity increased.
$10 000 -   Constructability: Installed foundation bolts for clarifier tank.
                                         (Wrong size and location.)

It is important that the explanation for variances be specific. It would be wrong to say "Cost went up by $60 000 because of changes made."

The above indicates, that incurred cost went up unexpectedly. A proper report however, should have foreseen early enough that the pump capacity had to be increased. Alternative courses of action should have been recommended at that time. The Project Manager may then have been able to affect other changes to reduce total project cost.

There is such a large variety of reports, most of them are not standard. Only *two* examples are shown below:

Integrated Cost/Schedule Reporting, sometimes called Earned Value Reporting is an effective tool in trending (see 4.4.4 -Cost/Schedule Integration). The report can be done in tabular form or as a summary in graphic form (Figure 5-03).

Performance Measurement and Analysis Reports can be depicted in a combination of tabular and graphic form (Figure 5-04).

In addition, Project Completion Reports (close-out reports) are essential as an information source for future projects or for bidding on another similar project. The owner would record the performance of suppliers and contractors. Work Item productivity data and delivery performance would be another item in the completion report. The assembly of this report can begin when the project is about 75% complete. It is not advisable to wait until the very end when all staff has been transferred to other jobs or left the company.

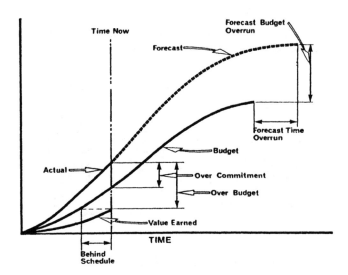

Figure 5-03
Performance graph.

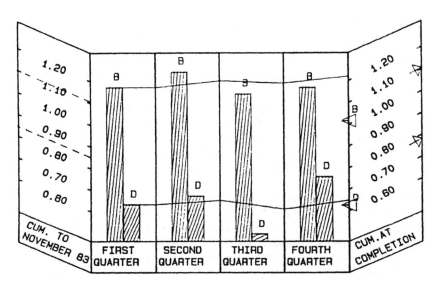

Figure 5-04
Labor performance index.

5.2 CHARTS AND GRAPHS

To translate the old saying: "A picture is worth a thousand words" into cost reporting, one could say that a graph is worth a thousand figures.

A well designed chart can be comprehended more quickly than abstract verbal or
numerical explanations. A graph can convey *effectively* the intended message in
an instant (25). When presentations are made in a descriptive manner at meetings
or during the presentation of cost status to management, it may take the audience
a long time to "get the picture." So why not present a graph in the first place?

The Cybernetic Aspect of Charting

Usually, the types of graphs and charts seen in literature are of relatively simple
nature and are displayed with clarity. This apparent simplicity is often deceiving.
We may know all about charts in theory, yet we are very often careless about
them (18).

In an age of persuasive advertising, image creation and a desire to be enter-
tained, it is easy to produce graphs for presentations that can gratify an audience.
Technology gives almost anybody the unprecedented power to alter or recompose
any picture at will. *(A photograph of people can be scanned by an image digit-
izer, retouched and edited to show different heads on different people or give
them pointed ears.)*

Graphs can now be produced that impress because of their good looks, having
a great impact on an audience. This may hide the fact that the message is wrong or
misleading. It is very important that graphs be produced to reflect reality con-
veyed in a manner readily understood by the reader.

The *flow chart* (Figure 5-05) shows the steps to be followed in preparing a
chart or graph. All efforts benefit from careful review and re-design. Quite often, a
critical examination of the visual presentation reveals new insights, aspects and
overtones of meaning that may, at first, not have been apparent.

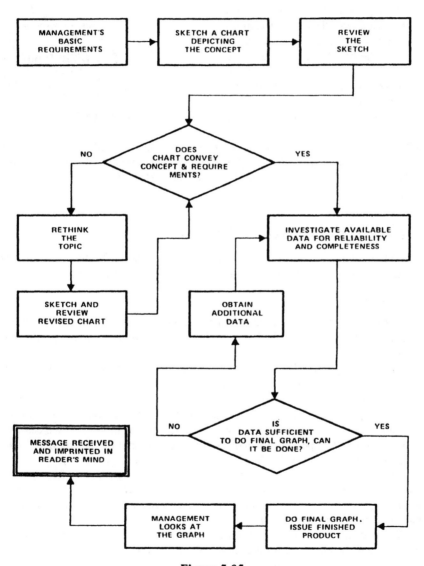

Figure 5-05
Cybernetic aspect of charting.

When all factors are considered, it will be found that the sheer mechanism of the process from conception development to the published product leaves one with many choices and possibilities of presentation.

A graph must be of standard size with a standard title block. Headings and foot notes should be provided to define, clarify and qualify data on which the graph is based. It is a common error not to provide enough explanation. Some of the basis for data may be obvious at the time but may not be remembered at a later date.

When a graph is manually produced, small grid paper should only be used for *preparation,* not for *presentation.* Lines should be **heavy** and obvious for visual effectiveness. The exception is background lines. They should be light and not necessarily continuous.

5.2.1 Types of Charts and Graphs

The following pages contain illustrations from a publication "Better Cost Reports through Graphic Presentation", permission to reprint by Ontario Hydro (18).

Bar Charts

The most elementary chart is the one-dimensional *bar chart.* Numerical relationship is easy to judge when the bars start from a common base, starting with zero (Figure 5-06).

The bar graph also lends itself well to negative and positive quantities and component classifications (Figure 5-07).

When percentages are used, the *component bar* chart is helpful This is essentially an apportionment of common totals. Budgets are often presented this way. This chart gives the user a clear picture of expenditure relationships. We can easily see how vacation periods affect department labor output or that computer costs are apparently not affected by the amount of direct work (Figure 5-08).

When actual values instead of ratios are used, the interpretation of component bar charts can be difficult because each separate bar is *added* to the next, thereby imprinting a picture of *relative* rather than *actual* sizes in the mind, i.e. comparing parts to their total (Figure 5-09).

When there are only *two* categories, it may be better to use *compound bar charts.* Here, the emphasis is on comparison of the parts with one another (Figure 5-10). The "overcrowded" look should be avoided.

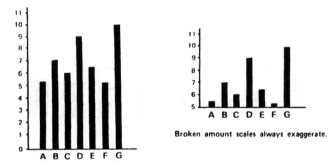

Figure 5-06
One-dimensional bar chart.

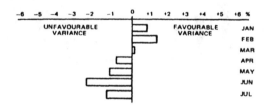

Figure 5-07
Plus/minus bar chart.

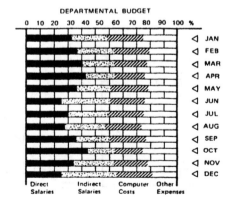

Figure 5-08
Horizontal component bar chart.

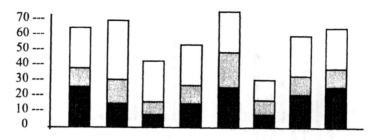

Figure 5-09
Vertical component bar chart.

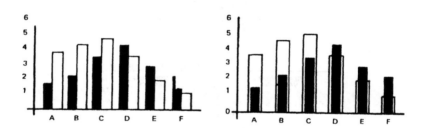

Figure 5-10
Compound bar chart.

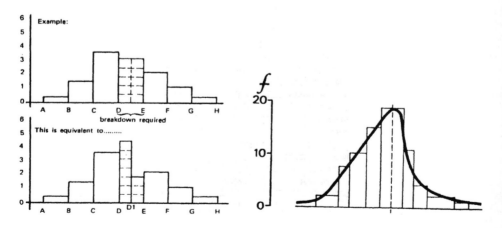

Figure 5-11
Histogram and frequency diagram.

Histograms

The histogram looks similar to the column chart but represents a frequency distribution. The histogram is an *area chart*. Its bars are joined. When class limits are changed, the *area* must be adjusted accordingly. It is desirable, that all classes in a frequency distribution be made the same width. In some cases, however, the classes have to be unequal to bring out a smooth and clear distribution curve (Figure 5-11).

Circle Chart

The circle chart is sometimes called the *pie chart*. It is used to cut totals into parts and is very effective with appropriate data because it highlights the relative magnitude of the various *portions of a fixed total* (100%). Even the simple production of a pie chart may cause problems. It is easier to "scissors cut" a circle and then present the assembled parts, but it is much neater to maintain the smoothness of the circle circumference (Figure 5-12).

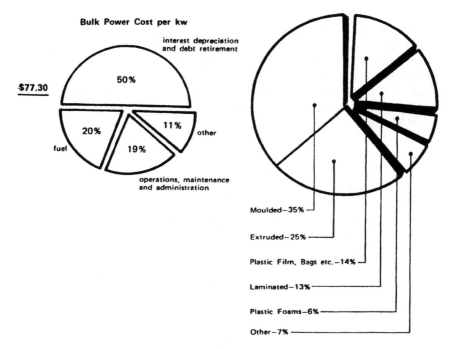

Figure 5-12
Percentage circle chart.

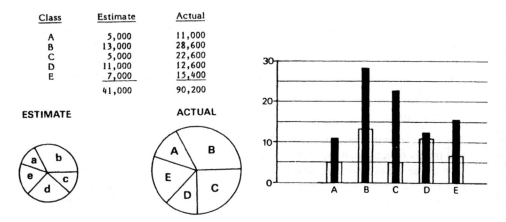

Class	Estimate	Actual
A	5,000	11,000
B	13,000	28,600
C	5,000	22,600
D	11,000	12,600
E	7,000	15,400
	41,000	90,200

ESTIMATE   **ACTUAL**

Figure 5-13
Pie chart - absolute data.

Pie charts must only be used with *relative* data, absolute pie charts must *never* be used. If we consider to use two or more circles to represent differing totals of *absolute* data, the areas must be made proportional to the data. But area judgments are difficult to make and we would be wise to avoid these charts. In the example (Figure 5-13), it is obviously difficult to tell that the bigger pie is more than twice as large in area as the smaller one. A better choice would have been to use a compound bar chart:

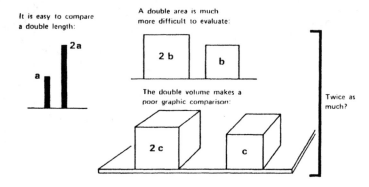

Figure 5-14
Area and volume charts.

Area and Volume Charts

These should strictly be avoided if at all possible. Few persons can judge areas, and hardly anyone can judge volumes with any degree of accuracy when displayed on a two-dimensional plane (Figure 5-14).

Line Charts

Line charts have the greatest application in cost reporting. They are two-dimensional in the sense that the viewer has to consider *two dimensions simultaneously*. This does not mean that areas are being judged. Areas are basically the *product* of two dimensions. In line charts, we are talking about the *location* of curves identifiable by two dimensions.

Lines are plotted on a vertical (Y) scale, the ordinate, and on a horizontal (X) scale, the abscissa. When the base of the line chart represents a time scale, while the vertical scale is *not* a frequency, we are dealing with a *time series*. Several lines may appear on the same chart *only* if there are useful and logical relationships between the various data which the lines represent. The most common are *arithmetical* line charts, where equal distances on the vertical and horizontal scale represent equal amounts (Figure 5-15).

When the equal distances on a line represent equal *ratios* or rates of change, those are *logarithmic* line charts (Figure 4-16). Logarithmic line charts are used when the line that is displayed is an exponential curve of the form

$$Y = a+bX+cX^2$$

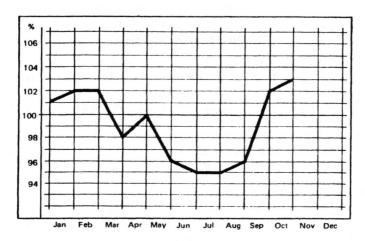

Figure 5-15
Arithmetical line chart.

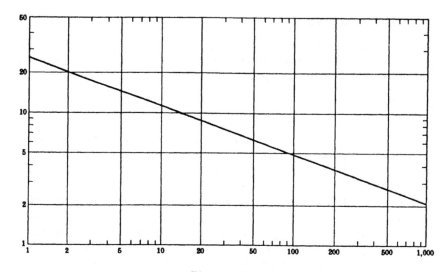

Figure 5-16
Logarithmic line chart.

Generally, the slopes of lines on an arithmetical chart depict the *absolute* amount of change per unit, not the "rate of change" per unit. Any change in the *unit* of an arithmetic scale changes the *slope* of lines charted to that scale!

It is therefore quite illogical to chart two or more differing categories of data in the same arithmetic line chart. *Absolute* comparison is meaningless!

The scale on a graph should be such that the slope of the curve(s) gives a good indication of fluctuations and trends. As a rule of thumb, the ideal curve occupies most of the space allotted.

In the example (Figure 5-17), the vertical scale in (a) should be reduced by one-tenth and in (b) increased by ten.

One can often see graphs that show heavy background lines (Figure 5-18 and 5-19). This should definitely be avoided! Any graph that is dependent on accurate scale reading should have figures in addition to thin lines.

The Ratio Chart

When we apply a percentage scale to the ordinate, the slope on the curve then indicates the *absolute amount of the percentage change* of the time ratio. Often, the base line is not zero, but the 100 percent line (Figure 5-19).

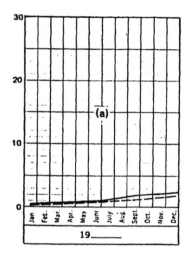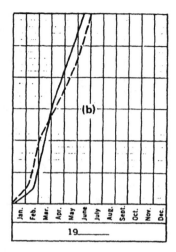

Figure 5-17
Scale on a graph.

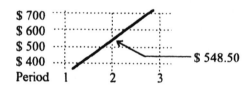

Figure 5-18
Background on graphs.

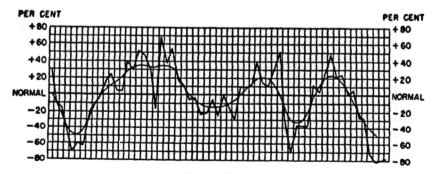

Figure 5-19
Percentage ratio chart.

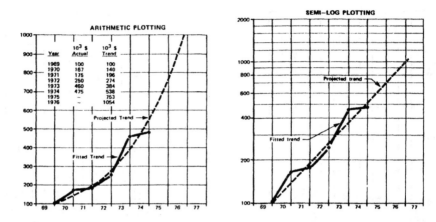

Figure 5-20
Logarithmic ratio chart.

In case of a growth curve, it is difficult to observe a trend when the graph is curvilinear. Graph paper with a grid, scaled in logarithmic increments is usually the best medium. Equal percentage increments plot in equal scale steps. Here, it is simpler to make a linear extension (Figure 5-20).

Miscellaneous Charts and Graphs

There is an abundance of miscellaneous graphic presentation to choose from. Some of the basic categories are:
- ♦ Flow Charts
- ♦ Calculation Charts
- ♦ Picture Graphs
- ♦ Step Series Diagrams

Flow Charts: A flow chart is a visual aid and a medium to clarify ideas. It usually does not represent numerical data, but may be useful in developing logical procedures, processes or concepts through a series of stages. The most common flow charts are:
- ♦ Process Chart
- ♦ Concept Chart
- ♦ Decision Flow Chart
- ♦ Multiple Activity Chart
- ♦ Gantt Chart
- ♦ Logic Diagram
- ♦ Procedure Flow Chart
- ♦ Information Flow Chart   etc.

The *Process Chart* is a pictorial representation of the activities of a process in which the symbols shown are used to represent standard activities The symbols are placed one below the other, in activity or event sequence. They may be described or numbered. They may consist of a main trunk, branches or loops.

Symbols may also be combined. (See also Applied Problem Solving - 7.2 for an example process chart). The main application is work study and the purpose is productivity improvement.

There is not always a need to have scale readings to convey a message. The *Concept Chart* is a good example for this. All it needs to indicate is the interrelationship between two curves. The chart here shows the diminishing influence of cost control on levels of capital expenditure during the planning and construction stages of a project (Figure 5-21).

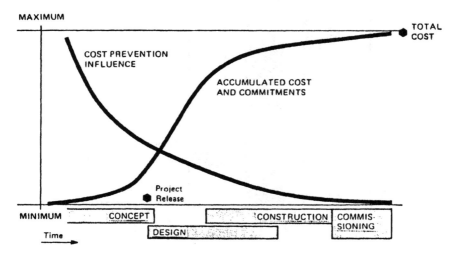

Figure 5-21
Concept chart.

The *Decision Flow Chart* is often used for the design of computer programs. The symbols used pertain to data processing systems. The use of templates make it easier and at the same time standardize the symbols. This type of decision flow chart is seldom used for reporting purposes. Most decision flow charts are prepared for procedures and instructions. An example of this is the "Cybernetic Aspect of Charting" under Figure 5-05.

When many activities are recorded against a common time scale, a *Multiple Activity Chart* is used. Since this chart presents a chronological record of several activities occurring simultaneously, the correct sequence and duration of each must be known (Figure 5-22). At the development stage, the chart can be an invaluable tool in the planning of team jobs. It can be used to compare alternative staffing schedules and to develop optimum utilization of men and machines. For an example of a similar chart, called the *Multiple Event Chart* see Chapter 7.2. - Applied Problem Solving.

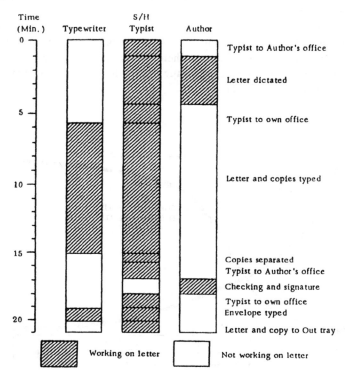

Figure 5-22
Multiple activity chart.

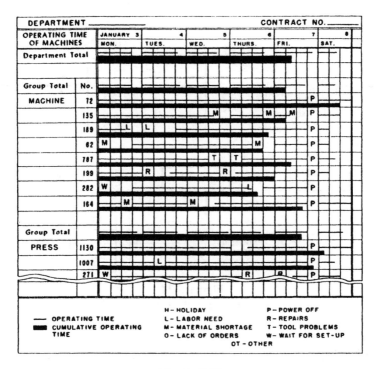

Figure 5-23
Gantt Chart

Well known is the *Gantt* Chart. This is a scheduling tool, where activities are shown on a time scale (Figure 5-23).

A *Logic Diagram* adds interdependencies and durations to activity flows in sequence and/or in parallel. This is also called the *Network Diagram*. For examples see 3.3.4 - Network Planning

The *Organization Chart* is well known and need not be shown here.

Procedure Flow Charts can take many forms. They can be drawn as shown in Figure 5-24.

Procedure charting is not standardized and a variety of symbols are used in practice. The chart in chapter 4 (figure 4-12) just outlines broadly how engineering expenditures get into a certain computer output system:

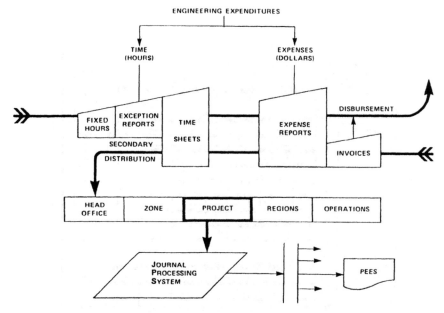

Figure 5-24
Procedure flow chart.

Document: GPP-663-10						
Date: 30 March 19….	Section: CONTRACT ADMINISTRATION - INSTALLATION					
Subject: EQUIPMENT ACCEPTANCE CONDITIONS						

○ Operation ▽ File ◄— Transmission			Manager Constr	Manager Engrg	Comm Supt	As Listed
□ Inspection ✱ Approve ◇ Decision						
ERF and list of reservations reviewed and corrective action taken when necessary....................................			▽ □			
The corrections are noted on the ERF...........................			○			
Copies of the approved ERF are issued to:						
Manager of Engineering Commissioning Superintendent			▽	▽	▽	
PART II - EQUIPMENT CLEARANCE AND TAKE-OVER CERTIFICATE						
1. COMMISSIONING SCHEDULE						
Commissioning schedules are prepared and copies issued to:					▽ ○	
Thermal Operations Manager Station Manager Manager of Engineering Manager of Construction			□	□		□

Figure 5-25
Interfunction chart.

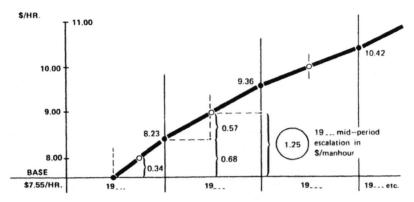

Year	½19...	19...	19...	19...	19...	19...
Escalation Index	1.09	1.24	1.38	1.48	1.58	1.61
Escalated $/hour	8.23	9.36	10.42	11.17	11.63	12.16
Increase per period	0.68	1.81	2.87	3.62	4.08	4.61
Mid-period escalation	0.34	1.25	2.34	3.25	3.85	4.35
Total hours / period	683	1188	1071	991	779	505
$ due to escalation	232	1485	2506	3221	2999	2197

Total expenditure due to escalation = $ 12 640

Figure 5-26
Calculation chart.

When there is more than one sequence in the process, the *Interfunction Chart* (Figure 5-25) is spread across several columns representing the functions concerned. This is a good way to present organization *and* procedure:

When it is advantageous to include calculations in basic graphic presentation, use can be made of *Calculation Charts* (Figure 5-26). These permit a ready overview without having to turn from graph to text and/or table.

A useful chart is the *Nomograph* (Figure 5-27). It is extensively used in engineering. When established for a certain purpose, the nomograph can provide approximate results that otherwise would take some effort to calculate. It is more and more being replaced now by mathematical computer models (what if). The nomograph was presented at the 1971 AACE annual meeting by representatives of the General Electric Company in a paper "Normalization of Power Plant Costs".

With the improvement of photographic techniques and desk top publishing and the increased importance of commercial communication, pictorial presentations are growing universally more popular. They can add greatly to the interest of what otherwise might be a dull subject. *Picture Graphs* (Figure 5-28) are eye catchers and make reading more interesting. Instant symbols are available through commercial firms.

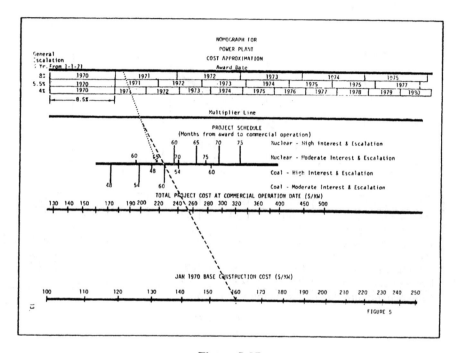

Figure 5-27
Nomograph.

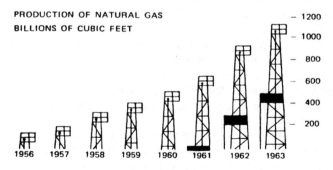

Figure 5-28
Picture graphs.

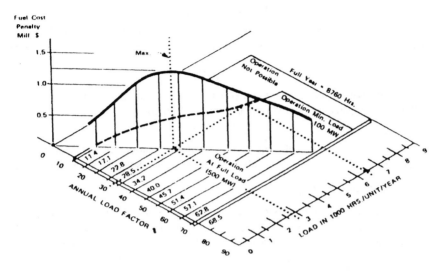

Figure 5-29
Step series diagram.

Background pictures can be produced photographically or by computer, forming the background of a chart.

In some cases, cartoons maybe used to liven up the image and, what is more important, leave a more lasting impression on the viewer.

When an idea is too complex to be grasped from a single chart, it is often desirable to design a series of step-by-step built-ups. This is called a *Step Series Diagram* (Figure 5-29).

5.2.2 Peak Reduction

When greatly scattered data is plotted, it is generally hard to read. For reporting purposes, it is a definite advantage to remove extreme peaks without affecting the arithmetic mean. A mathematical method of doing it was explained previously (see 1.1.2 - The Estimating Line). As shown in the example, this estimating line can somewhat be improved by the use of second or higher degree equations.
The equation for a second degree polynomial is

$$Y = a + bX + cX^2$$

From this, three simultaneous equations can be obtained:

$$\Sigma Y = Na + c\Sigma X^2$$
$$\Sigma XY = b\Sigma X^2$$

$$\Sigma X^2 Y = a\Sigma X^2 + c\Sigma X^4$$

from which a, b, and c can be found. The previous straight estimating line will now be a curve. When a constant percentage change is expected, the best curve is the exponential curve:

$$Y = AB^x$$

Here, Y changes at the constant rate B per unit change in x. Taking logarithm on both sides and substituting a = logA and b = logB, then

$$\log Y = a + bX$$

which indicates that a straight line will be fitted when plotted on semi-logarithmic paper. There are *other methods* available to reduce the peaks of a scatter diagram.

Exponential Smoothing

Business often uses trend-based forecasting models that have a long historical trend. The least square approach does not work in all cases even though the exponential regression curve gives a fairly good representation of widely scattered data. Regression looks at input data as having equal importance. The latest data is usually of more importance for trending than long gone past history. For data that is not scattered too much, exponential smoothing can be used.

Single exponential smoothing assigns weights to the most recent observation by assigning a smoothing constant "a". This constant declines as we move farther into the past. Assuming we have the following scattered data:

X	3	5	10	14	16	20
Y	15	12	6	5	4	4

Using single exponential smoothing, the equation we use to forecast scattered data is

$$F_{t+1} = aY_t + (1 - a)F_t \qquad\qquad\qquad (1)$$

where F_{t+1} is the forecast for period t+1 and Y_t is the actual value for time t.

The smoothing constant "a" depends on the importance we assign to the present data as compared to past history. The closer "a" is to zero, the less influence it has on current observations. Let us use t = 20 and a = 0.3 (30% of the latest observed value Y_t and 70% of prior average F_t-1). Since we have no actual data for the starting point, we will set $F_3 = Y3$ to start the curve. The curve is now fitted as follows:

$$
\begin{aligned}
F_{t+1} = \quad F_5 &= \quad (0.3 \times 15) + (0.7 \times 15) &= 15.0 \\
F_{10} &= (0.3 \times 12) + (0.7 \times 15) &= 14.1 \\
F_{14} &= (0.3 \times 6) + (0.7 \times 14.1) &= 11.7 \\
F_{16} &= (0.3 \times 5) + (0.7 \times 11.7) &= 8.2 \\
F_{20} &= (0.3 \times 4) + (0.7 \times 8.2) &= 5.7 \\
F_{21} &= (0.3 \times 4) + (0.7 \times 5.7) &= 4.0
\end{aligned}
$$

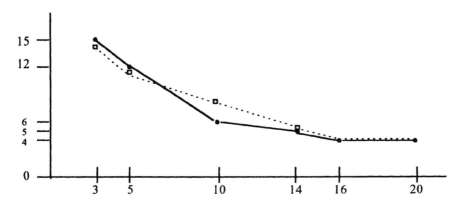

Figure 5-30
Exponential smoothing.

The forecast seems to level out at 4.0. This type of presentation is far better than a scatter diagram even though the curve is not very smooth. Some visual fitting may be acceptable. A decreasing trend has the tendency to exceed the actual values. If the trend were increasing, the tendency is to lower the actual values.

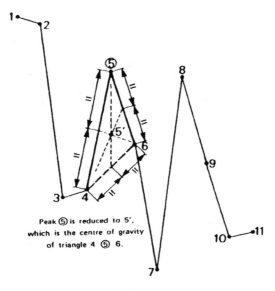

Figure 5-31
Graphic peak reduction

The single exponential smoothing can be somewhat improved by introducing another constant "b", which will flatten the trend in the previous calculation. It is called *Double Exponential Smoothing*. The same input is used as before. The new forecast is

$$F_{t+1} = S_t + T_t \qquad\qquad ....(2)$$

where

$$S_t = aY_t + (1 - a)(S_{t-1} + T_{t-1}) \qquad ....(3)$$

and

$$T_t = b(S_t - S_{t-1}) + (1 - b)(T_{t-1}) \qquad ....(4)$$

The Graphic Method

Here is another method available to graphically reduce the peak of a scatter diagram (Figure 5-31).

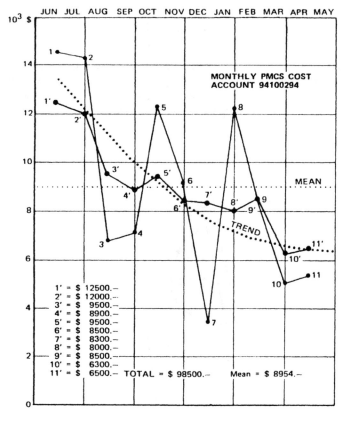

Figure 5-32
Graphic peak reduction example.

It was used by draftsmen in the old times. It is doubtful that anybody would use that method any more now, but the knowledge how this used to be done does not hurt. The computer can now calculate the peak reduction points by using the co-ordinates of the scattered data and then plot the results.

The center of gravity of a triangle is used to reduce peaks (Figure 5-31). On the first diagram, peak 5 is reduced to peak 5' by bisecting the sides of the triangle 4 5 6 and drawing lines to opposing peaks. Where those lines intersect, is the center of gravity of 4 5 6 To reduce peak 6 the triangle 5 6 7 is used the same way. Point 4 is reduced by using triangle 3 4 5. The end points 1 and 11 are reduced by projecting one additional point outward at the intersection of the arithmetic mean. See example Figure 5.32

Rolling Averages

The method of rolling averages involves little computation and is often sufficiently accurate for most practical applications.

Table 5-01

Rolling average cost data

PMCS* COMPUTER COST

June 70 to April 71

Item No.	Month		Monthly $
1	Jun	70	14,516.00
2	Jul	70	14,348.00
3	Aug	70	6,785.00
4	Sep	70	7,062.00
5	Oct	70	12,316.00
6	Nov	70	9,148.00
7	Dec	70	3,498.00
8	Jan	71	12,139.00
9	Feb	71	8,304.00
10	Mar	71	5,065.00
11	Apr	71	5,300.00
		Sum	98,481.00
Arithmetic Mean			8,952.84

*PMCS = Project Material Control System

An *uneven* number of periods are chosen and the arithmetic mean of their values is plotted at the median point. This is then repeated by moving one period forward at a time.

Example for monthly computer cost data (Table 5-01). The total amount over 11 months is $ 98 481 for an average of $8 953/month (mean). This average now becomes both the start and finish point of the three-month rolling average calculation (Table 5-02).

Table 5-02

Rolling average calculation

Item	Month		Dollars/Month	Average Dollars/Week	Adjusted Dollars/Month
1	Jun	70	(14,516 – 5,000)	2,379	10,333
2	Jul	70	14,348	2,870	12,460
3	Aug	70	6,785	1,696	7,364
4	Sep	70	7,062	1,766	7,667
5	Oct	70	12,316	2,463	10,693
6	Nov	70	9,148	2,287	9,930
7	Dec	70	(3,498 + 5,000)	2,125	9,226
8	Jan	71	12,139	2,428	10,542
9	Feb	71	8,304	2,076	9,013
10	Mar	71	5,065	1,266	5,497
11	Apr	71	5,300	1,325	5,754
TOTAL			98,481	22,680	98,479

Table 5-03

Rolling average - adjusted data

```
8,953 (mean)
14,516    14,516
14,348    14,348    14,348
37,817    6,785     6,785     6,785
  3      35,649     7,062     7,062     7,062
  =        3       28,195    12,316    12,316    12,316
12,606     =         3       26,163     9,148     9,148     9,148
          11,883     =         3       28,526     3,498     3,498     3,498
                   9,398      =         3       24,962    12,139    12,139
                             8,721      =         3       24,785     8,304
                                       9,509      =         3       23,941
12,139                                           8,321      =         3
 8,304     8,304                                           8,262      =
 5,065     5,065     5,065                                          7,980
25,508     5,300     5,300
  3       18,669    8,953 (mean)
  =         3       19,318    For a Total of    $ 97,845.
 8,503      =         3       And an Average of  $ 8,895.
           6,223      =
                    6,439
```

The result is quite close to the graphical method. It will be noticed, that the total for the rolling averages (\$ 97 845) is fairly close to the true total (\$ 98 481).

Had we used a five-month rolling average, the result would have been

10 711, 10 333, 11 005, 9932, 7762, 8833, 9081, 7631, 6861, 7952, 7315
Total = 97 416 ; Mean = **\$ 8856**

It should also be noted, that extreme values will influence the arithmetic mean and thereby the slope of the estimating line. Before any data are taken at face value they should first be analyzed carefully.

Monthly values include both, four-week billings and five-week billings. Establishing an average week over the 11 months period and an error adjustment that had to be made in April, an average dollar/week was calculated. In addition, an error of \$ 5 000 was made in the first month and credit for this error given in the following December billing. Those adjustments are made ahead of applying new rolling averages. The adjusted data now reads as listed in Table 5-03.

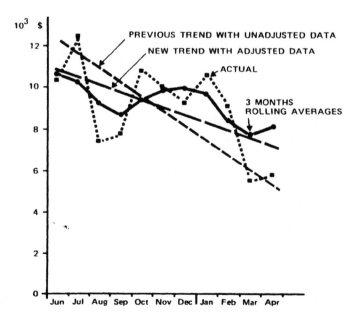

Figure 5-33
Rolling averages -trend graph.

Arithmetic moving averages used for trend lines can have disadvantages because they can remove or reverse some periodic variations or produce cycles that are not in the data. Actual data moves down from October to December in the graph (Figure 5-33) while the three months rolling average shows an upward trend. Even though the total picture seems quite adequate, the validity of presentation could be questioned. Assuming we have the following cost flow

Jan.	Feb.	Mar.	Apr.	May	Jun.
200	400	800	500	100	400 = 2400/6 = 400 mean

The three-month rolling averages are

$$333, \ 467, \ 567, \ 467, \ 333, \ \text{Average} = 458.$$

That means the forecast for July is 333.

The estimating line will give us a forecast of 380 for July with a downward trend. The weighted moving averages over three months are

$$
\begin{array}{lll}
\underline{Y_{t-1}} & \underline{Y_t} & \underline{Y_{t+1}} \\
400(1/6) + 200(2/6) + & 400(3/6) = & 333 \\
200(1/6) + 400(2/6) + & 800(3/6) = & 567 \\
400(1/6) + 800(2/6) + & 500(3/6) = & 583 \\
800(1/6) + 500(2/6) + & 100(3/6) = & 350 \\
500(1/6) + 100(2/6) + & 400(3/6) = & 417 \\
100(1/6) + 400(2/6) + & 400(3/6) = & \mathbf{350} \quad \text{July Forecast}
\end{array}
$$

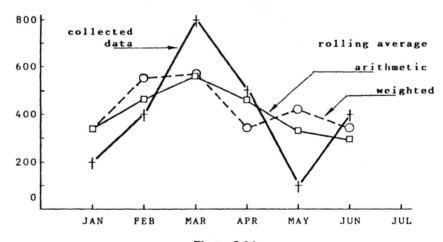

Figure 5-34
Weighted moving averages.

5.2.3 Graphic Distortions

A good costing system must display and highlight problems in the right perspective so that they can be recognized and corrected with good judgment. Distortions of true data will cause management to draw improper conclusions which, in turn, can result in incorrect decision making.

Distortions and illusions in graphic presentations have many causes:

- ♦ Thoughtlessness
- ♦ Lack of knowledge
- ♦ Artfully "shading" the truth
- ♦ Cleverly planned distortions
- ♦ Outright cheating

There is very little reason to believe that any cost engineer or those reporting professionally to management would fall into the last three categories. Those three occur, however, quite frequently in magazines and articles, and even government reports that are directed toward the unsuspecting public.

There are five basic rules that apply to recognize and avoid the production of misleading graphs:

RULE #1

The time sequence of a time series diagram must be *equally spaced*.

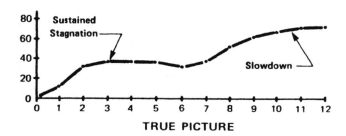

TRUE PICTURE

Figure 5-35
Distortions rule #1 - true picture.

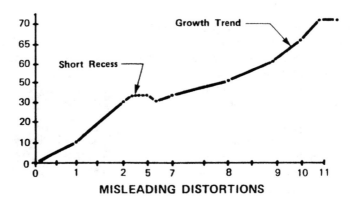

Figure 5-36
Distortions rule #1 - misleading picture.

Any change in scale in the time series will affect the slope of the curve. When the true display shows a sustained stagnation (Figure 5-35), the distortion indicates a recess (Figure 5-36). A slowdown is distorted into a growth trend, which is just the opposite from the truth.

' It is quite common to see a single line chart in which the horizontal scale unit for a time series has been enlarged for the last part of the series, such as changing from annual to quarterly data. This gives a much gentler slope for the quarterly data.

Month	A	B	Total
Jan	6	1	7
Feb	6	1	7
Mar	1	1	2
Apr	5	2	7
May	6	0	6
Jun	6	1	7

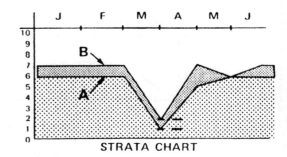

Figure 5-37
Distortions rule #2 - strata chart.

RULE #2

When extreme fluctuations are shown on a cumulative strata chart, the picture can give misleading impressions.

In Figure 5-37 "B" is constant for Jan., Feb., and March, and doubles in April. The shaded line, however, appears *smaller* during March and April:

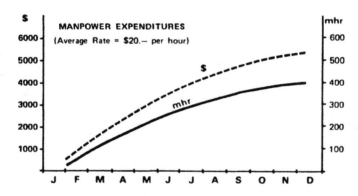

Figure 5-38
Distortions rule #3 - invalid slope.

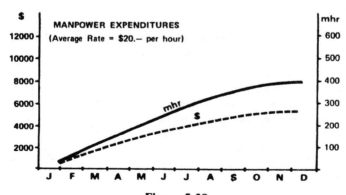

Figure 5-39
Distortions rule #3 - valid slope.

RULE #3

The slopes of lines plotted to different arithmetic scales cannot be compared with validity unless the ratio of the scales is truly related.

When workhour and dollar expenditures are combined on one sheet, the scales cannot be arbitrarily chosen just for the graph to "look good" (Figure 5-38).

In our example, it gives the wrong expression, that labor cost have increased, when in fact they have decreased. The dollar scale should be 20 times the Whr. scale, i.e. 600 Whrs. × \$20/hr = \$ 12 000. This results in an acceptable relationship (Figure 5-39).

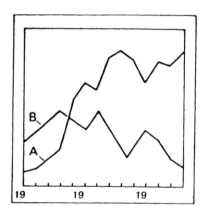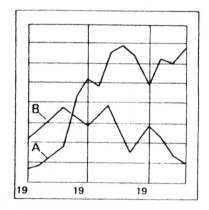

Figure 5-40
Optical illusion.

RULE #4

Optical illusions can occur, intentionally or unintentionally, when preparing charts.

When comparing the length of the vertical lines a_1 and a_2, to the right, it appears that a_2 is longer than a_1. This is an optical illusion. When we compare the variance between curves A and B for the years 1955 and 1960 by just looking at the first chart, we probably assume, that the variance in 1960 is larger (Figure 5-41).

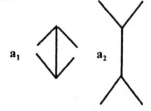

Figure 5-41
Optical comparison

To overcome this optical illusion, it is good practice to show guiding lines.

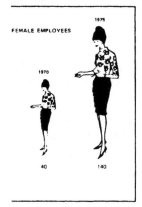 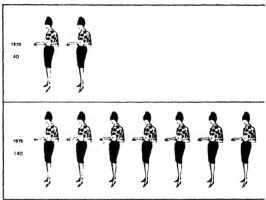

Figure 5-42
Compatible quantities.

RULE #5

When comparing alternatives use constant dollars and compatible ratios or quantities.

This is the proverbial comparison between apples and oranges.

In addition to the above, it should be mentioned, that errors are often made when displaying picture graphs (Figure 5-42). Each symbol must be treated as *a unit of measure* and occupy the same space so that larger numbers are indicated by *more* identical picture symbols, not by a *larger* picture. As mentioned previously under the heading "area and volume charts," it is difficult to compare the magnitude of areas.

5.3.1 Examples

The rules above have been devised to display and highlight information in the right perspective, unbiased and contributive to proper interpretation. The common expression "buyer beware" also applies to graphs:

"Observer beware!"

A survey was made once to find out if people would line up at the pump selling gasoline for 59.[99] cents/liter or would use the pump selling for 60 cents/liter. The vast majority opted for the first pump. The bill is the same in either case up to 50 liters. The *visual* effect made people decide spontaneously. The large 59 is not considered "shading the truth." We have become accustomed (conditioned?) to this deliberate creation of an image. To many of us it has become "good salesmanship," not misleading advertising.

It is therefore of utmost importance to be able to analyze the graphic pictures we see based on elementary rules of graphic presentation.

This holds true whether a graph is produced by a draftsperson or a computer. We just have to look at financial pages or advertisements to see plenty of violations to those rules.

Let us analyze a government publication printed several years ago (Une nouvelle direction pour le Canada - 1984). The graphs that support many arguments look good (Figures 5-43 to 5-50). No wonder, they were produced by *professionals*. There is absolutely no doubt that the sources used to produce the graphs supplied the proper statistics. Could it be, that they were accidentally presented in a wrong way or designed to sell the program proposed by the Government to the voter? We will never know.

Let us look at the first chart (Figure 5-43). The first impression one has is to see a frightening increase of the net public debt. The shaded area between 1982 and 1984 appears smaller than that between 1979 and 1982 and then it really widens very much after 1984. Almost like a huge mouth opening up wide into a terrible future. What could the message be? (We have inherited a huge debt load from the previous Government?) There are two deceptions here and three basic rules are violated:

Rule #2: Severe slopes on a cumulative strata chart.

Rule #4: Optical illusion.

Rule #5: Use constant dollars for comparison

By plotting the Net Public Debt Charges as a time series instead of a strata chart, indicating that the dollars are *escalated* dollars, one gets an entirely different picture (Figure 5-44). The time series shows a typical, less dramatic growth curve of approximately 15%. The forecast is a simple projection of the present trend.

The Figure 5-45 chart violates rule #1. The slope of a curve is a function of the *two* scales, horizontal vs. vertical. Equally wide columns are shown for six-year, five-year and one-year durations. This has a tendency to accentuate later percentage increases relative to the earlier part of the graph. Growth is emphasized by varying the time scale. Figure 5-46 shows adjusted data. The message? (Our exports are on a steep increase when in fact the rate is pretty well maintained.)

Furthermore, assuming a continuous annual time scale, it would have been better, not to separate the columns and show their true widths (Figure 5-46).

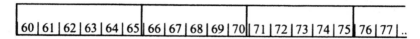

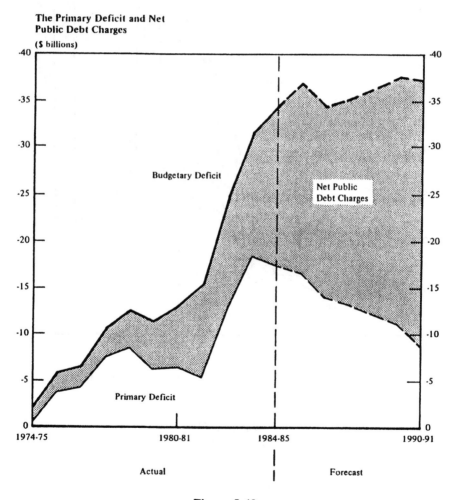

Figure 5-43
Public debt charges "A".

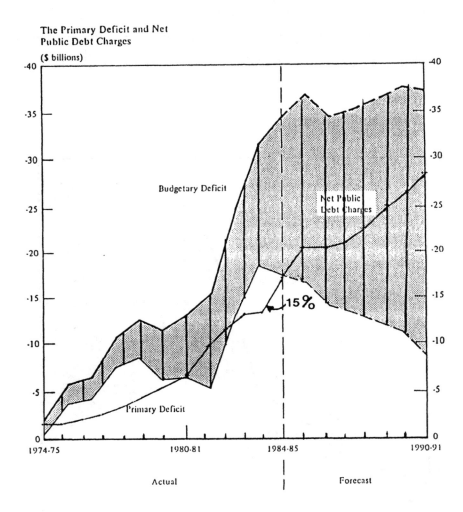

Figure 5-44
Public debt charges "B".

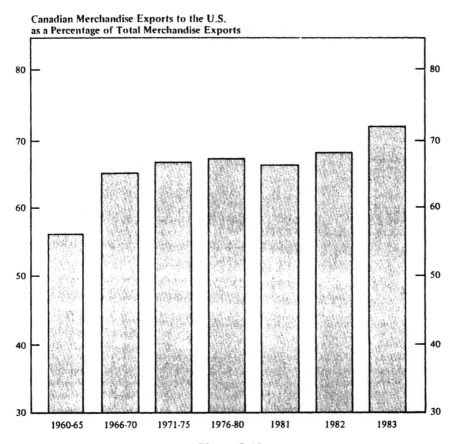

Canadian Merchandise Exports to the U.S.
as a Percentage of Total Merchandise Exports

Figure 5-45
Merchandise exports "A".

Another graph (Figure 5-47) shows the increase in elderly benefits and child benefits paid by the Canadian Government during the period 1979 to 1985.

This shows the money *spent* on the *program* during the period *not* benefits *received* per Capita. Furthermore, it does not factor in the benefit *loss* due to inflation. When including these factors, the *true* benefits (buying power) have almost stayed the same for the elderly and have actually gone down from 2.8 in 1979 to 2.6 in regard to child benefits. Even that figure is not very meaningful unless we know the *number* of children or elderly receiving the benefits.

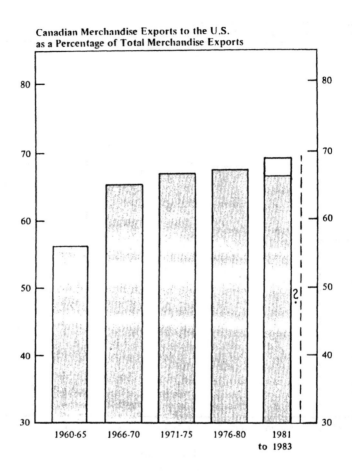

Figure 5-46
Merchandise exports "B".

The graph of Figure 5-47 violates rule #5. Benefits per capita such as M$/100 000 elderly, where *constant dollars* are related to quantities would be the only way to find out if they are better off. The Consumer Price Index was used as a deflator (Figure 5-48). This is the information we are actually after.

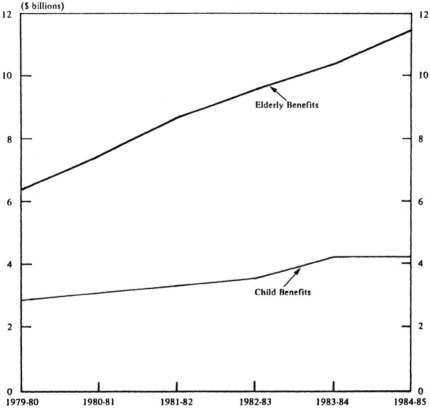

Elderly Benefits

Include Old Age Security (after tax), Guaranteed Income Supplement, Spouses Allowance and Foregone Tax Revenue from the age exemption and the pension income deduction.

Child Benefits

Include Family Allowance (after tax), Child Tax Credit and Foregone Tax Revenue from the Child Tax Exemption for children under 18. For the credit it is assumed that in respect of taxation year 1978, payments were made in 1979/80 and so on.

1984/85 values are estimates.

Figure 5-47
Elderly benefits "A".

Figure 5-47 shows the increase in elderly and child benefits paid by the Canadian Government from 1979 to 1985. It shows the money *spent* on the program during the period, **not** benefits *received* per capita. Furthermore, it does not factor in the benefit *loss* due to inflation. When including those factors, the *true* benefits (buying power) have stayed almost the same for the elderly and have actually gone down, from 2.8 in 1979 to 2.6, in regard to child benefits.

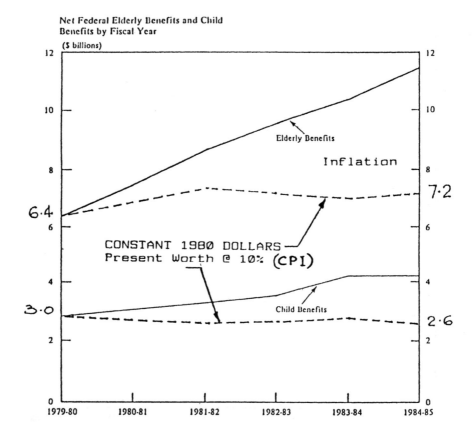

Figure 5-48
Elderly benefits "B".

This graph violates rule #5. Benefits per capita such as M$ per 100 000 elderly, where *constant dollars* are related to quantities would be the only way to find out if they are better off.(Figure 5-48).

Figure 5-49 gives the impression that there is a rapid downward trend. What is the message here? (We can not continue subsidizing increasing unemployment insurance cost. We are moving drastically into an ever increasing deficit situation!)

If we just change the word "cumulative" into "annual," we can see that the surplus situation shows some hopeful signs during 1983 and 1984 (Figure 5-50).

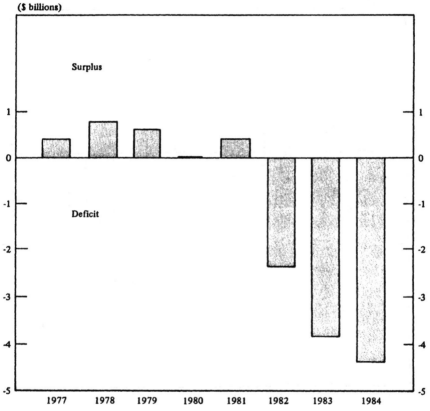

Figure 5-49
Unemployment insurance "A".

It is the intention here to demonstrate the pitfalls of graphic presentations, *not* to discredit the originator of these graphs. When the member of parlament was contacted to point out the discrepancies, the explanation was:

...quite interesting...thought-provoking...

...there is much room for difference of professional judg(e)ment on how best to present graphical information.

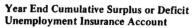

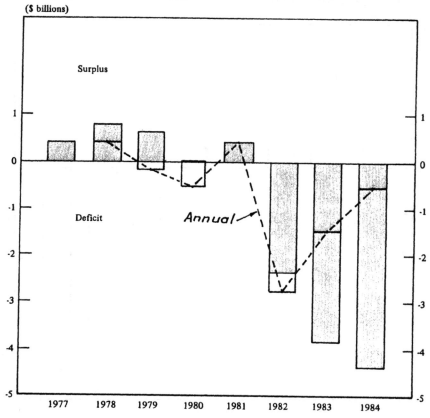

Figure 5-50
Unemployment Insurance "B"

Q #5.01:
What principles for reporting should a corporation adopt?

A #5.01:
♦ Source data elements should be standardized for all projects.
♦ Index coding should be consistent for all cost documents, both for what is done and for who is responsible for doing it and who does it.
♦ Summary report content and format should be common.

Q #5.02:
Should all project reports be standardized?

A #5.02:
Internal to the project the report content and format should not be in conflict with Management Summary Cost Reports. However, Analysis Reports can be locally developed as needed to communicate laterally among contributers to the project. It is important that those who report follow the rules of timely, reliable and accurate reporting. Those reports are input, but not part of the hierarchy of formal cost reports and must be identified as such.

Q #5.03:
There are five basic rules that apply to recognize and avoid the production of misleading graphs. Is there anything misleading about the graph below (Figure 5-51) if the hourly rate is $ 26.67?

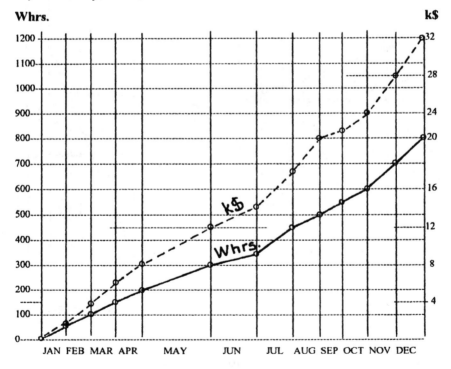

Figure 5-51
Graphic rules example.

A #5-03:

Rule #1 is violated. The time sequence must be equally spaced. May, June and July are stretched, thereby reducing the slope of the curve. This results in an unrealistic view of a "quiet" three-month period.

Rule #3 is also violated. The ratio of scales is related, but the k$ line is plotted using a different ratio. It appears that the workhour costs are going up, when in fact they are constant. The graph should be shown with one line only.

Q #5-04:

What is misleading in the picture puzzle below ?

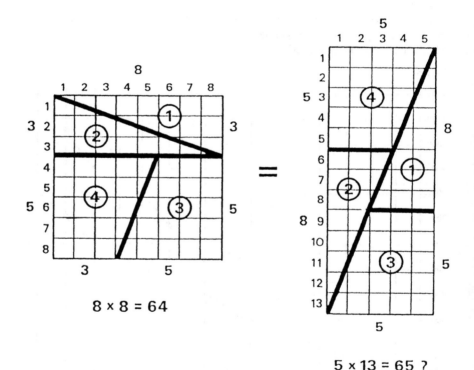

$$8 \times 8 = 64$$

$$5 \times 13 = 65 \ ?$$

A #5-04:

The answer is hidden behind those heavy lines. Calculate the proportion, using the triangle ① and ③ .How long is the dividing line? $X = (5 \times 8)/13 = 3.08$, not 3.

CHAPTER 6
PRODUCTIVITY

The objectives of this chapter are:

1. to define the terms "productivity" and "production".
2. to describe performance standards and adjustments to a uniform base.
3. to identify factors that affect productivity.
4. to describe a system to measure and improve performance.

The term productivity has different meanings for different individuals. Most people associate productivity with hard work. An economist measures productivity as a percentage yield on capital investment. The Department of Commerce measures dollars of output over workhours of input. Dollars per kilowatt is another measure used by industry. Whatever definitions we use, it is the objective to improve total performance and to become more effective within the competitive world economy.

Introduction

The harder I work, the behinder I get.

Unknown

Productivity is directly related to competition and living standards. This chapter deals with the identification and measurement of factors that affect productivity, the optimization of human resources and the outline of a system that measures performance against productivity standards and the resulting reports.

6.1 DEFINING PRODUCTIVITY AND PRODUCTION

Many of us who have walked a construction site felt an active air of efficiency. The trades looked busy and everything seemed to harmonize.

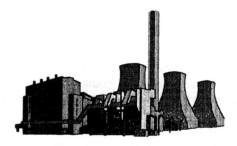

On another site, we may have encountered the opposite. Groups of workers stood around or were in each other's way. Tools and materials were scattered all over the place.

We have no doubt in our mind, which site is more productive. It is a *subjective observation,* however, the apparent efficiency was not defined or measured. We assumed, for example, that all direct work done was necessary. An example of a very efficient and highly productive operation was the construction of the Manufacturer's Life Building in downtown Toronto several years ago. The deep excavation took in a whole city block bordered on two sides by the busiest streets in Toronto. The only access to the site was from one of the side roads. There was very little room for material storage and equipment movement. Yet, in spite of all the complicated logistics, the construction activity went on like clockwork. The proper amount of material arrived at the right time keeping the trades constantly occupied.

We all know, that the beautiful execution of the work is not just accidental. Thoughtful and thorough preparation for all phases of design, procurement and construction and the coordination between various disciplines *prior* and during the installation of permanent facilities is essential for success.

Even a highly motivated tradesman will not be able to contribute to increased productivity if materials do not arrive on time or a poor layout of facilities causes delays in construction activities. This, of course holds true also in manufacturing, office work and other areas of utilization of human resources. In order to plan, measure and control operational efficiency we should first define the terms

Productivity and Production.

6.1.1 Production

Production is making things. All construction estimates are based on production.

Production is the quantities installed over a fixed time period.

The expression "500 cars are produced daily at plant C" is a typical expression for reporting production. It is the *output* of a production process.

The measurement for production gives no indication how *efficient* the job of producing 500 cars was performed. If the corporate requirement was 500 cars/day, then the production manager reached the goal.

The basic system model (Section 4.2.1) can be used to demonstrate the *Production Flow Process* (Figure 6-01).

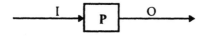

Figure 6-01
Production - single input.

where I = Input, P = Production Process, O = Production Output
If there is a multiple input that depends on each other and a mutual production process, we call it an interdependent system (Figure 6-02)

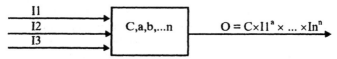

Figure 6-02
Interdependent production.

where C is a constant that, together with "a...n" will indicate increasing, decreasing or constant return to scale (amount of output per unit I of input). In construction, several inputs may be independent of each other. In this case,

$$O = K1 \times I1 + K2 \times I2 + ...+ Kn \times In + C \times (I1)^a \times (I2)^b \times .... \times (In)^n$$

The constants Kn are independent of each other.

When there are several production processes with *inputs in series*, then the input into the next process is the output of the previous process (figure 6-03). Because of the interdependency, if either of the production processes P break down, the output O = 0. A typical example is the assembly line.

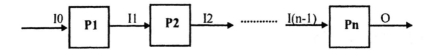

Figure 6-03
In-series production.

A totally independent input production process can be shown as in Figure 6-04.

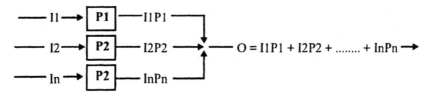

Figure 6-04
Totally independent production.

P represents the transformation of inputs into outputs. The production function shows that the total product of an input is transformed into the total output of the system.

Without going into many details and complicated mathematics, let us assume that there is an *interdependent* construction function of a welding activity. The welder is dependent on materials and the materials cannot be installed without a welder. We have a simple situation of two interdependent inputs I1 and I2. If we now estimate the output for various realistic combinations of inputs (fictitious data in Table 6-01).There will be several combinations resulting in the same interval of output, e.g. 300 to 399. All outputs within equal production intervals can be tabulated in the form of an array (Section 1.2.0). This can be visualized by equal production curves (Figure 6-05):

Table 6-01

Equal production curves

Input I1	Input I2	Output O
0	0	0
1	1	30
2	1	55
1	2	50
.....	.....	....
20	10	300
15	15	310
18	16	300
22	12	320
28	15	390
.....	.....	....
50	40	1000 etc.

We can now tabulate the outputs for various inputs I1 for a fixed input I2. If we fix I2 = 30 for example we will be able to calculate both, the average production process and the marginal process as shown in Table 6-02 (not real data, just for demonstration). Since this is an interdependent function, the following formula applies.

$$O = C \times (I1)^a \times (I2)^b \times \ldots \times (In)^n$$

To determine the constant C, we look at $(I1)^a$ and $(I2)^b$, which in our example are

$$(I1)^a = 1 \text{ and } (I2)^b = 1.$$

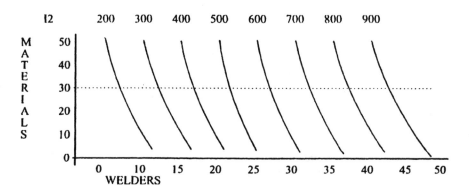

Figure 6-05

Equal production curve.

Table 6-02
Average and marginal production

I1	O	Average P	Marginal P
0	0	0/0	-
1	10	10/1 = 10	10 - 0 = 10
2	30	30/2 = 15	30 - 10 = 20
3	50	50/3 = 17	50 - 30 = 20
4	60	60/4 = 15	60 - 50 = 10
....	...	....	....
48	850	850/48 = 18	....
49	900	900/49 = 18	900 - 850 = 50
50	950	950/50 = 19	950 - 900 = 50

OUTPUT

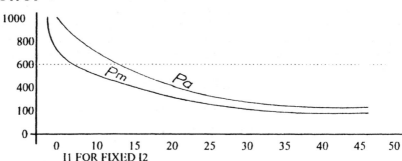

Figure 6-06
Average and marginal production.

No matter what the values of a or b are,

$(I1)^a = 1$ and $(I2)^b = 1$, and from $O = C \times (I1)^a \times (I2)^b = C(1)(1) = 30$

The model in our case is mathematically presented as

$$O_{30} = 30 \times (I1)^a \times (I2)^b$$

The values of a and b are found mathematically through substitution with any combination of inputs and corresponding outputs. This gives us a function of average and marginal product (Figure 6-06).

Looking at the cost of various inputs and determining minimum cost ratios, i.e. the dollar amount of one input in relation to any other, we obtain the optimum welder/material combination for any required output.

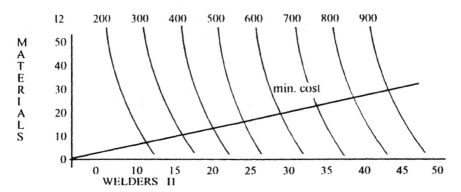

Figure 6-07
Minimum production cost.

With approximately parallel lines for equal production curves (Figure 6-05), the minimum cost ratio is a straight line (Figure 6-07).

The above is only a generalized version of production criteria. It is mentioned only to explain the difference between the terms production (quantitative) and productivity (qualitative).

> ***Production* rates are used to measure *progress*.**
> ***Productivity* rates are used to measure *performance*.**

6.1.2 Productivity

Productivity has many definitions in literature.

The Oxford Dictionary:	Capacity to produce, effectiveness of productive effort.
Collins Dictionary:	Fertile, creative, efficient (productive).
Reader's Digest:	Efficiency in industrial production.
AACE Terminology:	Relative measure of labor efficiency, either good or bad, when compared to an established base or norm as determined from an area of great experience. Alternatively, productivity is defined as the reciprocal of the labor factor.
PMI Terminology:	The measurement of labor efficiency when compared to an established base. It is also used to measure equipment effectiveness, drawing productivity, etc.
Samuelson's Economics:	The net productivity of a capital project is that annual percentage yield which you could earn by tying up your money in it (real interest yield).

Obviously, there is not a clear consensus of what productivity is. The past success of the American industry has been based on products.

Productivity emphasized how best to produce those products and how efficient tasks are performed. It is not only the efficiency or the measure of efficiency of production but also the relation to rational business goals.

Quality and standardization may be more important than efficiency of output to stay competitive worldwide.

In a broader sense it means the best use of any resources - human, capital, material, land - with due regard to the environment. It means productivity of *all* factors of production, leading to long term viability and competitiveness of the business.

Total factor productivity has multiple inputs. The productivity can be likened to an onion with many skins. The core may have strict single input labor units. The second layer is the immediate environment in which that labor is performed. This could be the equipment used, the local organization unit that produces and converts the goods and services. The third layer is the next level environment, which may be the corporate business. There could be many layers before we come to the total factor productivity.

As an example, a typist is capable of typing a record number of words per minute. We can call this "very high labor productivity". The typewriter is an old-fashioned impact machine. This is the second layer of productivity. Unfortunately, the "tool productivity" (equipment) is very low. The letter that was dictated by the manager will have little impact on additional sales or increased production, therefore resulting in a medium "supervisory productivity", and so on and so on. Any one of those separate "onion skins" are connected and have actually a real time simultaneity rather than sequential stages. Alvin Toffler (44) in his book *Powershift* writes:

"We are, in fact, discovering that "production" neither begins nor ends in the factory. Thus, the latest models of economic production extend the process both upstream and downstream - forward into aftercare or "support" for the product even after it is sold, as in auto repair warrantees or the support expected from the retailer when a person buys a computer. Before long, the conception of production will reach even beyond that to ecologically safe disposal of the product after use.. Similarly, they may extend the definition backward to include such functions as training of the employee, provision of day care, and other services... Hence, productivity begins even before the worker arrives at the office."

From this follows, that the best method to increase productivity is teamwork, involving all employees in problem solving and quality control.

Even though we should be aware of this "total factor" productivity, for the purpose of "project cost management" in the broad sense and "productivity and cost control" in a narrower sense, the emphasis will be on labor productivity, mainly in construction. Those "partial" productivity measures are important because labor constitutes a large part of the cost of construction.

6.1.3 Work Items

If we want to monitor and control the actual installation of equipment and materials at the construction site in detail, it requires an itemized breakdown of the work performed and quantities installed at the *commodity* level. This includes the installation of pipe, the pouring of concrete, the length of cable pulled or the number of light fixtures installed. Those are *work items*. Each work item is coded and subdivided into sub-items that are also appropriately coded. With quantities estimated and measured against those work item codes, we are able to obtain a unit rate of performance. By applying workhours to quantities and/or dollars to both, we can measure productivity and compare trade performance with historical standard cost.

Table 6-03

Work items - data collection

INPUT FROM ESTIMATOR:	HISTORICAL DATA ELEMENTS:	WORK ITEM RECORDS:
(10) Effective Wage Rate	Identification	
	(1) Item Number	(1) Item Number
(11) Unit Cost for Labor	(2) Work Description	(2) Work Description
		(12) Quantities
(4) × (10)=(11)		(13) Labor $
	Unit Data	= (11) × (12)
(12) Quantities	(3) Unit of Operation	(14) Material $
	(4) Workhour/Unit	= (9) × (12)
		(15) Construction Equipment $
	Labor Data	= (8) × (12)
	(5) Crew Composition	(16) Production Days
	(6) Crew Output/Day	= (6) × (12)
		(5) Crew Mix
	Constr. Equip. & Material	(7) Construction Equipment
	(7) Type	No./Type
	(8) Cost per Equipment	
	(9) Material Cost/Unit	

It is important, that the content of each work item be described in detail in order to apply a correct rate for future estimates Inclusions and exclusions should be defined (i.e. reinforcing included in the concrete code?; how about formwork, cleanup, etc?)

The work item represents a building block of data which has been quantified and priced according to the location and configuration of the building site. There are three major data input values to the system:

1. Quantity
2. Material Cost
3. Labor Cost

Table 6-03 shows a schematic depiction of data collection for work items (6).Work item records can be assembled into groupings according to the construction plan. These groupings can be time framed on the schedule and separated from other project cost. Available data can be processed to assist in the preparation of installation work schedules and resource planning.

Work item codes are used extensively in the building industries. The Work item principle offers a well organized and efficient method for computerization. Escalation factors can be included to adjust from base year data to reference year. The above mentioned work item record is sometimes called "The Cost Analysis Table" or CAT (12). In addition to CAT, a Rate Table or RT can be set up to collect historical data separately for a specific period in time and particular geographical location. This table can be used for several projects in the organization and updated periodically by the functional departments to help future projects.

Q #6.01: For purposes of *resource allocation*, the estimator uses company records of previous jobs which indicate that it takes 40 Whrs to install five sections of cable pans four years ago. It took a crew of one welder and two electricians with a wage rate of $ 25.00/hour and $ 30.00/hour respectively to install those five sections. The welding equipment rental costs were $ 50.00/day and the materials cost $ 75.00/section. Average inflation was 3%/year. Assuming an 8-hour work day, set up a work item table (estimate work sheet):

A #6.01:

Date: May 1995   (2) Description: Cable Pan - straight section

(1) Item #:   5386 - CP1   Escalation: $1.03^4 = 1.1255$

(12) Quantities: 100 (each)   (5) Crew mix: 1 welder; 2 electricians

(10) Wage rates: welder 25 × 1.1255 = $ 28.14; electrician. = $ 33.77

(13) Labor cost: (4) × (10) × (12) = 8 × $ 95.67 × 100 = $ 76 536

(14) Material cost: 75 × 1.1255 × 100 = $ 8444

(16) Production days: (6) × (12) = (40/5) × 100 = 800 Whrs = 100 days

(7) Equipment type: Welding equipment, rented.

(15) Equipment rental cost: 50 × 1.1255 × 100 = $ 562.75

     Total Cost of Item : **$ 85 540**

Q #6.02: An estimator uses company records from previous jobs which indicate that it takes 40 Whrs (5 days normal time) to install five sections of cable pans with a crew of one welder and two electricians with a wage rate of $ 25.00/Whr and $ 30.00/Whr respectively. One hundred sections need to be installed. The welding equipment costs $ 50.00/day, material costs are $ 75.00 per section. The WBS account is G53 and the work item number is CP-1.

  a) Set up a listing of historical data elements.
  b) Show input to the work item record.
  c) What is the total estimated cost of the work item?

A #6.02:

HISTORY		RECORD	
Identification	(1) G53-CP-1	(12)	100
	(2) Straight Section	(13)	$ 68 000
Unit Data	(3) each	(14)	$ 7 500
	(4) 24 Whrs./section	(15)	$ 5 000
Labor Data	(5) 1 welder, 2 electrician	(16)	100 days
	(6) 1 section/day		
Equip./Material	(7) Welding equipment		
	(8) $ 50.00/day	Total Cost:	$ 80 500
	(9) $ 75.00/section		

If there is a viable historical work item data bank, the updating to present worth labor rates and material cost will have been done automatically. Work Item *"Baskets,"* which contain several activities, can be established on historical data. For example, the cost for the installation of a service water pump based on past performance will indicate how well the crew performs now.

6.1.4 Measuring Productivity

The basic concept of productivity is a ratio which relates some volume of output to some volume of input. It measures the use of resources or the degree of their use.

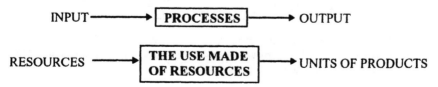

$$\text{PRODUCTIVITY} = \frac{\text{OUTPUT (units of products)}}{\text{INPUT (all types of resources)}}$$

Input and output can be quantified in different ways. Assuming the process is *learning by reading this text.* You are putting into this process your *existing knowledge.* Your output would then be *enhanced knowledge.* Obviously, by just reading the text, you will retain only a part of the total knowledge contained in the text. We can measure this by writing an examination on the subject. This would be the *feedback.* Examination questions are established through experience and previous observation of performance. Those examination questions will become

standards against which we measure performance.

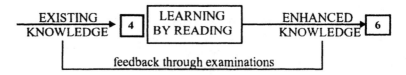

We now need units of measurement from *no* knowledge to *total* knowledge, say on a scale from 1 to 10. Assuming you have a knowledge ratio of 4 out of 10 when you start reading the text. The examination results indicate that you increased your knowledge ratio to 6 out of 10. This now leads us to the *efficiency* of your reading retaining skills, measured as output/input = 6/4 = 1.5 or 50% increase in knowledge. How good is this? We do not know unless we compare it with an established standard of average performance, which may be 1.6 for all we know.

The most common concept of productivity expresses output in physical units such as tons of steel, m^3 of concrete, number of houses, km of road built. The difficulty here is that we cannot aggregate different units of output. We may, however, equate different units of quantity to dollars. In this case, *constant dollars* must be used to measure changes in real terms over a period of time.

Input resources are mostly measured in hours of human labor because this is our prime economic resource. In this case, we are referring to a *single input* measure.

$$\text{Single Input} = P_s = \frac{\text{OUTPUT}}{\text{ONE INPUT}} \quad \text{or} \quad \frac{m^3 \text{ concrete / day}}{\text{daily workhours}}$$

This measurement gives no indication how well craft workers perform.

Sixty workers with wheelbarrows will probably put in place less concrete (200 m^3) than six workers with a creter crane (1000 m^3).

For *example*:

   Wheelbarrows: 200 m^3/60 workers × 8 hours = 0.42 m3/Whr.
   Creter Crane: 1000 m^3/6 workers × 8 hours = 21 m3/Whr.

Therefore, the *condition* under which the work is done does not show in the INPUT/OUTPUT formula. Those influences are hidden from our view. For a fairer picture, we should factor in the equipment used. We now establish a *multiple input* measure.

$$\text{Multiple Input} = P_M = \frac{\text{OUTPUT}}{\text{LABOR AND EQUIPMENT}} \quad \text{or} \quad \frac{\text{m}^3 \text{ concrete / day}}{\text{dollars / day}}$$

Assuming that a laborer with a wheelbarrow is paid $ 15/hr and the laborer working with a creter crane $ 20/hr. To rent a crane costs $ 150/hr. We now convert the input into dollars:

     Wheelbarrows: Creter Crane:
200 m^3/60 × 8 × 15 = 0.028 m^3/$ 1000/6 × 8 × 20 + 8 × 150 = 0.46 m^3/$

Obviously, the effective use of equipment is more productive than the extensive use of labor in this case. Of course there are cases when the use of labor in place of sophisticated equipment is more cost effective.

    The term 0.028 m^3/$ as a productivity ratio is seldom used in construction. Those figures can be very small. The *economic* productivity concept outlined above is therefore changed to

$$\text{UNIT RATE} = \frac{\text{INPUT}}{\text{OUTPUT}}$$

In regard to our example, we have now $ 36/m^3 for the wheelbarrows and $ 2.16/m^3 for the creter crane, a more familiar expression.

Cost Effectiveness

If we ignore the *resources* as a input measure, and state the input in terms of *time*, it is called

$$\text{PRODUCTION} = \frac{\text{QUANTITY}}{\text{TIME}}$$

A single wheelbarrow would have an average production rate of 200 m^3/60 = 3.33 m3 per day, whereby the creter crane has a production rate of 1000 m^3/day in our example.

> **Overall performance must properly balance between measures related to efficiency and those related to production.**

Remembering the production example previously mentioned at the beginning of this chapter, when producing 500 cars/day at plant C was meeting the goal. Just meeting the goal is not good enough for cost optimization. Meeting the goal *effectively* is more important. Assuming there is a group of workers that are highly skilled, experienced and motivated. They are working efficiently for 4 hours on the production line. The other 4 hours are wasted with "make-work" activities and stand-by. Only 4 hours are productive. Those, the productive workhours and the others must be *separately recorded* and reported in order to control the effectiveness of the operation. There is a *cost performance indicator* combining productivity and production:

COST PERFORMANCE = PRODUCTION × UNIT RATE

For the creter crane, that would be 1000 m³/day × 2.16 $/m³ or $ 2 160/day, and for the wheelbarrow, 200 m³/day × 36 $/m³ or $ 7 200/day. This now gives us a production rate in terms of dollars. Using the crane is more effective.

Here is a word of caution when using dollars as a measure of production or productivity. Dollars are variable, they change with time. When comparing performance at different dates, this should be taken into account. Below are three examples on labor productivity:

$ $ $ $ $ $ $

T I M E ⇒

Q #6.03: An identical job calls for 120 m³ concrete to be poured in Canada and Country "X." The following information is available

Canada	Country "X"
5 workers for 12 hrs @ $ 20.00/hr.	50 workers for 45 hrs @ $ 4.00/hr.
One creter crane rented for 2 days	Wheelbarrows - negligible cost.
for a cost of $ 300 per day.	Concrete mix on site = $ 20.00/m³
Cost of redi-mix = $ 50.00/m³	
Working hours = 8 hrs/day	Working hours = 10 hrs/day.

For both, Canada and Country "X",

a)   What is the single input labor productivity?
b)   What is the single input unit rate?
c)   What is the daily production rate?
d)   What is the overall (multiple input) unit rate productivity?
e)   How do Canada's total costs compare with the other country?

A #6.03:

	CANADA	COUNTRY "X"
a)	$P_s = 120/(5 \times 12) = 2.00$ m³/Whr	$P_s = 120/50 \times 45 = 0.053$ m³/Whr
b)	$P_s = 1/2.00 = 0.50$ Whrs/m³	$P_s = 1/0.053 = 18.75$ Whrs/m³
c)	$P_n = 120 \times 8/12 = 80$ m³/day	$P_n = 120 \times 10/45 = 26.67$ m³/day
d)	$P_m = (5 \times 12 \times 20 + 600)/120 = \$15/m²$	$P_m = 50 \times 45 \times 4/120 = \$75/m³$
e)	$\$15/m³ + \$50$ redi-mix $= \$65/m³$	$\$75 + \$20$ (concrete) $= \$95/m³$
f)	or 0.0154 m³/$	or 0.0105 m³/$
g)	$120(15 + 50) = \$ 7\ 800$	$120(75 + 20) = \$ 11\ 400$

Q #6.04: 15 welders work on project A for 10 weeks to install 500m of pipe. It needs 10 welders for 12½ weeks to install the identical length of pipe on project B. The work week is 5 days at 8 hours per day.

a) Calculate the productivity and the production rate for A & B

b) Which project has (i) a higher labor productivity? Which has (ii) a higher production rate?

c) The labor cost of welding for project A is $ 150 000 and for project B it is $ 125 000. What is the hourly rate of pay for the welders?

d) Assuming project management decides to buy modern automated welding equipment for project B for each welder, adding $ 60.00 to the hourly equipment rate. This will reduce the labor requirement to four welders doing the job in five weeks. What is the new production *unit rate* in m/hr for project B?

e) The *total* installation cost for project B with automated equipment is now:

f) Applying multiple input P_m, what does it cost now to install one meter of pipe in project A and project B?

g) With automated equipment for project B, what is the *production rate* per day on each project in terms of quantity?

A #6.04:

a)

$$\text{Productivity A} = \frac{500m}{15 \text{ welders} \times 10 \text{ weeks} \times 40 \text{ hours/week}} = 0.0833 \text{ m/Whr};$$

$$\text{Productivity B} = \frac{500 \text{ m}}{10 \times 12.5 \times 40} = 0.1 \text{ m/Whr};$$

$$\text{Production A} = \frac{500 \text{ m}}{10 \text{ weeks} \times 40 \text{ hrs/week}} = 1.25 \text{ m/hr};$$

$$\text{Production B} = \frac{500 \text{ m}}{12.5 \text{ weeks} \times 40 \text{ hrs/week}} = 1.00 \text{ m/hr}$$

b)

(i) Project B   (ii) Project A

c) Input stays the same as above, output changes to dollars.

$$\text{Project A} = \frac{\$\,150\,000}{15 \times 10 \times 40} = \$\,25/\text{hr}; \quad \text{Proj. B} = \frac{\$\,125\,000}{10 \times 12.5 \times 40} = \$\,25/\text{hr}$$

d)

500 m/5 weeks × 40 hrs/week = 2.50 m/hr

e)

4 welders × 5 weeks × 40 hrs/week × ($ 25/hr + $ 60/hr) = $ 68 000

f)

Project A = $ 150 000/500 m = $ 300/m;
Project B = $ 68 000/500 m = $ 136/m

(Project B without equipment would have been $ 150 00/(500/m) = $ 250/m)

g)

Project A = 500 m/10 weeks × 5 days/week = 10 m/day;
Project B = 500 m/5 weeks × 5 days/week = 20 m/day

Q #6.04: A piping job requires 1200 joints to be welded. There are 6 welders on the job, working a normal 40-hr week. They are able to weld 4 joints *each* per 8-hr day. Their rate of pay is $ 20/hr _

a) What is   - the production rate per hour?
                         - the duration of the job in days?
                         - the productivity unit rate?
                         - the total labor cost?
                         - the unit labor cost?

b) It has been decided that the welders will have to work 60 hrs/week to get the job done in time. Overtime pay is 1½ the normal rate. The overall efficiency drops by 14%. How does this affect the productivity, total cost and the unit cost? Traditionally overtime for trades are calculated on a daily basis. Assume a normal 5-day week at 8 hours per day.

c) The perceived benefit to advance the schedule was estimated to be $ 2000/day. How much was saved per day by working overtime?

A #6.04:

a) Production   = 6 welders × 4 joints/8 hrs = **3 joints/Whr**
    Duration   = 3 joints/hr for 8 hours = 24 joints/day or 1200j/24 = 50 days
    Productivity   = 6 welders work 50 × 8 hrs doing 1200 joints or
                       2400/1200 = 2 Whrs/joint
    Labor cost   = 6 welders for 10 weeks @ 40 hrs/week, which equals
                       2400 Whrs or 2400 × $ 20 = **$ 48 000**.
    Unit cost   = $ 48 000/1200 joints = $ 40.00/joint.

b) The productivity drops to

$$3 \text{ j/Whr} \times 0.86 = 2.58 \text{ j/Whr}.$$

Job duration is

$$1200/2.58 \text{ j/Whr} = 465 \text{ Whrs or } 465/60 = 7.75 \rightarrow 60\text{-hr weeks.}$$

For 5 days/week, it is $7.75 \times 5 = 38.75$ days or 38 days and 9 Whrs. This is 38 days with 4 hrs overtime plus 1 day with 1 hr overtime. Split into normal and overtime hrs:

> normal hrs $= 7 \times 40 + 32 = 312$;
> overtime hrs $= 7 \times 20 + 13 = 153$ hrs.

Productivity: 6 welders $\times$ (312 + 153) Whrs/1200 joints = 2.32 Whrs/joint.
Total cost $= 6 \times 312 \times \$20 + 6 \times 153 \times \$ 30 = \$ 64\,980$.
Unit cost $= 64\,980/1200 = \$ 54.15/\text{joint}$.
c) Overtime job duration:

$$465 \text{ hrs with } 12\text{-hr day} = 38.75 \text{ days.}$$

Difference in total cost:

$$64980 - 48\,000 = \$\,16\,980$$

with a schedule deduction of 50 minus 39 days (rounded) = 11 days, costing

$$2000 \times 11 = \$ 22\,500.$$

Therefore, the savings are

$$22500 - 16\,980 = \$\,5\,520, \text{ or } 5520/39 = \$\,142\,/\,\text{day.}$$

6.2 BASE PRODUCTIVITY

It is very difficult to measure a country's productivity in one set of figures. At present the overall productivity in the USA is still larger than in most nations, but the others have increased productivity at a faster pace, narrowing the gap. The average annual increase in productivity in industry between 1985 and 1993 were

Great Britain	+4.1%
Japan	+3.7%
Italy	+3.4%
Sweden	+3.4%
USA	+2.9%
France	+2.5%
Norway	+1.9%
Germany	+1.8%
Canada	+1.2% (Source: US Department of Labor

A breakdown of the above will show that there are marked differences between the sectors in industry. The manufacturing sector has a larger growth than the service industry and construction. The average output per employee can change to a more favorable statistic without an actual increase in productivity by simply closing factories and cutting the work force. The result is not a higher output per *existing* worker, but a balanced optimum output with less workers. Those statistics strongly reflect economic conditions, not necessarily the efficiency of the work force.

In order to measure whether there has been an increase or a decrease in a *project's* productivity, we need a different yardstick to measure against. Labor is only one component of the total project. It is not the total cause for good or poor performance. However, the interdependence of labor, machines and materials make labor a fairly reliable overall indicator of project performance.

6.2.1 Creating a Standard

Base productivity for labor represents the workhour rates that can be *expected* at a location under "normal" or under "average" conditions. It is either the published industrial standard or the company standard that has been established over a period of time. How do we obtain

$$P_B = \text{Base Productivity?}$$

There are several ways, some of which are listed below:

Historical Data

These are company records of past performance. Those records are usually categorized by work item. It is very important, that a uniform coding system be used from job to job to establish a desired unit rate to measure future performance.

There are basically two types of historical data, the estimator's record and the accounting-based standard. The accounting record is easier to obtain, but it has some shortcomings. It is based on pay items and does not make allowance for varying physical environments, design differences and working conditions. It is often produced at a remote central location by staff unfamiliar with field conditions.

The estimator's record takes into account actual job conditions which include physical measurements on the job and even unavoidable nonproductive time. This standard would also make adjustments to the "normal" base productivity standard when necessary.

Government Labor Surveys

Governments keep a record of labor performance by surveying a random number of companies. Average unit rates for various areas within the industry are then published (STATCAN, BLS - Bureau of Labor Statistics, USA etc). Average unit rates must be carefully analyzed as to work item content (inclusions, exclusions).

Data Catalogues

Those are publications in magazine or book form. Many of those are readily available such as Means, Craftsman, Walker, Engineering News Record. AACE have listed major sources in their Cost Engineers' Notebook (2).

AACE Notebook

The Cost Index Committee of the AACE collects and publishes unit rates for various areas and industries in the United States and other countries. This is periodically updated by membership input.

Company's Own Records

From all of the above, a company's own records are by far the best rates to be used as a standard to measure performance against from one job to another.

The most common way to measure performance against the standard is a ratio:

$$\text{PERFORMANCE RATIO} \quad P_R = \frac{\text{ACTUAL}\ \dfrac{\text{OUTPUT}}{\text{INPUT}}}{\text{BASE}\ \dfrac{\text{OUTPUT}}{\text{INPUT}}} = \frac{P_s \text{ or } P_M}{P_B}$$

and for

$$\text{COMMON INPUT (WORKHOURS)} \quad P_R = \frac{\text{ACTUAL OUTPUT}}{\text{BASE OUTPUT}}$$

Indexes which relate performance between two operating periods and/or between similar projects are useful management tools. The formulas above do not show unit rates, used often in construction. The workhours for common input would then be

$$\text{COMMON INPUT (WORKHOURS)} \quad P_R = \frac{\text{BASE OUTPUT}}{\text{ACTUAL OUTPUT}}$$

Unit rates and productivity rates can be converted easily from one to the other. When the output is in dollars, the wage rate of each individual worker must be taken into the conversion when the input consists of a group of workers with different wage rates.

6.2.2 Base Adjustments

Normal or average conditions do not exist all the time. Published base rates may have to be adjusted for one or more of the following reasons:

Location: Especially on international jobs, the local climate, the national characteristics, topography, different labor laws and workers' training have an impact on productivity. On the positive side, there may be considerably lower labor rates and lower cost housing in some countries. For companies that specialize in foreign projects, *base* productivity levels can be established for a given location over the years instead of adjusting existing local base levels.

Working Conditions: There is a difference whether a new plant is being built or an old plant is being rehabilitated. "Hot" work (radioactivity, chemical fumes) may have to be done or the plant configuration may result in crowded conditions. New safety rules may impose restrictions on access or movement. The type of contract may impact on working conditions, such as delays, interference and access restrictions.

CUMULATIVE EFFECT OF OVERTIME ON PRODUCTIVITY:
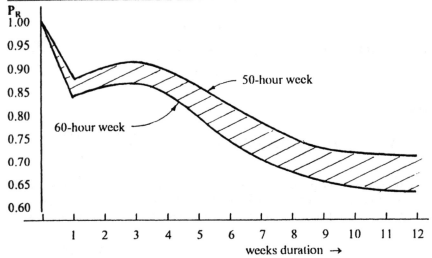

Figure 6-08
Impact of overtime.

Impact of Overtime: Longer than normal hours reduce productivity. The Business Roundtable (1) published statistics which shows the effect of reduced efficiency of a 50 hour week and the premium cost of overtime. When a job is scheduled for a 50 hour week, there is a reduction of productivity for the total 50 hours, not just for the ten hours of overtime.

As can be expected, the productivity loss is even more pronounced for a 60 hour week. This should be taken into consideration especially on "fast track" projects.

The graph (Figure 6-08) reflects the averages of many observations from project operations under normal conditions, i.e. good labor relations and management capability.

Below is an example how the decreased efficiency affects the cost of a job:

Q #6-06:

What are unit labor cost for a six-day week at 12 hours per day as compared to a normal 40-hour 5-day work week? How many days do we shorten the schedule? Assume:

Standard wage rate:	$ 20/hour
Overtime wage rate:	$ 30/hour
Normal Production:	16 units/hour
Productivity reduction:	from 1.0 normal to 0.8 incl. overtime
Crew size:	8 workers
Installation requirement:	90 000 units.

A #6-06:

Normal Work:

Eight craftpersons install $8 \times 16 = 128$ units/hour, which requires $90\,000/128 = 703$ working hours job duration, or $703/40 = 17$ weeks, 2 days, 7 hours.

Normal unit labor cost = ($ 20.00/16$ units) = $ 1.25/unit.

Total labor cost = $ 1.25 \times 90\,000 = $ 112\,500

Overtime Work:

Production: 16 units/hour $\times 0.8 = 12.8$ units/hour

Six-day-week working hours: $40 + (5 \times 4) + 12 = 72$ hours/week

Units installed per week: $72 \times 12.8 = 921.6$ units/week/worker

Job duration: $90\,000/(921.6 \times 8) = 12$ weeks, 1 day, 3 hours, or approx. 5 weeks saving in time.

Total labor cost:

$8[(40 \times $ 2o + 32 \times $ 30) \times 12$ hours/day] + [$ 20(8 + 3) + 4 \times $ 30] = $ 171\,680$ or $171\,680/90\,000 = $ 1.91/unit, a 53% increase in cost.

The Learning Curve

A worker becomes more productive when a task is performed repeatedly (5). The conscious application of this fact is an important part of cost optimization. Practice improves performance. The learning curve has major applications in the manufacturing industry. Construction projects are also affected, but to a lesser degree. A project which would not suffer to a large degree if the schedule is stretched or, in other words, the peak of the manpower loading curve can be flattened, will have an extended learning period. It has to be calculated, that the debit in schedule extension is less than the credit in increased productivity due to the learning curve.

Calculations for the learning curve are based on a so-called doubling effect. If a worker takes 10 minutes to perform a task for the first time, and 8 minutes the second time, there is a 80% improvement in performance. We may now say that the fourth performance time could be 80% of the second performance time or 6.4 minutes. Plotting curves for the actual performance by various workers (mainly in the manufacturing industry) have shown an exponential behavior. The calculations are not complex. The formula used is

$$E_N = KN^s \qquad \qquad \text{......................(1)}$$

where E_N is the expected time for performance class N. K is the performance time for the first performance and s is the slope constant on the curve. The slope constant can be negative when the effort per unit decreases with production.

Taking logarithm on both sides, the formula becomes

$$\log E_N = s \log N + \log K \qquad \qquad \text{......................(2)}$$

which is the equation of a straight line. The relationship between E_N and KN^s for N = 1, 2, 3, ...n can be tabulated with the following result:

$$E_2/E_1 = (K2)^s/(K1)^s = 2^s \text{ and } E_N/E_{\frac{1}{2}N} = E_{2N}/E_N = 2^s$$

that is, when production is doubled, the effort per unit is 2^s of what it was before. E_{2N}/E_N is the decimal learning ratio To obtain the percentage learning ratio L_P, we obtain

$$(L_P/100) = 2^s \qquad \qquad \text{......................(3)}$$

taking logarithms,

$$s = \frac{\log(L_P/100)}{\log 2} = \frac{\log L_P - 2}{\log 2} \qquad \qquad \text{......................(4)}$$

and to avoid negative logarithm, it is general practice to use the formula

$$-s = \frac{2 - \log L_p}{\log 2} \qquad \qquad \text{......................(5)}$$

This is best demonstrated with an example:

Q #6.07:

Welders are to install a total of 6940 m of piping in a process plant. The original schedule calls for a duration of 24 weeks. The workpower is loaded such that there is a rapid increase, then a high peak, followed again by a rapid decrease in the labor force. Under this scheme, a welder can install 20 m of pipe per week and no learning advantage is assumed.

As an alternative, it was considered to flatten the peak after five weeks and maintain a steady workpower level for a period of 12 weeks before a rapid decrease. The production rate of week 12 would then be maintained to the end of the job. From previous records, it was estimated, that welders will install 316 m during the eighth week and 360 m during the tenth week while maintaining their manpower level.

a): Find the percentage learning ratio and the quantities installed during the 12 week learning period. Use the relationships

$$E_N = KN^s \quad \text{and} \quad -s = \frac{2 - \log L_p}{\log 2}$$

b): Plot the learning curve.

c): In practice, the workpower loading is a bell curve. For this exercise and for the sake of simplifying the calculation, assume a linear distribution (triangle). What is the delay in schedule?

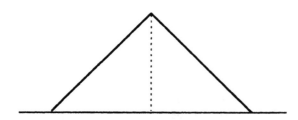

Figure 6-09
Learning curve - linear distribution.

d): If the unit rate for welders is $ 25.00/hr and there is a 40-hr week, what are the cost implications if a delay in schedule would cost $ 60 000/week?

A #6-06:

a):
$$E_3 = 316 = K(3^s); \quad E_5 = 360 = K(5^s)$$

$$E_3/E5 = 316/360 \quad = 0.87778 = (3/5)^s = 0.6s$$
$$\log 0.87778 = s \log 0.6 \quad \text{or}$$
$$(9.94338 - 10) = s (9.77815 - 10)$$
$$s = (-0.05662)/(-0.22185) = 0.25522$$

$$-0.25522 \log 2 \qquad = 2 - \log_{Lp}$$
$$\log_{Lp} = 2 + 0.25522 \times 0.30103 = 2.07683$$

$$L_p = 119.35 \ (20\% \text{ increase})$$

solving for K, using 9th week data

$$316 = K(3^{0.25522})$$
$$\log 316 = \log K + 0.25522 \times \log 3$$
$$\log K = 2.49969 - (0.25522 \times 0.47712) = 2.37792$$
$$K = 238.73$$

for production during week 17:

$$E_{12} = K \times 12^{0.25522}$$
$$\log E_{12} = 2.37792 + (0.25522 \times 1.07918)$$
$$\log E_{12} = 2.65335$$
$$E_{12} = 450 \text{ m installed in week 17}$$

similar,

$$E_1 = K \times 1^{0.25522} = K = 239 \text{ m in week 5}$$

b) All weeks can now be tabulated (Table 6-03) and plotted as a typical learning curve (Figure 6-10).

c) *Original Schedule:*
The installation of 6940 m of pipe in 24 weeks is distributed as shown in Figure 6-11.

Table 6-03
Learning curve example

	log(n)	x(n)	x(n) + K	E(n)
E_1	0.00000	0.00000	2.37792	239
E_2	0.30103	0.07683	2.45475	285
E_3	0.47712	0.12177	2.49969	316
E_4	0.60206	0.15366	2.53158	340
E_5	0.69897	0.17839	2.55631	352
E_6	0.77815	0.19860	2.57652	377
E_7	0.84510	0.21569	2.59361	395
E_8	0.90309	0.23049	2.60841	406
E_9	0.95424	0.24354	2.62146	418
E_{10}	1.00000	0.25522	2.63314	430
E_{11}	1.04139	0.26578	2.64370	440
E_{12}	1.07918	0.27543	2.65335	450
	Total length installed =			**4 448 m**

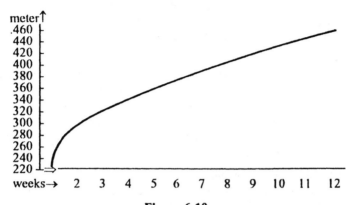

Figure 6-10
The learning curve.

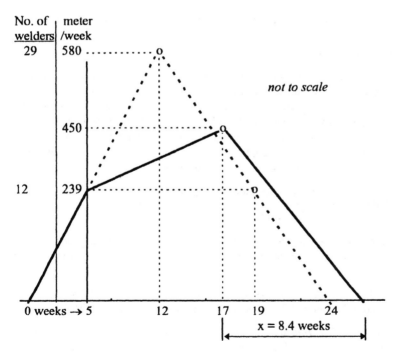

Figure 6-11
Flattened learning curve.

Number of welders at peak: $12y$ $= 6940/20$; $y = 29$ welders
They install 29×20 $= \mathbf{580}$ m of pipe per week.
After 5 weeks there are $y = (5 \times 29)/12 = \mathbf{12}$ **welders** on the job
They install $5 \times 6 \times 20$ $= \mathbf{600\ m}$ of pipe.

Flattened Schedule:
With the flattened schedule, $600 + 4448$ $= 5048$ m of pipe are installed
in $5 + 12$ $= 17$ weeks.
Remaining to be installed is $6940 - 5048$ $= 1892$ m of pipe.
Fixing the production at 450 m/week, the time remaining is
                                    $(450x)/2$ $= 1892$; $x = 8.4$ weeks
Therefore, the schedule is delayed by $17 + 8.4 - 24 = 1.4$ weeks.

d) The labor cost for the conventional schedule is
        (24 months/2) $\times$ 29 welders $\times$ 40 hrs $\times$ \$ 25/hr = \$ 348 000
and for the flattened schedule:
        12 welders for 12 months + 12/2(5 + 8.4) $\times$ 40 $\times$ 25 = \$ 224 400
Cost of schedule delay is   60 000 $\times$ 1.4 = \$ 84 000

It is more cost effective to delay the schedule.

Other Base Adjustments

Economic Activity: Recruiting workers is more difficult when there is a high economic activity in the area. Due many activities, the quality of labor that is available tends to move below standard and the turnover is greater. From the cost point of view, incentives to keep good workers may add to the base rate.

To import craftspersons from outside the local area will also have a declining effect on productivity. Those people have to be housed (moving or travel expenses) or camps will have to be provided. If the ratio

$$\text{Activity Index} = \frac{\text{Number of trades locally available}}{\text{Trades normally required}} = 1.0$$

we have a situation where the base productivity does not need any adjustment. Anything *less* than 1.0 indicates a surplus of trades in the area and a *positive* base adjustment, e.g.

Activity Index = 0.8; Base productivity = 1.0 + 0.08 = 1.08
Activity Index = 2.5; Base productivity = 1.0 – 0.25 = 0.75

Direct Hire vs. Subcontract: Statistics show, that generally subcontractors are more productive than direct hire labor. This is mainly due to the specialization of subcontractors. Many subcontractors specialize in a particular type of work and have a core of steady experienced craftspersons who work more efficiently.

It does not mean that the cost are lower when a subcontractor is hired. There are some duplications in the use of facilities and equipment. High contingencies are usually put on fixed-price contracts.

Job Size: Smaller jobs are usually more productive than large jobs. Very large scale projects usually have communication and motivation problems. The larger an organization, the more difficult is the coordination between the various groups. There seem to be more changes due to design reviews and interference between subcontractors, often resulting in a large number of claims.

Other Adjustments: There could be justifiable adjustments to the base rate such as the first job using metric dimensions and materials based on ISO standards. The unfamiliarity and opposition by less enlightened personnel with the new measures could initially affect the base rate (the Darlington Generating Station near Toronto was built mainly to SI dimensions with very little problems).

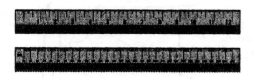

The adjusted base should now become the standard to measure against.

Q #6.07:
Why should we apply the learning curve when performing time- and cost estimates for labor performance?

A #6.07:
The critical path method in scheduling assumes that activity times remain constant. The estimator applies established rates which are based on the records for an experienced worker performing at a normal pace. This does not include a learning-by -doing effect which improves with experience.

It has been well documented over the years (5) that a constant percent reduction in performance time can be expected with each doubling of the number of performances. This doubling effect is not an assumption, but a statistical reality. For example, if it takes a welder twenty minutes to do a joint the first time, and eighteen minutes the second time, there is a ten percent improvement over that time period. We can now expect another ten percent improvement at the end of the fourth performance time, i.e. 16.2 minutes.

This will have an effect on the estimate for long duration jobs and should be taken into account. The principle of the learning curve is that

the rate of improvement remains constant for a laborer.

6.3 FACTORS AFFECTING PRODUCTIVITY

When we discussed base adjustments under 6.2.2, we already described several factors that affect productivity such as overtime, working conditions and others. Those were specifically identified as measurable or quantifiable by calculation. There are other factors affecting productivity that are not added to the base rate because they are often intangible. Below is an outline of some of those factors we should be aware of while working on a project.

6.3.1 Ergonomics

Ergonomics is the science that is concerned with the relationship of a person
to the workplace. This applies not only
to the construction environment, but
also to the office. It is sometimes called
biomechanics or human engineering.

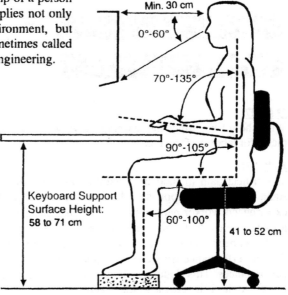

The goal is to in-
crease productivity
by providing physical
conditions that give
the worker the means
for optimum per-
formance on the job.
It starts with the lay-
out of facilities, job
methods, tooling and
type of equipment to
be used. There are
usually engineers in-
volved in conjunction
with behavioral sci-
entists.

The objectives of Ergonomics are to achieve functional effectiveness of the equipment and tools used by people for cost-saving aspects of safety, health and job satisfaction. Ergonomics are generally involved in the following areas:

Negative Stress

Stress is very much an individual matter. Stress for one person can be a welcome challenge for another. *Negative* stress adversely affects the health of the

employee. Proper ergonomic measures can minimize stressful situations and thereby improve productivity. The types of stress workers are exposed to are

♦ Fatigue from repeated motion.
♦ Mechanical vibrations. This can result in tenseness, nervousness, or difficulty in hearing.
♦ Assignments exceeding the capability of the employee.
♦ Unreasonable expectations can result in depression, or loss of confidence.
♦ Monotony of the job can lull employees into careless situations.

Tools

The design of handtools resulted in many improvements because of recent ergonomic studies. Modern tools use damp- ening and cushioning devices. When a tool is well balanced for example, it is easy to handle and the chance of an accident is greatly reduced. Especially in North America and Europe, tool designers have made innovative improvements in recent years.

Noise Control

High noise can reduce a worker's ability to think, it can cause high blood pressure and muscular tension. Sound absorbing materials are used to combat noise.

Work Station

Productivity will increase when surroundings are comfortable. Color, temperature, ventilation, lighting and ergonomic seating arrangements play a large part in office or plant layout.

6.3.2 The Environment

Construction workers are not as fortunate as those in the manufacturing industry because they are exposed to the outdoors and must endure heat, cold, wind and rain. There is no doubt that this has a marked effect on productivity. Contractors must take environmental factors into consideration when bidding on a job.

Noise

Loud and unwanted noise interferes with work and communication. We may recall from school physics that sound is a compression and rarefaction of air. The level of energy that produces the sound is measured on the decibel scale (dB). High amplitude sound is damaging to the ear. The noise level should not exceed 90 dB. Sound is also measured by Hertz, the unit of frequency (cycle per second). It is commonly referred to as pitch. Very high pitch even with relatively low dB can be very annoying, especially when the high pitch does not vary (compressors, transformers).

Temperature

There are areas on our globe where there is little variation in temperature. Seasonal changes can be extreme in the far north. Temperatures between 5°C and 20°C have little effect on performance unless the humidity and winds are high. If the windspeed at the freezing point (0°C) is 11 m/sec, the equivalent temperature is −11°C. Technically, windspeed is measured in m/sec, not km/h. (Because we are used to the expression miles/hour, publications in metric conversion often translate into km/h). Below is a table showing how temperatures are affected by wind velocities ranging from 10 km/h to 60 km/h:

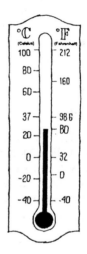

| | degree Celsius | | | | | |
km/h	0	-5	-10	-15	-20	-25
10	0	-7	-12	-17	-22	-27
20	-2	-13	-19	-25	-31	-37
30	-7	-17	-24	-31	-37	-44
40	-11	-17	-24	-34	-41	-48
50	-15	-22	-29	-36	-44	-51
60	-16	-23	-31	-38	-45	-53

Good contractors will include the above information when they prepare labor estimates.

Humidity

Humidity reduces the body's ability to evaporate skin moisture, thereby cooling the body when temperatures are high. Fatigue results under hot, humid conditions.

Wind

As mentioned above, the velocity of wind can have a severe cooling effect on exposed skin when the temperatures are low. Cool fingers and toes reduce efficient work. Under dusty conditions, the wind can pick up particles that impair vision and make it difficult to open and read drawings.

Precipitation

Rain, sleet, snow, hail and dew have an impact on construction activity. During adverse weather conditions, some work may have to be canceled. Even after the rain has stopped, muddy conditions at the job site affect equipment and labor activity. Precipitation is difficult to predict for weeks or months ahead. Meteorological records can be consulted to forecast the chance of precipitation in the job area.

6.3.3 Motivation Factors

Managers must achieve results through the efforts of other people. This calls for a motivated work force. How do we motivate people? There are two opposing views:

A)   A strong leader that concentrates power and responsibility at the top office. Workers are motivated by coercion such as extra pay or punishment (carrot or stick). Management believes that employees dislike work, that they avoid responsibility whenever they can and are influenced by physical desires only.

B)   Managers encourage employee participation. Workers make work place decisions, including job structure, job layout, methods and processes and the work environment. They allow workers to express ideas on how to achieve higher quality, more efficiency and lower cost. This is sometimes called "democracy in the work place." Basic physical needs allowed a worker is now expanded to include fulfillment, esteem and belonging. It is hoped that this will increase productivity.

What is the better way, "A" or "B"? "A" is the traditional view. Modern management is more often than before leaning toward "B". However, that could change in the future. "A" looks down at the workers, evaluating *their* productivity. "B" says: "It is not only the management of productivity that matters, but also the productivity of management!"

The question therefore is: How successful is management in motivating the worker? It is probably not a question of either/or, but the makeup of the working force. Not all workers can be treated under "A", neither can all perform under "B", which is far more difficult to manage.

Let us look at the many factors that bear on the productivity of an individual. These are both, physical and psychological (Figure 6-12). *The background of the individual* has a bearing on the attitude of both, worker and manager. This in turn affects motivation. Upbringing and place of *origin* has a large impact on how an individual can be motivated to contribute to a high level of productivity. Managers must select staff on the basis of *ability, not appearances.*

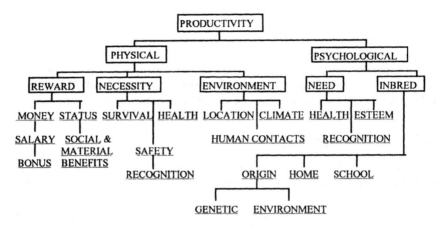

Figure 6-12
Motivation factors.

Below is a condensed listing of the most common techniques that increase productivity through improved motivation(1):

Goal Setting

Acceptance of a carefully set goal is more probable if workers participate with management to establish it. Goals should be set at a high but attainable level. The goal should be specific, allowing direct comparison with actual performance. Disclosing workers' performance in relation to the goal will give them direction and an incentive to compete with other workers.

Incentives

Incentives are tangible rewards given to those who perform at a given level. The rewards can be monetary or in kind such as golden rings, vacation trips, day off with pay etc. Other incentives are better working conditions (larger office), recognition by announcement or certificate. Minor monetary or small item rewards usually do not motivate. Profit sharing schemes seem to be much more successful. Incentives must be very carefully planned and reviewed regularly. They should be clearly linked to performance, which means productivity, not just output.

Worker Participation

Active involvement in managerial decisions gives workers stronger feelings of loyalty to a firm. Almost any changes that affect the working conditions are apt to be more readily accepted if the workers have been consulted in advance in more than a perfunctory way. Worker participation can have several forms:

- ◆ Suggestion plan
- ◆ Labor/management committees
- ◆ Quality control circles
- ◆ Consultant orchestrated
- ◆ Worker/top management meetings

All work is equal, there are no inferior jobs.

Most participation plans have proven to be successful. Workers who are actively involved in decisions that affect them are more receptive to change, work harder as they develop more enthusiasm, become more loyal to the employer, do not suffer from job alienation, experience greater job satisfaction, and show increased morale and creativity.

Employee participation programs seem to be the most effective way to boost productivity, especially in construction.

Positive Reinforcement

It is also known as behavior modification, where desired behavior is rewarded by praise and/or compliments. The company must remind supervisors that a pat on the back, when deserved, should not be forgotten. This instills a sense of pride in the job. It appears that positive reinforcement works best for construction jobs. The reason may be that trades usually receive little or no recognition for their efforts (Why praise them, they get paid, don't they?)

Work Facilitation

Most workers can best be motivated through the satisfaction inherent in the work itself. Employers must provide benefits such as a good social environment on the job (recreation facilities), insistence on good workmanship, salary reviews linked to the company's profit, making the work interesting, improving work content.

New employees should be given all the information necessary, including an introduction to co-workers in order to become productive employees in a short period.

6.3.4 Management and Organization

While North America's productivity has declined in the past, the Japanese for example, are highly productive. Are they doing anything different? Not really, they apply some of the most basic principles of organization, marketing, and manufacturing, in short, recognized principles of management. But those principles are not new, in fact, North America was the *originator* of most of them. That is why many of us seem surprised that this could have happened. It is probably caused by a certain smuglyness and a reliance on past momentum.

We became information gatherers, form designers, managers of data, i.e. top heavy. Our "systems" isolated us from the "workers." The result was a certain alienation of the big boss from the lower ranks. It is not unusual, that the industry still has 12 layers of management. We should compare this with 7 in Japan. It appears that we became an over-supervised, over-managed society.

The functional manager usually came up "through the ranks" and accomplished results through "*position power.*" This can be an individual who takes "his job upstairs with him", often indulging himself in facets of a job he enjoys the most. Promotions were often a matter of years of service (see *The Peter Principle*).

Not enough attention has been paid to *managerial productivity.* Productivity is just as much important for the executive office as it is in the factory, the construction site, the warehouse or the typing pool. We now recognize that in addition to the management of productivity, we need to

increase the productivity of management.

How do we do this?

Managers must stay in much closer touch with their employees and their customers. All must become aware of the demands tough competition places on them. Only people can make companies more productive, people using computing for competitive advantage within a multi-disciplinary environment. Information technology has "democratized" the work force, leading to the empowerment of workers. Informed decisions can therefore be made at all levels of the organization. Middle managers can largely be eliminated, as they no longer provide a pathway for information flowing up and then back down. More work can be accomplished by fewer people. However, we should make a distinction between the functional manager and the project manager.

Direction to a project team is provided by a project manager who will have to be supported by the team in order to be successful. This requires a great amount of "*knowledge power,*" including superior communication skills. It needs a manager who can delegate and therefore will trust his/her team. In order to reach a high level of managerial productivity, an inspiring manager must learn all the pertinent skills. Organizations such PMI and AACE are very effective to assist in obtaining those skills. R.M. Wideman, Past Chairman of the PMI Standards Board introduced the "Project Management Body of Knowledge" (PMBOK) as follows:

"*...there has been a growing recognition that management, and particularly project management, is a special skill that can be codified and learned. Project management skill is quite different from the technical skills that are so often associated with most projects. Indeed, there are aspects of all projects which are outside the scope of these technical areas, yet which must be managed with every bit as much care, ability and concern...*"

This indicates that there are skills beyond the pure technical or financial aspects of a project, such as
- problem solving
- understanding PM software
- stress and time management
- information handling
- decision making
- communication and presentation skills
- human resource management

Training managers in these areas will enhance their contribution to increase productivity.

Organization

Traditionally, organizations are structured to represent a network of vertical and lateral relationships. Specialized parts of a business are in sealed compartments. The optimum hierarchy consists of

1. Specialization
   subdivision of work to develop higher competence.
2. Unity of command
   each person to be accountable for his performance to only one superior.
3. Delegation
   defined responsibility and requisite authority at sufficiently allocated levels in the hierarchy.
4. Span of Control
   for efficiency, the number of individuals reporting to one supervisor must be optimized (about 8).

"Structural" thinkers emphasize that one must start with organizational logic and design the best possible network of functions and positions with little consideration how it will suit people. In reality, an organization is a human complex, it is a social phenomenon.

Drucker (*The Practice of Management*) says

".... the starting point of any analysis of organization cannot be a discussion of structure. It must be the analysis of the business. The first question must be: What is our business and what should it be? Organization structure must be designed so as to make possible the achievement of the objectives of the business for five, ten, fifteen years hence..."

A productive organization is therefore a framework within which persons are expected to contribute to the making of a product, the provision of a service, or the realization of an ideal.

The members of an organization must participate in the decision making process. The mindless "following of orders" is not very effective any more.

<u>Following orders</u>

The structure of an organization should be used as an implementation tool for its overall strategy. Tailoring a strategy to fit an organizational structure is far less productive. *Method Study* is an excellent tool to design the optimum organizational structure.

6.3.5   Standards

The good thing about standards is that there are so many of them to choose from...

Unknown

Standardization is a process that began untold centuries ago and let us along the road to civilization. Society as we know it, we may say, got its start with the setting of standards. The first measuring device known to man was man himself.

For centuries, size was determined by a hand, a foot, a finger or an outstretched arm. King Henry I (1496) established the yard as the distance from the tip of his nose to the end of his thumb (had he looked sideways, it could have been a meter). Since men are not created physically equal, the need for standards became apparent.

Rapid development in standardization took place during the industrial age. Standards were developed not only for weights and measures, but also in regard to fire and building codes, safety regulations, food and drugs, clothing, furniture, utilities, appliances, tools and vehicles. Those standards, however, were mainly developed on a national or regional basis and included often only minimum requirements.

Standardization can be a strong force toward the simplification of production procedures and of everyday life. Thoughtfully and intelligently applied, standardization limits the *unnecessary* variety of products and components. This greatly reduces costly duplication without sacrificing quality.

Standardization also creates meaningful communication between the designer, the manufacturer and the consumer. Known standards of which the buyer is aware *before* purchase will build confidence in the product.

We are surrounded by a great number of standards which we take for granted and which form an invisible but important part of our daily lives. The plug will fit the electrical wall outlet, traffic light signals show green for go everywhere and so on.

On the international level, the absence of mutually agreed upon standards of production and performance can be as much a barrier to world trade as any tariff laws. International standardization provides a stimulus to world trade. This is of mutual advantage to all countries concerned. Standards use in North America is essentially a *voluntary* matter.

There are three major levels of standards as shown Information on standards flow up and down. The acceptance and use of the standards depends on the individual parties in the process. Unless a standard is cited in the law, regulation, building code, or contract, no one is compelled to use or comply with it.

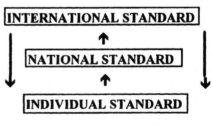

Many companies develop their own standards *on purpose*. Parts are designed such as to fit only *their* product. Any replacement needed has to be purchased from *that* company, and for a higher price. A national proliferation of parts is the result. For example, approximately two million different fasteners are being produced in the USA whereby DIN standards in Europe specify only 300 000 or 15% as many. Obviously, this multiplication of effort is expensive and unproductive.

To add to the problem, the USA has been reluctant for decades to join the world in converting to international units of measures (SI). International standards are based on those units. It is the opinion of the author (shared by many others) that the reluctance by some companies to convert to "metric" has hurt our international trade.

A report to the Congress by the Comptroller General (CED 78 128) rejected metric conversion on the basis of 1977 statistics which indicated that only 7% of the GNP are exports. The manufacturing of products in metric units were considered to be of minor significance. They did not seem to realize what impact their rejection may have on future world trade. Since then, the world has become even smaller and new markets, especially in the East have opened up. The European Community has agreed to prohibit the purchase and the sale of non-metric products.

Fortunately, the USA have been active in the International Standard Organization (ISO) through input by the American National Standards Institute (ANSI). However, most metric hardware is produced to DIN (Deutsche Industrie Norm) standards first developed in Germany long before ISO was created in 1947. The Industrial Fastener Institute of the USA (IFI) is not recognized word wide. So far, their fasteners, even though produced in large numbers, can only be secured in the USA and Canada.

Japan also plays an important role in terms of metric hardware standards. The Japanese Industrial Standard Institute (JIS) standards are mainly based on DIN. The Standard Council of Canada (SCC) is encouraging international standardization in Canadian industry.

ISO International Standard Organization

ANSI American National Standards Institute

DIN German Institute for Standards

JIS Japanese Industrial Standard Institute

IFI Industrial Fastener Institute — Some other standard associations

Some other standard associations:

UNI	= Italy
AFNOR	= France
BSI	= Great Britain
SN (VSM)	= Switzerland
SIS	= Sweden
NNI	= Netherlands
SCC	= Canada
DGN	= Mexico

To summarize:

> **Standardization leads to higher productivity, broader markets - especially international markets - more time to devote to innovation, and less expensive products for the consumer.**

It should be noted, however, that we should not equate standardization with *uniformity*. Standardization will rationalize the production of parts that make up a finished product. We should still maintain a variety of cars, food, dresses, etc. We will never standardize human taste.

6.3.6 Quality

Put Quality in first, Productivity will follow.
                    Dr. Edwards Deming

How does quality relate to productivity?

The ratio output/input should be maximized, should it not? If we add quality to the processes, does it not make the output more expensive? It may seem that way at the lowest stage of production. If we speed up the assembly line, we will produce more.

INPUT X $\longrightarrow$ FAST PROCESSES $\longrightarrow$ OUTPUT (Y-R)

where R is the number of rejects in the form of broken parts, nonfitting assemblies, buyer's dissatisfaction etc. Therefore, output needs to be defined as *good reliable output*.

Quality is hard to define. One can make a very high quality product and still not sell it on the market. The customer will always let you know what products are desirable over others. In some cases quality means *durability* and reasonable cost, in other cases the product has to look good or is disposable after short use.

Why do we see so many Japanese made cars on the road? The main reason seems to be that they are reliable, easy to handle, in short: They are of good quality. This is what the customer seems to appreciate. Today, products and services are often exported to many parts of the world. Understanding exactly who the customers are, and what they want is the first step in productivity improvement.

The American Society for Quality Control (ASQC) is an organization that promotes continuous quality improvement. It is their understanding that there is a definite relationship between quality and productivity.

We have, in the past, always believed that quantity and mass production (cheaper by the dozen) are the driving forces of economic affluence. We were a land of plenty. Appliances and cars broke down after a short period. We had cheap energy and easy access to major markets. To make it right would slow things down. We then found out, that we could not sell our inferior products on the world market. The competition from Japan and Europe taught us that

Improving quality also improves productivity.

A lean and efficient manufacturing process with downstream quality requirements for suppliers' materials and "doing things right the first time" can be more cost effective than the traditional mass production method without quality control. Through higher quality, manufacturers are now creating a more loyal, satisfied, and growing customer base.

How does this apply to management? The quality improvement principles do not seem to have yet caught on enough in the service industry which has grown enormously with the event of the computer. With huge investments in information technology we must be able to deliver the output necessary to increase productivity. We are becoming more conscious of the advantages quality can have on the world market. The demands in achieving world class quality can be more formidable than may have been anticipated.

The US Congress passed a law in 1987 mandating an American national award, the Malcom Baldridge Quality Award. On the international scene, the European Economic Community has adopted the ISO 9000 series which are basic quality assurance standards. These standards must be met by any company seeking to do business in Europe. ISO 9000 4-2, a standard for all service industries, was scheduled to be finalized by 1996. The trend seems to be that sooner or later all companies around the globe will adopt these standards, even though they may not do business in Europe. This is because ISO 9000 is the best attempt by a governing body to establish universal standards.

The implementation of the ISO 9000 program must be carefully planned to make it effective. It can only work when the top executive becomes fully responsible for the program and requires management to take an active role in its development and control.

I.S.O. 9001 CERTIFIED

Jim Clemmer, President of the Achieve Group, presented a paper at the 1991 ASQC convention in Toronto, outlining the difference between traditional thinking and quality-oriented thinking of modern management. Below are excerpts from notes taken:

TRADITIONAL THINKING	QUALITY ORIENTED VALUES
☹ Management is the brain, employees are the hands	☺ Trust the expert to do the job.
☹ Management plans, directs, and controls all aspects of daily operations	☺Front-line performance teams run daily operations with the full support of management.
☹ All operational problems are solved by management	☺ Management makes sure that performance teams have the proper skills and are given the necessary information and tools to solve operational problems.
☹ It is the sum of the efforts of the individuals that result in overall organizational performance	☺ Organizational performance depends mainly on its systems, processes and structure.
☹ Snappy decision making is the mark of a good leader	☺ Building broad consensus and having a stake in the outcome of decision making improves implementation.
☹ Lock in operations after defects are modified and fixed permanently	☺ Improvement is a continuous process, there is always the potential for a "better way of doing things".

The Department of Defense has introduced the term "Total Quality Management" (TQM). Their definition:

TQM is the management application of methods and human resources to control all processes with the objective of achieving continuous improvement in quality. Under the TQM strategy, we are seeking a cultural change from defect correction to defect prevention; from quality inspected into products to quality designed and built into the process and product; from acceptable levels of defects to continuous process improvement. To achieve the goal of TQM, we intend to encourage implementation of concepts such as Statistical Process Control and Continuous Process Improvement, and to emphasize the use of sound engineering design and manufacturing practices.

Excerpt from the Master Plan for DoD, TQM, Jan.12, 1987.

6.3.7 Computers

"Existing processes are left intact and computers are used to simply speed them up."
(Michael Hammer - Harvard Business Review - Jul/Aug 1990)

Have computers contributed to increased productivity? Perhaps. This question is difficult to answer. Overall improvement in national economic productivity in the USA had been less than that of some other nations (see 6.2 Base productivity). Is it in spite of computers or do computers have little influence on overall productivity?

What areas are more productive due to computerization and automation, and what areas have hardly been affected? Computers have added speed to the way we are doing things. But has the way we are doing things improved *because* of computers?

The first computers supported existing bureaucracies in large corporations and governments. (Small companies could not afford those huge machines). There were two major players involved: Top management and information officers. The main frames were centrally controlled. Huge data bases were meant to put order into information systems. Highly paid computer experts installed management information systems (MIS) that would give CEOs all the information needed to manage and control a company. With increased memory more "fields" were opened up in the data base with huge amounts of information flowing up the channels of the organization to top decision makers.

For a while it looked as if the traditional "structured" organization was improving productivity. But soon, information overload set in. The people on top did not always know *how to specify the information* needed. Neither did the departments who reported to them. Then came desk top computers who by-passed the giant brain. Each department had its own budget for those smaller machines with related software. But there was little standardization until IBM cornered the market. This decentralization was counterproductive. The mainframe was still very important for payroll, employee information systems, corporate finances, etc.

Word processing, spread sheet estimating, and other functions were handled locally. But the micros were not integrated with the main frame. Corporate data processing regained control again by specifying and thereby standardizing the hardware and software of the micros and even minis. They devised rules and regulations to restrict information to those employees that were included within the upward moving information channel.

The old hierarchy started to blossom again - for a while - until the micros became almost as powerful as the mainframe computers used to be. This now gave more initiative to middle managers. "Network Management" was then devised to re-centralize information flow.

The great difficulty was: How do you set up rules to police the network? On the one hand, management wanted to delegate, on the other hand it wanted to have security over information. Companies became extremely concerned about information leaking out to competitors. They started to tighten up, increasingly restricting information even to their own employees.

Can this information "policing" help to increase productivity? Definitely not! The more structured, channeled, and policed a system is, the less likely is the generation of innovative ideas. A company cannot keep information and knowledge secret forever and stay competitive. Secrecy leads to spying. If no creative ideas are generated within, data are accessed by penetrating other, more successful companies' data base. They in turn spy on others. This is an expensive, short sighted undertaking. It does not give us a competitive edge in the world market. It is self defeating.

Top management alone cannot solve all problems. We should realize that "mind workers" can be creative only if given freedom of access to information. Employees have contacts with the outside world, they have friends and neighbors working for other companies. They are constantly importing and exporting information. Companies that support individuals who join professional associations for example, gain a great amount of *input* for the benefit of the company.

If researchers from different companies, universities, and associations could share more of their knowledge instead of keeping it secret, we would probably see less economic stagnation and higher productivity. In the evolution of computerization, the technology usually runs counter to controls from the top.

When micros challenged main frames, the organizational hierarchy was challenged. The threatened bureaucracy was re-established by "networking". This resulted in a tight security conscious organization devoting much of its resources to intelligence and counter intelligence instead of innovative research and development.

Computers once again challenged the structured bureaucracy. There was more use made of "relational" data bases that add flexibility to stored data. Instead of the vertically structured data base, the fields could be laterally arranged by the user. This allowed new relationships of information to emerge. The computer now became the catalyst for new ideas.

Similar to the discussion under Chapter 7.1 "Creative Thinking," one can access knowledge modules, without knowing at the time what the end result of the search for information will be. This relational "hypersystem" reminds one of brainstorming, where the random access will stimulate the creative brain.

It will probably not be too far into the future when we will have information access as individuals we can now only dream about. The construction of the National Research and Education Network (NREN) has begun (at the point of writing) research and development in high performance computing (up to 3000 Mbps fiber optic). The aim was to promote more effective scientific research, technological development and university/industry interaction.

This could mean the breakup of rigid bureaucratic information channels, directing our efforts toward organizational creativity and a communication network that moves information across, not just up and down. It links users nation- and worldwide. As we all know, the changes in the field of computerization occur so rapidly, that it would not be wise, to predict the future. It appears that the benefits of computerization (return on investment) do not derive from the machines themselves alone but from the changes in the organization and mode of operation, taking advantage of the computers' strength. This will hopefully result in "flexible" organizations and thereby cause a rise in productivity.

Computer Expressions

When we want to drive a nail into the board, we cannot do this with our fists, we use a hammer. A screwdriver is used to fasten a hinge to the door. Our fingernails would not be strong enough for this. Tools are there to extend our bodily capability. If we want to have a fast answer to a mathematical problem, we use a calculator.

Why is it, that the computer is considered by many as being more than just a tool? We tend to mystify this machine by giving it almost human qualities. Should we not be glad that the computer can outperform in speed and logic? We do not want the hammer to be as soft as our first or the computer as emotional and irrational as the human brain often is!

We talk about "Artificial Intelligence," and the computer "brain" that can get sick by having a "virus." The word "neurochip" is derived from the human brain cell. "Expert" systems may lack common sense, so we want to develop computer "consciousness." Those expressions can be misleading. We may expect performance from a computer that it is not designed for. As the hammer is a tool extending the use of the arm, so is the computer a tool exceeding the brain in areas of poor brain performance.

Most users do not know why a computer acts the way it does. Not knowing causes some users to feel that computers operate by "magic" or what computers do is almost "human." In spite of giving computers human terminology, computers deny the irrational and intuitive by reducing experience to step-by-step programs that admit no contradiction or ambiguities. We should be glad that it is so or the computer would become a less reliable tool.

On the one hand there are many positive aspects of computerized technology. It is technically very effective, but on the other hand, every computer program is a mass of congealed logic calculated in advance to replace the human presence. Computerized devices run on "automatic." Everything that once followed from a meeting between two people such as needing a bank loan or expressing a grievance, becomes a mere "transaction" based solely upon standard, on-line data. We must recognize that only people can make companies productive by using computing for competitive advantage.

 Computer Accuracy

Engineers strive for computer assisted optimal designs. CAD information is then used at the construction site where robots do most of the building jobs. This would make use of the full productive capability of computers. Can this scenario be dangerous? What if the structure falls down? Who is to blame? At present, we are demanding low-cost user-friendly programs and assume that calculated results are unquestionably correct. It is very dangerous to use unverified results without the thought that they can have significantly wrong figures. An article in Scientific American (1992) claims that as much as 50% of all software systems may contain errors which are virtually uncorrectable.

One can read of many instances where fatal accidents resulted in "computer errors" (airplanes crashing, buildings collapsing, hospital treatment devices malfunctioning etc). Quality control of software is still lacking in spite of ISO 9000 requirements. We still seem to rely on vendor demonstrations and unreliable claims. There is no license required for software developers. To avoid the risk of bridges collapsing, we must develop backup systems that are completely independent of the main program.

An admirer looked at the beautiful bridge over the river and asked: "Wonderful! Who was the architect?"
When the bridge fell down, he asked: "Who was that incompetent engineer?"
The newspaper reported that the computer was to blame. The computer was charged with negligence.

 Effect on Productivity

The word "user productivity" or "productivity software" is often used in the promotion of products. Many of those products fall short of the claim. Have we seen much improvement in white-collar productivity over the past several years? Office software does almost everything, word processing, spreadsheets, communication, and maintains data bases. This integrated software is used to type letters, schedule appointments, keep historical file data, and play games, etc. Correspondence can now use hundreds of different fonts of different sizes (points) and can be lavishly embellished with graphics. Expensive laser printers and color copiers are used to take advantage of the software capabilities.

The time saved is being offset by the additional effort to include all these extras, many of which are absolutely unnecessary. *To be productive, we need to choose software which enhances the thought process and stimulates creativity.* Often, a simple note may do instead of a formal laser printed letter to the person sitting in the next office. Just speeding up routine office work with more sophisticated software does not seem to be the answer.

Neural Networks

A newspaper recently announced: "Bioelectrics will make a computer think!" It went on to say that a silicon chip has been created that behaves much like a human brain cell. Instead of working digitally, it works in the *analogue* mode, similar to our brain or the electric clock on the wall. Digital information is represented by zeros and ones, everything must be chopped up in pieces, pixels are either white or black. For most applications, this is not only acceptable but also desirable. *Analogue* is looking at subtly graded contours and continuously varying shades of gray. Neural networks can juggle many tasks simultaneously. Artificial neural nets (ANN) are trainable models that can either "learn"(human expression) to pick out relationships between cause and effect or are self-organizing, e.g. filtering noise from radio transmissions.

Neural Network Applications to Cost Engineering (42) uses the basic system model

INPUT ⟶ PROCESSES ⟶ OUTPUT

to describe the network.

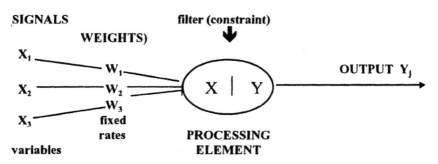

McKim uses an example for estimating the costs of pumps using the parameters
of flow and head. Application at this time is still not as common as conventional
computer applications. The Swedish carmaker Volvo uses an artificial neuronet to
grade the paint finishes of its automobiles and diesel knock in its engine
(Engineering Dimensions, Nov. 1993).

 Fuzzy Logic

Fuzzy logic is not a new idea. Lofty A. Zadeh published an article on "Fuzzy Set
Theory" in 1965 (Berkely CA) for the first time in America (it originated in Japan). Fuzzy logic exploits the tolerance admissible in systems control and behavior. It captures the subtle judgment humans make as compared to the exactness of
traditional information systems which exclude contradictions and inaccuracies.
Mathematical theories deal with either/or and equations. Fuzzy logic evaluates
expressions such as "the door is not quite closed or it is partially open". Humans
usually do not think in precise terms. Fuzzy logic translates inaccurate categories
into mathematical terms. Unfocused quantities have connecting factors that are
precise figures which can be used to program machines (robots for example).

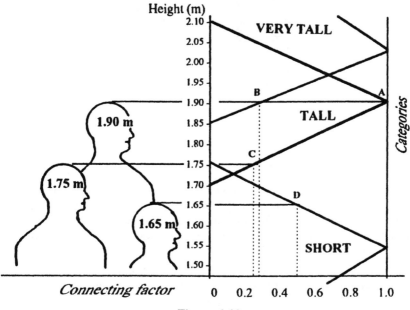

Figure 6-13
Fuzzy logic.

Assuming we have three categories of tallness for individuals: Very tall, tall, and short. This can be depicted on a sliding scale (Figure 6-13). An individual who measures 1.90 m in height describes himself as being tall. The horizontal line of 1.90 m crosses the very tall category at the connecting factor 0.25, meaning that he belongs actually to the very tall with a relatively low factor. The individual who measures 1.75 m considers himself not very tall. His connecting factor is 0.21 in the tall category. The one who measures 1.65 m says: "I am not tall". This not precise expression falls into the small category with a factor of 0.47. We now have categories and factors for all three men.

Fuzzy chips have been developed to speed the computer processing time. The application is mainly in robotics. Much research will still have to be done to convince the industry of its value.

On the humorous side, here is a fuzzy graph:

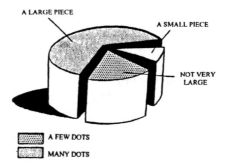

Figure 6-14
Fuzzy graph.

6.3.8 Value Engineering

Productivity is much more than just work efficiency. The objective of optimum performance is *effectiveness*. Even though it is extremely important for owners to minimize construction cost, they should not neglect to evaluate the total life cycle cost for total value and optimum return. That means recognizing the client's special needs such as quality, aesthetics, or marketability.

> **No matter how productive the workforce is and how low the
> capital cost are held, the project will fail when worth and cost
> are not matched.**

It all starts with the job plan. There is the temptation to take its usual structure for granted. Especially in regard to buildings, we often assume that there is little change in layout and design necessary (that's how we always do it). With the definition of the project scope we should ask ourselves: Is there anything else that can do the job? How can we improve the conventional design to better meet the needs of the client, add *value* to it and get a minimum of ten dollars for every one invested?

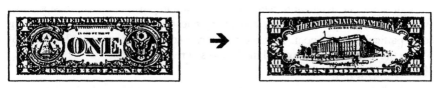

The combination of problem solving, method study and cost analysis performed in a creative way (see Chapter 7) is called *value engineering*. Value Engineering (VE) is specifically applied to project development, early design and product improvement. It is a formal step-by-step process with distinct problem solving phases interacting with each other (Figure 6-15).

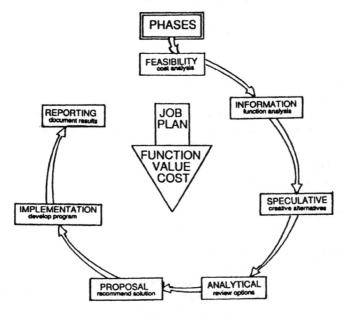

Figure 6-15
Value engineering.

Its application is most cost effective if a team of multidicipline members apply value engineering efforts against the highest cost components of a facility. This will yield the highest dollar savings. The required function of each of those components are identified, its value established, and optimum cost identified by analyzing alternatives that improve performance. The phases of the job plan are generally

📖 *Feasibility*

It is an economic evaluation to determine if value improvements and cost savings will result in adequate return by appointing a team performing value engineering on a particular component. This is essentially a cost/benefit analysis. The costs of doing the study must not exceed the benefits derived from it.

📖 *Information*

This includes setting up lines of communication and contacts. Information is then obtained from designers and estimators for current plans and specifications, design restrictions, codes and standards. The component is then broken down into functional areas for life cycle cost analysis. The "worth" of performing the basic function is the determined by expert opinion.

📖 *Speculative*

Creative thinking is necessary to develop alternative methods of performing the basic function. The most difficult task for engineers is the "lateral" thinking process (see 7.1.2) required to come up with new creative ideas.

📖 *Analytical*

The new ideas that were generated are now evaluated for each option and ranked in the order of life cycle cost savings and other advantages.

📖 *Proposal*

A convincing argument must now made why a certain option should be implemented. The proposal should include all the background information which leads to the improvement of the standard function.

📖 *Implementation*

When the decision is made to approve the proposal, design changes need to be made, and cost estimates need to be updated and documented.

📖 *Reporting*

Standard cost reporting procedures apply.

The importance of value engineering is recognized by the government. Bills HR133 and HR2014 were introduced to the 103rd US Congress requiring the application of VE in all federal agencies. They provide an incentive to do so in the form of an increase in monetary support to those states that can document positive VE results. This was done in the spirit of improving quality of services, minimizing the cost of construction programs, and reducing operating cost.

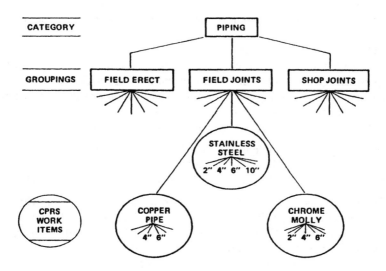

Figure 6-16
Productivity system breakdown.

6.4 PRODUCTIVITY CONTROL

To control productivity, we must establish a system that measures productivity as defined, then compares it with productivity standards and reports deviations from this standard.

6.4.1 Designing a System

Productivity measurement can be applied to
- Operating Labor
- Construction Labor
- Engineering Performance
- Manufacturing Operations

Even though there are considerable differences in the application, the same principles apply:
1.   Break down the work into categories such as
Work Items (Commodities)
Engineering Systems
Production Units

2.   Describe the content of each category. State inclusions and exclusions. Similar to the work breakdown structure, each category's content must be mutually exclusive.
3.   Content must be quantitative.

On a large project, each category can be broken down into groupings. Each grouping can again be broken down into several work items (Figure 6-16).

Assuming we are calling the system the Construction Productivity Reporting System (CPRS). The next step in designing the system would be to code each work item (or activity) in such a way that it can be summarized, e.g.

Code:	Category:	Grouping:	Work Item:	Activity:	Quantity:
4	P	FJ	CM06	JT	E
Piping	Pipe	Field Joint	6" Chr. Moly.	Weld Joint	Each

The Superintendent will now describe each work activity in detail with activity code, quantity code, inclusions and exclusions. Exclusions will show a reference to other activities. Those activity descriptions will be issued to the foremen concerned. For a viable cost control, the work breakdown structure code will also be shown under the activity description. This is a handy cross reference to another cost collection system. Table 6-04 shows an excerpt from a Superintendent's Work Activity Description.

The system now requires the collection of actual quantities installed and workhours expended on each work item. This is done at the working level by filling in a time distribution sheet.

Table 6-04

Activity description

Activity:	Description:	Quantity:	Reference #
SWGR	Installation of 4.16 KV Switchgear	Cell	53200500

Inclusions:	Exclusions:
1) All costs of installation	1) Cable and terminations
2) Material handling from ground floor to work location	2) Additional relays
3) Positioning leveling and securing of cells	3) Checking into service
4) Cell interconnections	4) Material handling to a prearranged drop off point (833051)
5) Breakers	5) Transformers

Table 6-05
Site productivity system.

1. Establish Cost Accounts ↓

Divide the project into logical blocks of work, say Account 2.640 - Sidewalks.

2. Quantity Take-off ↓

Measure the quantity of work to be performed for each cost account.
Estimate: 210 m sidewalk; 1.5×0.1 m ; 21 MPa strong concrete; $210 \times 1.5 = 315$ m^2

3. Establish a Budget ↓

Use historical data to develop the budget for each account. Budget for account # 2.640 = 315 m^2 $\times$ 0.44 Whrs/m^2 = 140 Whrs.

4. Measure Progress: ↓

As the work progresses, record the completed work, i.e. 90m x 1.5 m = 135 m^2

5. Collect Actual Cost: ↓

Charge labor hours to the cost accounts. As of..(date).. account 2.640 has been charged with 70 Whrs.

6. Report Project Status: ↓

Account 2.640: Original quantity = 315 m^2 Completed quantity = 135 m^2 Percent completed = 43 %. Actual hours used = 70 Whrs. Budget hours/m^2 = 0.44 Whr/m^2. Actual Whrs/m^2 = 70/135 = 0.52. Forecast (predicted) at completion = 70 + 94 = 164 Whrs. Over budget = 164 − 140 = 24 Whrs. Low productivity due to break-down of power screed.

7. Review Status Report: ↓

Management reviews and takes action. Repair of screed is time consuming. Decision: Rent a screed until the other is repaired.

8. Revise the Prediction: ↓

Required to complete is 60 Whrs. New forecast at completion = 70 + 60 = 130 Whrs

←9. Return to Step 4 to Measure Progress under New Plan

All this data is processed by a computer, which includes selection, calculation, comparisons and summarization at the various levels. Table 6-05 shows nine steps for a site productivity system (1):

6.4.2 Productivity Reporting

<u>Maintaining</u>   <u>Clean - up</u>   <u>Training</u>

Of all the workhours collected, only a certain percentage can be captured by the construction productivity reporting system. Certain jobs have to be performed on a construction site such as clean-up, job preparation, training, and many more that are excluded from performance monitoring. And not all of the remaining work is quantifiable. For example, on a particular nuclear station only 65% quantified Whrs applied (Figure 6-17).

On that job, there were over 240 work items quantified with 20 groupings and 11 main categories. The reports showed

♦   a target (or standard) man-hour unit rate
♦   an achieved workhour unit rate
♦   a performance index and
♦   allowed man-hours (quantity × target).

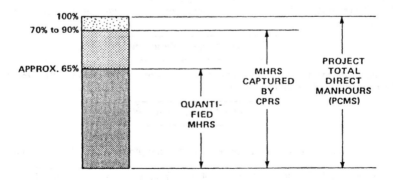

Figure 6-17
Quantified workhours.

The basic calculations needed to produce the reports:

(U_A) = Achieved Workhour Unit Rate

$$= \frac{\text{Workhours Captures by CPRS (M)}}{\text{Quantities Reported to CPRS (Q)}}$$

(T_D) = Divisional Target = A Standard Workhour Unit Rate

$(P.I.)$ = Performance Index = $\dfrac{Q}{M} \times T_D$

$(M_A) = Q \times T_D$

These reports were produced in tabular and in graphic form (Figure 6-18).

COMPARATIVE PRODUCTIVITY REPORTING SYSTEM
CUMULATIVE QUANTIFIED CPRS MANHOUR SUMMARY
PERIOD ENDING NOVEMBER, 19

CPRS NO.		DESCRIPTION	CUMULATIVE DIRECT MHRS CAPTURED BY CPRS	TOTAL DIRECT MHRS AS PER PROJECT COST STATEMENT	PERCENTAGE OF CPRS MONITORED MHRS
QT1XX	P	EXCAVATION	186,801	265,426	70.37%
QT1XX	B	EXCAVATION	158,925	344,472	46.13%
QT1XX	D	EXCAVATION	115,856	200,708	57.72%
QT2OX	P	BACKFILL	149,416	154,374	96.78%
QT2OX	B	BACKFILL	259,975	300,728	86.44%
QT2OX	D	BACKFILL	156,788	159,509	98.29%
QT21X	P	FORMWORK	1,901,847	2,201,695	86.38%
QT21X	B	FORMWORK	2,135,506	2,757,443	77.44%
QT21X	D	FORMWORK	2,695,108	3,488,701	77.25%
QT22X	P	CONCRETE	578,031	766,250	75.43%
QT22X	B	CONCRETE	631,809	937,926	67.36%
QT22X	D	CONCRETE	598,231	669,443	89.36%
QT3XX	P	STEEL	126,483	1,346,359	9.39%
QT3XX	B	STEEL	1,420,638	1,485,054	95.66%
QT3XX	D	STEEL	670,740	680,214	98.60%
QT4XX	P	PIPING	1,678,226	2,828,185	59.33%
QT4XX	B	PIPING	2,558,729	3,829,691	66.81%
QT4XX	D	PIPING	211,901	317,926	66.65%
QT5XX	P	ELECTRICAL	1,193,950	1,692,374	70.54%
QT5XX	B	ELECTRICAL	1,940,317	2,712,586	71.53%
QT5XX	D	ELECTRICAL	78,718	94,109	83.64%

Figure 6-18
Work item summary report.

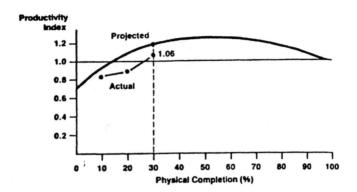

Figure 6-19
Productivity profile.

CPRS NO.		DESCRIPTION		MANHOURS	QUANTITY	ACHIEVED MIIR UNIT RATE
210	P	CONVENTIONAL FORMWORK	CUM	1,011,482	199,423 m^2	5.072
210	P	CONVENTIONAL FORMWORK	PER	0	0 m^2	0.000
210	B	CONVENTIONAL FORMWORK	CUM	1,212,511	206,586 m^2	5.869
210	B	CONVENTIONAL FORMWORK	PER	4,382	2,231 m^2	1.964
210	D	CONVENTIONAL FORMWORK	CUM	1,761,519	154,634 m^2	11.391
210	D	CONVENTIONAL FORMWORK	PER	157,924	16,036 m^2	9.848
211	P	PANEL FORMWORK	CUM	790,469	310,326 m^2	2.547
211	P	PANEL FORMWORK	PER	0	0 m^2	0.000
211	B	PANEL FORMWORK	CUM	798,510	208,273 m^2	3.833
211	B	PANEL FORMWORK	PER	1,207	0 m^2	0.000
211	D	PANEL FORMWORK	CUM	765,666	171,168 m^2	4.473
211	D	PANEL FORMWORK	PER	60,548	15,329 m^2	3.949
! 212	P	FORMWORK OTHERS	CUM	251,054	N/A	N/A
! 212	P	FORMWORK OTHERS	PER	0	N/A	N/A
! 212	B	FORMWORK OTHERS	CUM	527,557	N/A	N/A
! 212	B	FORMWORK OTHERS	PER	5,830	N/A	N/A
! 212	D	FORMWORK OTHERS	CUM	593,270	N/A	N/A
! 212	D	FORMWORK OTHERS	PER	66,580	N/A	N/A
! 214	P *	MOVING FORMWORK SYSTEM	CUM	N/A	N/A	N/A
! 214	P *	MOVING FORMWORK SYSTEM	PER	N/A	N/A	N/A
214	B	MOVING FORMWORK SYSTEM	CUM	47,973	34,018 m^2	1.410
214	B	MOVING FORMWORK SYSTEM	PER	0	0 m^2	0.000
214	D	MOVING FORMWORK SYSTEM	CUM	20,124	19,294 m^2	1.043
214	D	MOVING FORMWORK SYSTEM	PER	394	0 m^2	0.000

Figure 6-20
Productivity achievement report.

CUMULATIVE PERFORMANCE INDEX DATA (BASED ON CPRS GROUPINGS)
PERIOD ENDING — NOVEMBER, 19

CPRS NO.		DESCRIPTION	C-MHRS TO DATE	C-QTY TO DATE		DIV TARGET	ALLOWED MHRS	PERFORMANCE INDEX
G530	P	CONDUIT	320,746	1,063,562		0.293	311,623	0.971
G530	B	CONDUIT	358,613	1,068,562		0.293	313,088	0.873
G530	D	CONDUIT	8,696	19,419		0.293	5,689	0.654
G600	P	INSTRUMENTS & DEVICES	82,338	37,731 Each		2.168	81,800	0.993
G600	B	INSTRUMENTS & DEVICES	95,932	53,940 Each		2.168	116,941	1.219
G600	D	INSTRUMENTS & DEVICES	0	0 Each		2.168	0	0.000
G620	P	S/S TUBING	554,354	679,236		0.845	573,954	1.035
G620	B	S/S TUBING	387,800	612,432		0.845	517,505	1.334
G620	D	S/S TUBING	249	757		0.845	639	2.568
G640	P	COPPER TUBING & PIPE	214,467	344,766		0.640	220,650	1.028
G640	B	COPPER TUBING & PIPE	87,662	242,372		0.640	155,118	1.769
G640	D	COPPER TUBING & PIPE	0	0		0.640	0	0.000
TOTAL P			6,913,656				6,812,993	0.985
TOTAL B			8,383,969				9,455,168	1.127
TOTAL D			4,679,381				3,679,187	0.786

Figure 6-21
Performance index data.

Productivity changes with the phases of job completion. This is especially true in civil work. In the beginning of a job, big machines dig deep holes and bulk concrete is being poured. Later, there is less access, more congestion, greater heights combined with smaller pours (beams, plinths, floors). This affects the unit rate and in turn the productivity index. Actuals rates must therefore be compared with a productivity profile that changes with physical completion (Figure 6-19)

Figures 6-20 and 6-21 show examples of other reports.

6.4.3 Production Studies

Productivity reports provide management with means to identify problems in productivity performance. This must now be followed by appropriate action to improve performance.

The identification of productivity problems is followed-up by what is sometimes called "Production Studies". It is actually some form of method study or work study (see 7.2.- Problem Solving). There are three basic phases in doing the study:

1. Investigate
    deviation from standard
    reason for production problem
    data collection problems
    organizational deficiencies
    work methods
    lack of incentives
    other problems

bright ideas

2.   Record Findings
         analyze present situation
3.   Recommend
         potential improvements
         how to implement

After implementation, the results are closely monitored to ensure that improvements have been made.

6.4.4 Activity Sampling

To supplement direct productivity measurement and derive at specific indications of site productivity, activity sampling (also called work sampling) is sometimes done. Activity sampling measures the time spent on work (and non-work) activities rather than on work output (25). Activity Sampling is based on statistical analysis of random observations. Various categories of activities are tallied through "snap shot" observations.

According to a survey made by Dr. John Borcherding (26), the average percentage of activities for all crafts on a typical nuclear power plant was:

Direct Work...............	32%
Waiting...................	29%
Moving about.............	13%
Lateness, personal breaks	11%
Receiving instructions	8%
Carrying tools, materials	7%

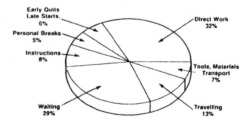

Direct work done on Ontario Nuclear Stations (25) were approximately 47% according to a study done by a consultant for Ontario Hydro (Thorne, Stevenson & Kellogg).

Statistical Basis

Working activities are regarded as being composed of "moments". A sample of these moments, taken at random, will reflect the activity pattern of the total work cycle. The more samples we take, the more accurate the result will be. (Similar to flipping a coin. If you do it often enough, you should get as many heads as tails.)

**We therefore assume statistically that random samples follow a
normal distribution**.

The formulas (Figure 6-22) show the standard deviation of the binomial probability distribution, i.e. the distribution associated with the occurrence of an activity (p) or its non-occurrence and the standard error (l) of the distribution.

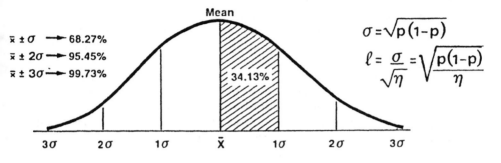

Figure 6-22
Activity sampling formulas.

This is the range of error by which the "study mean" deviates from the theoretical (true) mean.

Assuming we are shooting for a confidence limit of 95% or two standard deviations from the mean (2σ),the number of observations needed are:

$$N_{2\sigma} = \frac{4P\left(100-P\right)}{L^2},$$

where "L" is the degree of accuracy or "precision" of the sampling. Assuming a 2% accuracy and 8 trade groups (classifications) to be observed, the number of observations to be made are 20 000, and assuming that work over non-work (p) is 50%, how many observers will be needed? Experience tells us, that an observer can make 300 observations per day under normal conditions. If the job has to be done in 15 days with a 75% observer utilization, we need 6 observers.

$$N = 8 \times \frac{4 \times 50\,(100 - 50)}{2^2} = 20{,}000$$

Needed at ±2% Accuracy:

$$\frac{20{,}000}{15 \times 300 \times 0.75} = 6$$

There are several approaches to activity sampling. The two best known are the "tour approach" and the "crew approach". The tour approach tallies *all* craftsmen whereby the observer walks a pre-determined route. This way, a large number of samples can be collected in a relatively short time. It gives a macro view of the work activities relative to the construction site as a whole.

The *crew* approach is more selective and rather applicable to larger projects. The total study is divided into "cells." Observations within each cell are randomized. A large project could have about 1200 tradespeople. Of those (nominal population) foremen, crafts working in shops, warehouses, and gangs working away from the plant area are excluded from the study. This now could leave less than 1100 tradespeople as a "population of interest." This population of interest is then stratified into trade groups such as carpenters, laborers, riggers, electricians, etc. (Figure 6-23).

The next step is to identify the foremen for each trade group and the number of tradesmen in the group. In our example, the team had 15 days to perform the study with six observers for eight trade groups. From this, the leader of the study group prepares a preliminary assignment schedule for the observers, considering the size of the sample population in each trade group.

This now leads to the preparation of a *specification* sheet which shows the work areas where specific trades have to be sampled. The example below (Figure 6-24) shows 37 pipe fitters working under three different supervisors. There are six observation areas the sampler has to cover and tally what he sees. He is alternating 1 to 6 at random in regard to time and path. This will reduce some bias that may exist by knowing that an observation is going to be made.

Functional Group:	1 Nominal Population			2 Population of Interest			3 Sample Size		
	Bruce	Pickering	Darlington	Bruce	Pickering	Darlington	Bruce	Pickering	Darlington
Piping	352	217	91	308	165	58	276	137	54
	(338)	(405)	–	(331)	(380)	–			
Structural/ Concrete	330	294	811	292	236	751	244	205	744
	(492)	(384)	–	(328)	(314)	–			
Instrumentation	313	239	N/A	278	214	N/A	244	190	N/A
	(195)	(319)	–	(183)	(300)	–			
Electrical	266	192	35	249	157	33	190	125	26
	(324)	(243)	N/A	(279)	(186)	–			
Mechanical	297	251	305	244	153	257	235	125	125
	(.527)	(159)	N/A	(366)	(156)	–			
Total	1558	1193	1242	1371	925	1099	1189	782	949
	(1876)	(1510)	N/A	(1487)	(1336)	–			
*Percentage of Nominal Population = $\frac{2}{1}$				* 88%	78%	88%	**87%	85%	86%
				(79%)	(88%)				
**Percentage of Population of Interest that was sampled = $\frac{3}{2}$									

Notes:
● Figures in brackets (. . .) are last year's statistics.

Figure 6-23
Population of interest.

The observer will be able to make several tours during a day. Up to six observers could be on site at any time, anywhere. In the example below, walk #3 was scrapped because 4 tradesmen worked on insulators outside the study area. In this particular case, the sample size was reduced from 37 to 24.

Many precautions have to be taken before the actual observations start. The sampling program must be well publicized in order to avoid labor unrest. Meetings must be held with the Personnel Department and labor union representatives. Facilities and services must be available at the time the study starts. Any interruption can seriously affect the flow of sample taking which in turn will bias the results. The computer must be ready to accept immediate input with at least a daily turn-around.

SAMPLE SUB-GROUP SPECIFICATION SHEET

DATE FOR: 23 SEP		PREPARED BY: KH	SHEET 1 OF 1	GEN.FOREMEN:	VIC WHALEN KEN COOPER AL ZOLDY
#	Task or Work Description	Work Site(s)/Location(s)	Trade(s)	No.Men	Supervisor(s)
1	Fab. Pipe	RB 3 (Beth)	PF	22 14	A. Beth
2	" "	RB 3 (Harris)			
3	Install const. steam line	TH3 , TH 2 , RB2 (Beth)	PF(i)	4̶ 4	" insulators are not studied
4	" serv. water piping	PH #2	PF	9̶ 8	J. Burke
5	Testing EPs	Water Treatment Area (Burke)	PF	2 ✓	B. Harris
6	Insulate steam line pipe	TH3 (Beth)			
				37→	population to be studied
				24	sample size
					OK

OBSERVATION TOUR ROUTE ALTERNATIVES	A	1 to 6 alternated	NOTES: OBSERVER: Reg. Norris (Beth) J.Burke: One PF sick, 3 out of area of study = 4, A. Beth: 2 PF night shift, 2 PF sent to shpp, - 4 exclude insulators = 4 /(12)+ 1 moved out
	B		
	C		
	D		
	E		
TOUR I.D. NUMBERS			

Figure 6-24
Sample specification sheet.

MANAGEMENT DELAYS
Wait for other trade
Wait for materials, tools
Wait for instructions

PRODUCTIVE ACTIVITIES
Direct work
Job preparation/Lay away
Travel and transport

PERSONAL ALLOWANCES
Taking a break
Personal needs
Prepare for work, wash up >15 min.

WORK ALLOWANCES
Discussing job
Balancing delay
Obtaining information
Incidental stoppages
Safety observers

NON-ACTIVITY
Observation not possible

Activity categories are designed such that the study objectives can be met, i.e.
- Plan the work better
- Give adequate instructions to the workforce
- Reduce personal and work allowances
- Improve supervision and communication
- Have tools and materials available when needed.

Based on those objectives, several categories can be established. Below are some sample observations and their interpretations (Figures 6-25 to 6-27).

Sub-categories can be reduced for ease of observation. For example, *Discussing Job* and *Obtaining Information* could be combined, as could *Taking a Break* and *Personal Needs. Management Delays* may not need any sub-categories. It all depends on the purpose of the study.

If there are too many categories, observations are subject to too many interpretations by the observer. Some of the above are very much subject to interpretation.

Travel-   **Figure 6-25**   *Balancing*
loaded.   *delay.*

Figure 6-26

Direct Work   *Obtaining*
assisting.   *information.*

Figure 6-27
Taking a break   *Job preparation?*   *Attending to personal need*

Job preparation could also be "obtaining information" or even "incidental delay" by talking to the individual who takes a coffee break. But then again, they may be "discussing the job" (Figure 6-27).

Sampling Tally Sheet

Observer:	Tour Number	1081		1081		1082		1082		1082	
T. McC.	Trade Group	CA		LA		CA		LA		CF	
	Day of Week	2		2		2		2		2	
	Shift	D		D		D		D		D	
	Date	9 Nov 82		9 Nov 82		9 Nov 82		9 Nov 82		9 Nov 82	
	Start Time	8:30		8:30		9:28		9:28		9:28	
	Finish Time	9:20		9:20		10:14		10:14		10:14	
	Sample Size	24		15		23 *		15		2 **	
No.	Activity Description	Tallies	Σ	Tallies	Σ	Tallies	Σ	Tallies	Σ	Tallies	Σ
Directs 101	Direct Work	ℍℍ ℍℍ ℍℍ	15	ℍℍ I	6	ℍℍ IIII	9	ℍℍ II	7	I	1
201	Job Preparation/Lay Away	IIII	4								
Work Allowance 301	Discussing Job	I	1			II	2				
302	Balancing Delay	II	2	III	3	III	3			I	1
303	Obtaining Information – By Self							I	1		
304	Incidental Delays and Work Contingencies										
305	Safety Observer							II	2		
Personal 401	Taking a Break on the Job			II	2	III	3	III	3		
402	Attending to Personal Needs										
403	Prepare/Wash Up at Start, End, Break > 15 min.										
Travel 501	Travel/Transp. – Loaded			III	3	I	1				
502	Travel/Transp. – Unloaded										
Delays 601	Wait for Other Trade					II	2				
602	Wait for Material, Tools, etc.										
603	Wait for Instructions, Superv.							II	2		
	Total Observations	22		14		20		15		2	
No Observ. 701	Observation Not Possible < 15 min. After Start										
702	Observation Not Possible – Within Time	II	2	I	1	III	3				
703	Observation Not Possible < 15 min. Before End										
	Total Tallied	24		15		23		15		2	

Notes:
* One carpenter went home sick at 9:30
** Included 2 cement finishers. They joined the group at 9:20

Figure 6-28
Sampling tally sheet.

It should be remembered that the observer does not linger on to find out, he makes fast "snap shot" observations that are entered immediately into a tally sheet. Below is a sample of a tally sheet (Figure 6-28), which observer "T.McC." filled during one tour, starting at 8:30 and observing two trade groups (# 1081 and 1082).

The result of the tally is fed into a computer right after the observations are completed (Figure 6-29). Good data processing services and fast turn around are very important because future tour assignments have to be prepared at the end of the working day for the following morning. The assignments are dependent on statistical results summarized by the computer.

The accuracy of observations is calculated for each trade group as shown in Figure 6-30.

It is shown in the Figure 6-30 example that, after five days of observation, the two groups (pipe fitters and boiler makers) have been tallied 600 and 100 times respectively. The ratio of total observation over work activity for the two trades is 43% and 50%.

Population Activity Estimates for Functional Groups and Total Sample

Group No. of Obs. Made	Piping 1591 (2588)	Structural 3050 (1895)	Instrument 2928 (2531)	Electrical 1312 (720)	Mechanical 2478 (1270)	Total Sample 11 359 (9004)
Direct Work	43.4 ± 2.5 (44.9 ± 2.0)	51.6 ± 1.5 (51.1 ± 2.3)	48.4 ± 1.8 (50.6 ± 2.0)	37.9 ± 2.7 (39.2 ± 3.6)	47.7 ± 2.0 (40.5 ± 2.7)	47.2 ± 0.9 (46.7 ± 1.1)
Job Preparation	5.6 ± 1.2 (6.3 ± 0.9)	4.5 ± 0.6 (5.0 ± 1.0)	4.9 ± 0.8 (5.1 ± 0.9)	7.0 ± 1.4 (4.5 ± 1.5)	6.4 ± 1.0 (4.5 ± 1.2)	5.5 ± 0.4 (5.3 ± 0.5)
Work Allowance	19.3 ± 2.0 (13.0 ± 1.3)	14.8 ± 1.0 (13.4 ± 1.6)	13.6 ± 1.3 (11.7 ± 1.3)	17.3 ± 2.1 (13.2 ± 2.5)	19.0 ± 1.6 (16.2 ± 2.1)	16.3 ± 0.7 (13.2 ± 0.7)
Personal Allowance	18.0 ± 2.0 (22.1 ± 1.6)	14.7 ± 1.0 (17.9 ± 1.8)	19.4 ± 1.5 (23.1 ± 1.7)	20.9 ± 2.2 (26.2 ± 3.3)	14.7 ± 1.4 (21.6 ± 2.3)	17.0 ± 0.7 (21.7 ± 0.9)
Travel	11.8 ± 1.6 (9.9 ± 1.2)	10.8 ± 0.9 (7.8 ± 1.2)	12.0 ± 1.2 (7.3 ± 1.0)	14.8 ± 2.0 (13.9 ± 2.6)	10.0 ± 1.2 (7.4 ± 1.5)	11.6 ± 0.6 (8.7 ± 0.6)
Mgmt Delay	1.9 ± 0.7 (3.8 ± 0.8)	3.6 ± 0.5 (4.8 ± 1.0)	1.7 ± 0.5 (2.2 ± 0.6)	2.1 ± 0.3 (3.0 ± 1.3)	2.2 ± 0.6 (9.8 ± 1.7)	2.4 ± 0.3 (4.4 ± 0.4)

Notes:

1. Results are expressed as percentages of observations made and are corrected to a 38-hour week.

2. Figures in brackets are last year's results.

Figure 6-29

Sampling - computer output.

MONITOR OBSERVATIONS

STUDY DAYS	TRADE GROUP	OBSERVATIONS		RATIO %	ACCURACY (PRECISION)
		WORKED	TOTAL		
5th	PF	260	600	43%	± 4%
	BM	50	100	50%	± 10%
7th	PF	680	1 440	47%	± 2.6
	BM	130	280	46%	± 6%
10th	PF	890	1 930	46%	± 2.2%
	BM	130	650	52%	± 4%
12th	PF	890	1 930	46%	± 2.2%
	BM	590	1 180	50%	± 3%
15th	PF	890	1 930	46%	± 2.2%
	BM	880	1 730	51%	± 2.4%

$$L_{2_r} = 2\sqrt{\frac{43 \times 57}{600}} = \pm 4\%$$

Figure 6-30
Monitoring observations.

Because the boiler makers have a lower accuracy (10% vs. 4%), the tour specification would have to put more emphasis on the touring of boiler makers. The result is a very close accuracy on the 15th day (2.2% vs. 2.4%).

Activity measurement, when used and interpreted properly, can be a valuable diagnostic tool for implementing productivity improvement programs. The technique is relatively simple and inexpensive. However, since it does not address work output per workhour it can only supplement, not replace, existing productivity measurement systems.

Detailed tables and graphs are produced at the end of the study period to analyze and evaluate the information. The time-of-day graph (figure 6-31) is an example which compares the productive activity with direct work plotted in 15-minute intervals. The horizontal axis shows percent of total time spent on an activity. The vertical axis is the time scale. It can be seen in this example, that work activities drop sharply an hour before quitting time.

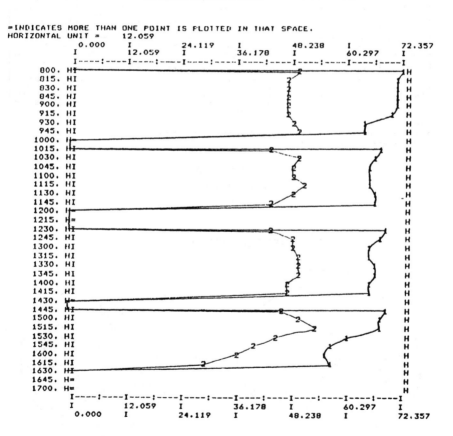

Figure 6-31
Time-of-day graph.

Population Activity Estimates for Functional Significant Trades

Group No. of Obs. Made	Carpenter 1083 (874)	Labourer 1736 (836)	Inst. Pipefitters 1549 (1443)	Inst. Elect. 1379 (1088)	Iron-Worker 1515 (876)	Boiler-Maker 515	Electrican 1312	Pipefitters 1591
Direct Work	54.2 ± 3.0 (52.8 ± 3.4)	48.3 ± 2.4 (47.7 ± 3.4)	42.9 ± 2.5 (48.6 ± 2.6)	54.5 ± 2.7 (53.4 ± 3.0)	47.7 ± 0.4 (34.4 ± 3.2)	44.5 ± 4.4 (51.2 ± 4.4)	37.9 ± 2.7 (39.2 ± 2.7)	43.4 ± 2.5 (44.9 ± 2.5)
Job Preparation	5.1 ± 1.3 (4.7 ± 1.4)	4.0 ± 0.9 (4.3 ± 1.4)	6.6 ± 1.3 (7.1 ± 1.4)	3.0 ± 0.9 (2.5 ± 1.0)	5.9 ± 1.2 (3.2 ± 1.2)	8.7 ± 2.5 (4.7 ± 1.9)	7.0 ± 1.4 (4.5 ± 1.1)	5.6 ± 1.2 (6.3 ± 1.2)
Work Allowance	15.4 ± 2.2 (12.4 ± 2.2)	15.4 ± 1.7 (16.5 ± 2.6)	12.4 ± 1.7 (14.6 ± 1.9)	14.8 ± 1.9 (7.9 ± 1.6)	10.4 ± 2.0 (17.8 ± 2.6)	10.7 ± 3.3 (11.7 ± 2.8)	17.3 ± 2.1 (13.2 ± 1.9)	19.3 ± 2.0 (13.0 ± 1.2)
Personal Allowance	15.5 ± 2.2 (20.2 ± 2.7)	14.5 ± 1.7 (17.0 ± 2.6)	19.9 ± 2.0 (18.4 ± 2.0)	18.9 ± 2.1 (29.1 ± 2.8)	13.9 ± 1.8 (24.0 ± 2.9)	16.4 ± 3.3 (20.5 ± 3.6)	20.9 ± 2.2 (26.2 ± 2.4)	18.0 ± 2.0 (22.1 ± 2.1)
Travel	6.5 ± 1.5 (5.4 ± 1.5)	13.8 ± 1.6 (10.8 ± 2.2)	16.4 ± 1.9 (8.4 ± 1.5)	7.1 ± 1.4 (5.8 ± 1.4)	11.1 ± 1.6 (8.2 ± 1.8)	11.1 ± 2.8 (7.0 ± 2.2)	14.8 ± 2.0 (13.9 ± 1.9)	11.8 ± 1.6 (9.9 ± 1.5)
Mgmt Delay	3.3 ± 1.1 (4.5 ± 1.3)	4.0 ± 0.9 (3.7 ± 1.3)	1.7 ± 0.7 (2.9 ± 0.9)	1.6 ± 0.7 (1.3 ± 0.7)	2.1 ± 0.7 (12.4 ± 2.2)	2.7 ± 1.4 (4.9 ± 1.9)	2.1 ± 0.8 (3.0 ± 0.9)	1.9 ± 0.7 (3.8 ± 1.0)

Notes:

1. Results are expressed as percentages of observations made and are corrected to a 38-hour week.
2. Figures in brackets are last year's results.

Figure 6-32

Activity estimates.

WORK ACTIVITIES – MAJOR CATEGORIES

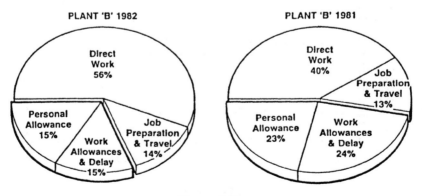

Figure 6-33

Sampling pie graphs.

Functional Group	Trade Group Abbreviation	Trade Group	Trade Colour
Piping	PF	Pipe Fitters	Orange
	PI	Pipe Insulators	Green/Red
Structural/	CA	Carpenter	Green
Concrete	LA	Labour	Yellow
	CF	Cement Finisher	Brown Wood Grain
	PA	Painters	Brown
Electrical	EL	Electricians	Grey
Mechanical	BM	Boilermaker	Black
	IR	Iron Rigger	Purple
	OE	Operating Eng.	Red

Figure 6-34

Trade group breakdown.

Figure 6-32 shows sampling results over a period of two years in tabular form. The various trade groups are shown and the number of total observations made for those trade groups. This is compared with total observations in the previous year. A percentage breakdown by activity is listed under each trade group. For example, a laborer did less direct work than a carpenter, but had almost twice as much travel time. Those statistics raise the question: "Why?" An investigation may show, that a laborer is required to do more clean-up work. Personal allowance is almost the same.

Figure 6-33 is a pie graph report used for a presentation to upper management.

It is obvious that the sampling from the previous year gave the managers and foremen an incentive to put in place a productivity improvement program. Direct work increased from 40% to 56% in one year. However, there may have been other reasons for the improvement, depending on the physical completion of the plant (see Figure 6-19). There is less congestion during the early stages of a construction project and thereby less interference with work activities.

Figures 6-34 and 6-35 are preparation sheets in the execution of the study. It gives a breakdown by trade and raw data that is recorded on a day-to-day basis.

This data is helpful to the analyst of the work activity study. It is of strategic value. The assignment of the next day's observations depend on it.

Week No.	Date	Tour No.	No. of Observations	Cumulative Total	No. of Observers
	15 Nov Mon	1250-1279(30)	693	693	3-1/2
	16 Nov Tues	1280-1319(40)	879	1,572	5
1	17 Nov Wed	1320-1347(28)	817	2,389	5
	18 Nov Thur	1348-1384(37)	1084	3,473	5
	19 Nov Fri	1385-1411(27)	780	4,253	4-1/2
	22 Nov Mon	1412-1444(33)	690	4,943	4-1/2
	23 Nov Tues	1445-1478(34)	899	5,842	5
2	24 Nov Wed	1479-1503(25)	760	6,602	5
	25 Nov Thur	1504-1543(40)	1382	7,984	5
	26 Nov Fri	1544-1569(26)	464	8,448	4
	29 Nov Mon	1570-1600(31)	928	9,376	5
	30 Nov Tues	1601-1637(37)	987	10,363	5
	01 Dec Wed	1638-1674(37)	868	11,231	5
	02 Dec Thur	1675-1691(17)	238	11,469	5

Figure 6-35
Sampling, raw day-to-day data.

6.4.5 Foreman Delay Survey

One of the major causes for low productivity are delays in the work flow. This is mainly caused by poor management at all levels. Insufficient planning and organization causes all kinds of delays. Even though activity sampling is less subjective, a Foreman Delay Survey can give a more detailed analysis of non-work activities (1). Each foreman will make a daily evaluation of hours lost by problems causing delays (Figure 6-36):

Foreman Delay Survey

CRAFT:

NAME OF FOREMAN: GENERAL FOREMAN:

(DATE) DAILY EVALUATION: NUMBER IN CREW:

PROBLEMS CAUSING DELAY

	MANHOURS LOST		
	Number of Hours	X Number of Men	= Manhours
1.a Waiting for materials (warehouse)			
1.b Waiting for materials (not received or not ordered)			
2. Waiting for tools or tools not available			
3. Waiting for equipment			
4. Equipment breakdowns			
5.a Changes/redoing work (design errors)			
5.b Changes/redoing work (prefabrication errors)			
5.c Changes/redoing work (field errors)			
6. Move to other work area			
7. Waiting for information			
8. Interference with other crews			
9. Overcrowded working areas			
10. Plant coordination/authorizations			
11. Other_____			

COMMENTS: _____

Figure 6-36
Foreman delay survey.

Q #6.08:

In order to develop a labor productivity standard, which would you prefer, unit cost data or productivity data?

A #6.08:

A given unit cost can consist of multiple inputs. Mixed crews made up of verying crafts with different wage rates can produce cost data that is hard to analyze. Historical productivity data are less sensitive to changes over time. It would therefore be preferable to use productivity data as a standard against which future performance can be measured.

Q #6.09:

Five trade groups with a previous record of 45% direct work activity are to be sampled at a construction site. How many observations are necessary to state with 95% confidence that the accuracy of the result is within ±2 percent?

A #6.09:

95% is close to two standard deviations, therefore

$$N = 5\frac{4 \times 45(100 - 45)}{2^2} = 12\ 375 \text{ observations}$$

Q #6.10:

It takes a crew of two sheet metal workers working eight hours and two laborers working four and one half hours to install 100 m^2 of 0.5 mm thick aluminum siding on a building. What is the productivity of the crew and what are the unit cost if sheet metal workers charge $26/hr and laborers charge $18/hr?

A #6.10:

Productivity is   (100 m^2)/(26 Whrs) = 4 m^2 / Whr
Cost: Sheet metal workers   2 × 8 × 26 = $416
         Laborers   2 × 4.5 × 18 = $162
                    Total cost per 100 m^2 = $578

                    Unit cost = $578/100 = $5.78/m^2

CHAPTER 7
PROBLEM SOLVING FOR MANAGEMENT

The objectives of this chapter are:

1. to explain the concept of "horizontal" thinking as compared to both, "natural" and "vertical" thinking for the purpose of increasing creativity on the job.

2. to outline the use of method study techniques to increase the effectiveness of systems and to improve the productivity of resources.

A model of the mechanism of mind will give us a better understanding of the process of creative thinking.

Applied problem solving uses a strategy consisting of six basic steps. Their concept will be explained to better understand the benefits of method study.

Introduction

If you think creativity is a mysterious gift, you can only sit and wait for ideas. But if creativity is a skill, you ought to learn it.

Edward de Bono

Our school system is very much based on the logical or vertical thinking approach. This approach is very effective in the sciences and engineering. However, it is rigid and built up on established fixed foundations. *New* ideas are often related to insight and humor which is outside the logical thinking process. A discussion on a "model of the mind" may help to understand how incoming information ingrains upon our mind, thereby blocking unconventional insights.

Method Study deals with the elimination of unnecessary effort and waste by critically examining existing conditions, generating and analyzing alternatives for improvement through imaginary thinking, and implementing *better ways of doing things*.

During the early stages of a capital project, a team of method study experts can generate ideas that can determine which initial concepts are worth developing into a project proposal. Even though this is the most cost effective time, method study techniques can be used. Those methods are also applicable to improve productivity and in solving other problems management may encounter at *all* stages of a project.

One of the indulgences of older people is to be amused when a new generation discovers the wisdom of the past. It was once profitable as well as faddish to have work measurement and work simplification programs imposed on workers of all description. The vicious snap of the clip board was heard throughout the land.

Efficiency experts were created by the thousands. But somehow it all ceased to be popular, the fad was over and all these good tools were sealed away and forgotten. Many efficiency experts became computer system analysts and we all became too busy with electronic technology to notice that a big pile of fast printouts and proliferation of data was not necessarily a panacea to our problems. Double digit inflation in the 1980s, followed by an economic recession seemed to trigger the key to salvation:

Productivity improvement.

However, productivity requires efficiency. Efficiency requires a methodology. But this methodology was there all the time. We have now seemed to re-discover the *systems approach* to problem solving. (Some specific applications are using different names and terminology for basically the same approach, i.e. Configuration Management, Value Engineering etc.)

It is not the old "clip board efficiency" motion study of the worker any more (this could be politically incorrect). *The technique has evolved into a creative integrated method study approach.*

The state of mind and the mode of thinking by decision makers are very important to success. Strict logic and the application of familiar procedures are sometimes not enough.

7.1. CREATIVE THINKING

Create comes from the Latin word *creare*, meaning to produce or to bring about something new. We often make the assumption that we only have to collect more and more information for it to sort itself into useful ideas. Traditional education is concerned with conveying established knowledge and ideas. Whenever we have reason to change those ideas we are in for a challenge. There is always opposition to change. The new idea may adopted, replacing the old one by the method of science and logical thinking. This usually will add to human knowledge and the old idea becomes stronger and more rigid.

There is another, more effective way of changing ideas. It is through insight and creativity. This way of changing ideas is not from outside by conflict, but from within. Information can not always be evaluated objectively. It sometimes needs intuition, imagination and insight to rearrange established ideas. This leads to creative thinking.

7.1.1 A Model of the Mind

When an engineer is required to make complicated designs based mainly on empirical data such as river diversions, it is very helpful to have a true scale model available. This model can be used to observe and measure water flows, leading to the development of formulas for calculating river structures and contours.

Because of the different size, the model can never be completely accurate. Every stone and sand bank in the river cannot be duplicated and reduced in the model. The physical properties of the water (moles) are not "scaled down" and allowance must be made for that (similitude). Other examples of models are wind tunnels, ship models, economic models, mathematical model, environmental models (greenhouse effect) etc.

> **It seems, whenever we want to understand or explain some-
> thing in the real world that is too complicated to be used as a
> prototype, we use the second best approach-the Model.**

To understand *creative thinking* and *creativity* better, we should consider building a model of the mind.

The Brain System

Cogito, ergo sum (I think, therefore I am.)
Descartes

It would be useful, if we could understand, how our brain system handles information. This would help us to make more effective use of it. Especially during the last 20 years, brain researchers have come a long way to give us a much better understanding of the "hardware" of the brain. Psychologists observed the effects of human thought processes to explain human behavior.

However, it is generally conceded that little is known of how the brain manipulates concepts, how the brain puts meaning into our thoughts, how fantasy, facts and emotions are intermingled in our minds.

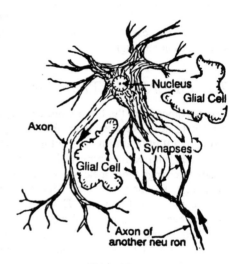

Figure 7-01
The neuron of the brain

Figure 7-01 shows a diagram (drawn by the author) on which a model of the brain can be based. There are about 10 billion nerve cells (neurons) in our brain with 200 000 entry ports (synapses) each. An output channel (axon) is attached to the body of the cell. The cell body has a nucleus where "decisions" are made. Input and output are like electrochemical flows, i.e. ions on a pulsating move. The nucleus decides when to release or *fire* ions down the axon. This can take place up to thousand times per second.

It is a simple decision:

```
              NO   ◇   YES
            ◄─────FIRE─────►
```

This decision is made only after all inputs exceed a certain threshold. It means, the **simple rule of additions apply.**

cogito, ergo am (I think, therefore I "sum".) Pun intended.

How can we visualize the functions? How does a neuron attractions? It is suggested that low threshold brain areas (ion-hungry neurons ☺) will attract ions to flow in their direction. This takes place at a higher level of memory. When a neuron is activated, a stream of ions flows down the axon which divides into several pairs of unequal branches. Therefore, one single pulse, originating at the cell body will split and each separate stream of ions will reach their destination at different times. The neuron needs a short recovery time before it can fire again.

This, in simple terms, is the lowest level organization of mind. Hofstadter (30) in his book Gödel, Escher, Bach calls neurons the brain's "ants" and discusses larger structures in the brain and the concept of higher level "signals" and "symbols." This, in a few words is his model:

Observing an ant colony, movement of individual ants seems random, but involving large numbers of ants, trends can emerge from that chaos. It seems that group phenomena which have coherence - trail building for example - will take place only when a certain threshold of ants get involved.

At any one moment, in any small area of a colony, there are ants of all types present, each specialized in its activity. The individual ant gets fired up when a task for which it is "designed" (load carrying, digging, fighting etc.) is in front of it. A "signal" will bring it to the place of the task. When ants receive the "signal," they form little teams and unform when the task is done. There is a delicate cast distribution among the ant colony which has evolved over millions of years due to the pressures of survival. Local matters of urgency, such as nest building and nursing will have the right composition of team members. This is a higher, second level of structure, which will receive another "signal" from the next, third level up, whose members are not ants, but teams on the lower level. Therefore, individual ants are not aware of a higher level than their own. There are several levels and the colony itself is at the highest level.

To summarize, it is a model that describes the thought processes as shifting from the *signal* level to the *symbol* level, which can be activated (triggered) from the outside. The interrelationship between symbols gives us a realization of concepts. In other words, individuals *react* to outside stimuli in accordance to the inherent network of symbols our brain possesses. Since the learning process during our life slightly modifies the inherent network, the input is *channeled into concepts that are solidified over time.*

Solidification of concepts can be considered the *natural* way or the *learned* way of thinking. Familiar information is processed to fit existing channels, further solidifying the mind.

Experiment 1: Do not look at the sign to the right. You already did ? Then look away now and tell what the sign said.

If you read: "PARIS IN THE SPRING", the familiar has been solidified, changing the incoming information.

Experiment 2: Please put you hand over your watch. Leave it there and describe what your [SIX] looks like. Is it 6 or VI or a line |, or blank?

One may have worn the watch for years and yet many are not able to tell. This is a case, where information is not consciously registered (threshold not reached) because it is not necessary for the final answer.

Is there a simpler model of the mind? Edward de Bono (10) developed an intriguing model of the brain which in turn let to his definition of lateral or creative thinking.

The following concept is one of several described by Edward de Bono. The short outline below can only give a cursory view of his fascinating "model of the mind".

NOTE: Edward de Bono is regarded as a leading authority in the field of creative thinking. His contribution has been to transform a mystical process into the behavior of a self-organizing information system. From this base he has developed the formal tools of lateral thinking for developing new ideas. See his book **SERIOUS CREATIVITY** *(Harper 1990)*

The Memory Surface

Consider a flat dish of ordinary table jelly as a model of the brain's memory. If we now use a fine spray of hot water randomly hitting this "memory surface", we may observe the following:

Individual tiny droplets that are still hot will hit the surface and combine with other droplets and in the process dissolve some of the gelatin. A small depression is formed. With another spray of hot water the surface of the jelly is sculpted into contours and channels as the collected water drains off the surface. After a while, droplets of water hitting a depression will make that depression and connecting channels deeper.

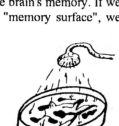

Eventually, patterns are firmly established and the memory surface develops contours. The water will not stay where it has been placed, but flow in the direction of adjacent depressions. It will now become very difficult to establish new patterns because the water tends to follow the old contours. What has this model in common with real memory?

Water will tend to flow toward the lower areas.

Neurons activate (fire) ions to flow toward low threshold units.

Once water has flowed through an area, it is more likely to do so again.

Activation lowers the threshold of a unit and makes it more likely to be reactivated again. Incoming information is processed by information that is already recorded.

This assumes the jelly has a property that makes it harden with extended water exposure. Similar to a new river flow, eating into the soil fast, then encountering stones and finally rock. It means that the erosion on the jelly surface would slow down. This would introduce a tiring factor to the brain model causing the pattern (pool of water) to shift to another area. This is known as *adaptation* in the nervous system.

There may also be some other property the jelly has. Its consistency may not be even. Water rushing over the memory surface may encounter lumps of harder-to-dissolve jelly. This would cause a staccato type of flow, equivalent to a succession of images that make up a thought pattern.

7.1.2 Thinking Processes

With the help of de Bono's memory surface model we will now be able to define the processes of natural, logical and lateral thinking which he described in his book *"Lateral Thinking."*

Natural thinking has all the errors and limitations of the memory surface. Flow is determined by the natural behavior of the jelly. With natural thinking, if something is said three times, then it is more correct than something said only once. Repetition gives dominance. In natural thinking, the thought flow is immediate, direct and basically adequate. There is no hesitation with natural thinking and alternatives are never considered.

☺My boss's mind was made up. He does not like me. This causes me deep concern, especially since he is doing a performance review tomorrow. No wonder I was dreaming of him last night. We were on a boat together, just 200 meters off shore. He had a black notebook in his hand labeled "Employee's Performance." I was going to impress him, climbed over the railing and - would you believe it - *walked* on the surface of the water to the shore and back.

I am sure that feat would affect his performance evaluation. Did it? He wrote in his black book: "This employee cannot swim!"

A typical case of natural thinking. Once a label is put on things or an image is created, there is no flexibility in thinking.

> ☺ "I came home late one evening and found my neighbor searching for a lost key in front of his house. I helped him look for the key without success. "Can you remember when you the last time had the key?" I asked. "Oh yes, I left the house by the rear door, that's where I must have lost it." "Why then are we looking for that lost key in *front* of the house?" "Because there is more light here and it is dark in the back!"

This is natural thinking with a trace of misguided logic in it.

Logical thinking is a deliberate attempt to restrain the excesses of natural thinking by selectively blocking natural flow pathways. The effect is like that obtained by a farmer who directs the water to his field by blocking up some irrigation channels in order to get the water to flow through some others. In logical thinking, each step must be justified or it is excluded from further consideration.

☺ Statement: Americans are liars.
Fact: I am an American.
Logical Deduction: Therefore, I am a liar.
Logical Conclusion: If I am a liar, my statement must be wrong.
Logical Conclusion: Americans are not liars.
Logical Deduction: As an American, my statement must be right.
etc.

Our education system is almost exclusively built up on logical thinking. It originated with the Greeks and is considered by many as the only correct way to solve problems. Logical thinking is tremendously effective in the processing of available information or the collection of more information. Logical thinking is sometimes called mathematical or *vertical* thinking because one proceeds carefully from one proven stage of knowledge to the next, like building a tower, selecting and placing one stone upon the other.

The sequential *decision tree* used by project management (see Chapter 2) is a typical example of the vertical thinking process. We come to a fork of branches and must decide which branch to choose. The most favorable branch will be selected, the others are eventually eliminated. This then makes for *one and only one* right solution. It is also considered a *practical* way of thinking:

> ⧖ ☺ A widow put her husband's ashes into an hour glass. This way, he continues working for her.

In the old times when engineers used slide rules and calculators, two or maybe three different schemes may have been used for a difficult design such as a bridge or city water intake. It took a team of engineers and draftsmen weeks or even months to develop very few schemes if any. Only the correct detail was chosen, usually based on precedence, with minor variations, to come up with the final plan.

Now we have developed a large variety of schemes, doing sensitivity analyses, using the Monte Carlo random method for optimum cost effectiveness, etc.

The first method is a logical, planned approach, the second method will produce something that comes out of a "black box" with a "what if" input. The result could never have been predicted. This is an example of a lateral input into a system.

Similarly, lateral thinking supplements vertical thinking. With logical vertical thinking one moves only if there is a direction in which to move. With lateral thinking one moves in order to generate a direction.

Lateral thinking is concerned with *restructuring existing* thought patterns and creating new ones. Lateral thinking can be learned. It is not an attack on vertical thinking, but a method of making thinking more effective.

Going back to the memory surface, we showed that incoming information establishes rigid patterns through repetition. This pattern is translated into any repeatable concept, idea, thought, image. A higher level pattern may also refer to an arrangement of other patterns which together will give us an approach to problems, opinions, how we see things, etc. We do not attempt to logically break down the complicated behavior of the brain, but put *together* the basic processes that identify a system capable of thinking creatively.

The purpose of lateral thinking is the generation of new ideas and the escape from old ones. Some sort of restructuring of patterns is necessary in order to make the best use of the information that established them. Old ideas must be scrutinized. Rigidity and dogma must be overcome. This is often not easy because it requires us to "unlearn" well established patterns.

How difficult the *unlearning* of old ideas is can best be demonstrated by looking at the metric conversion issue. The SI is a very simple system. It has only seven base units and 16 derived units. They are all related to each other by unity. And yet, it took over hundred years to implement this system worldwide. This in spite of knowing that *19 units* for energy for example (BTU, calorie, erg, therm, horsepower etc) are replaced by *one unit*, the joule, and 30 units for pressure (atm, psi, bar, ton, inch Hg, etc.) also are replaced by one unit, the Pascal.

This is a typical example how deep the "grooves" and knowledge patterns are ingrained into the memory surface. The main impediment for *unlearning* the old is, that incoming information is processed by information already perpetuatively stored in established patterns on the surface. When somebody in the USA is being told: "It is 200 km to Toronto," the immediate reply is often: "How many miles is that?" Miles are established and the incoming information (km) is processed in terms of miles. Since they are not compatible, confusion and rejection result. Miles are deep grooves in the memory surface.

This is not to say that ingrained knowledge is bad. It may in most cases be a great advantage. It makes us comfortable to do things automatically without thinking, e.g. to turn on hot water, we use the left knob, for cold water the right one.

Knowledge, that is stored explicitly is said to be *declarative* knowledge with long term memory. It is said that the imprint on the memory surface has been caused by a slow chemical change over a long period of time. Short term knowledge is due to electrical activity (first drops on the surface) which sorts things out, before a long-term record is made.

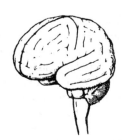

De Bono (10) came up with an illustrative example of the lateral thinking process as compared with vertical thinking, whereby the sequence of arrival of information plays an important role.

Cutoff pieces of cardboard, representing information, are given to someone who must put them together in such a manner, that a simple geometric shape results. The information (cardboard pieces) are given one at a time. The action of assembly is the logical conclusion at the time. This is a very crude example of a sequential decision tree, where only the best alternative is chosen before we proceed to the next step (Figure 7-02).

The cumulative memory system depends on the sequence of information that must make sense at each moment. Since piece 6 does not fit, we have to go back to restructure the situation even though we have been correct at each step up to piece 6.

Lateral thinking assumes that the existing arrangement of information on the memory surface is always *less* than the best possible arrangement.

Obtain the first piece of information in the form of piece #1:

$$\boxed{\qquad 1 \qquad}$$

Add pieces #2 and #3:

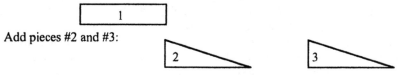

The information is now gathered and assembled for best results:

We now receive more information as shown by pieces #4 and #5:

Again, we are using the added information to the best of our advantage. This results in the following shape:

For the final new information, add piece #6:

It does not fit into the scheme of things. **?**

Alternative:

Add #2 and #3 to piece #1:   Add #4 and #5:   Add piece #6:

 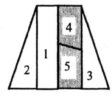 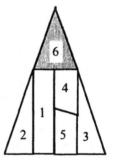

Figure 7-02
Lateral thinking illustration.

7.1.3 Creativity

If you lose the power to laugh, you lose the power to think.

                                            Clarence Darrow

The Knowledge Module

Lateral thinking is a tool which we use to become creative. We know, that we can learn lateral thinking as a skill through understanding of the method and through practice.

Declarative knowledge with long term memory can be considered the *first* stage of thinking. This is usually taken for granted and often assumed to be correct anyway or not to be there at all.

The *second* stage is the one we are conscious of. It is the stage of mathematics and logic (Figure 7-03). The computer is an excellent second stage type device. (Our expressions that humanize the computer, such as brain, artificial intelligence, memory, or virus are very much misleading. To compare the computer hardware to our brain is similar to comparing a pair of pliers to our thumb and forefinger. As a pair of pliers is much superior in strength to our fingers, so is the computer superior in speed and mathematical calculations to our brain.)

Laughter is a fundamental characteristic of the brain system. With laughter goes creativity. It means our mind is open to any alternative ideas even when they sound silly and are considered incorrect by the characteristic logical mind.

If we assemble knowledge into small packages and call it a *"Knowledge Module"*, then we have a large number of these modules in our memory. Typical modules are:

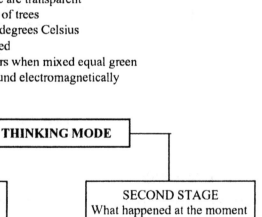

(We have yet to see a computer laugh.)

- ◆ we can multiply numbers
- ◆ glass and clear plastic are transparent
- ◆ furniture is made out of trees
- ◆ water freezes at zero degrees Celsius
- ◆ rubber can be stretched
- ◆ yellow and blue colors when mixed equal green
- ◆ we can reproduce sound electromagnetically
- ◆ etc.

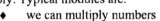

THINKING MODE

FIRST STAGE
What happened in the past

Experience, knowledge

SECOND STAGE
What happened at the moment

familiar   unfamiliar

MEMORY SURFACE
**Alteration or rejection
to old patterns**

new ideas, experience

Figure 7-03
Thinking mode.

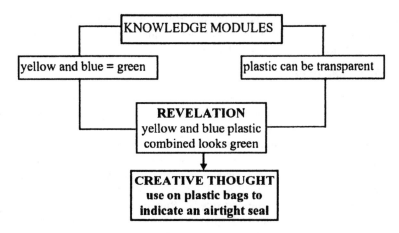

Figure 7-04
Knowledge modules.

Some of the knowledge modules are easily accessible, others need *stimulation* to be made aware of. There are thousands of those modules in our mind. If we were asked to write down only one hundred of them in a row, we will have a hard time doing so. This retrieval, however, can be trained.

We can also expand and combine those modules. This is even more difficult. It would certainly be a revelation to us, if we came up with a combination that creates an entirely new and innovative idea.

This combination is creative thinking.

As an example, let us choose from the modules listed above *two* that are familiar to us (Figure 7-04):

Broadcaster James Burke in his "Connections" TV series describes how a series of seemingly unrelated events have often accidentally been connected to create something new and unique. The greatest inventions are made that way.

As was mentioned earlier, humor plays a great part in lateral thinking. Most humor is based on images that are out of the ordinary, often impossible (cartoons) and outright silly. Much of it is based on the effect that words with different meanings have. The element of surprise, the unexpected, is also important.

- ☺ I went to the restaurant the other day and asked the waiter: "Do you serve prawns here?" "We serve anybody, Sir, please sit down!"
- ☺ A doctor once received a telegram: "Mother-in-law at death's door. Can you pull her through?"

These are some examples where the knowledge module is just a word or short expression. How do we get perfectly legitimate words, apparently unrelated, out of deep grooves of the memory surface and combine them into uninhibited expressions? Through *random* stimulation.

Random Stimulation

Random stimulation is a method which deliberately generates *external* input that triggers new ideas. With random stimulation one uses any information whatsoever, no matter how silly it may sound. While vertical thinking deals only with what is relevant, random stimulation accepts and even welcomes random inputs.

A practical application is "brainstorming," which has been used with success in method studies. To generate random input, any idea which comes to mind is accepted, not necessarily as being valid, but as a *catalyst* to stimulate the mind for better ideas. This is done at the *first stage* of thinking, which is the idea stage, the approach or perception stage. To help in this process, we may use random words, taken from a dictionary or a "word drum" designed for this purpose.

There is a Windows-based software available called IdeaFisher that allows the "brainstormer" to access word processors. The two main features are the Randomizer and the Synchronizer which enhance brainstorming sessions. Through the use of a special algorithm different combination of ideas are listed randomly. The synchronizer will juxtapose two or more lists, making unique connections between knowledge modules. This is meant to stimulate our brain.

For example, the brainstorm idea that cars should have square wheels is an illogical expression. But instant expressions of ideas should never be rejected outright or ridiculed. We should avoid negative reaction:

- ♦ "That would never work!"
- ♦ "It is well known that ..."
- ♦ "This is obviously wrong!"
- ♦ "This would be much too expensive!"
- ♦ "What is so original about that?"

Those statements would only inhibit free expressions of ideas. Generally, managers must *listen* to employees expressing new ideas and not cut them short with an instant judgment. Since we are not allowed to reject outright what is not logical, we go to the *second stage* and accept the statement for further processing.

Looking at it positively, we may find it easier to brake with this car. If we need a car, that climbs stairs, this would be a good solution. An undesirable effect would be an up and down movement. But we could have an axle which can vary in height. That means the up and down movement is caused by the differing dimensions between road and axle. Is it between road and axle or between road and car body? Obviously, it is both. Considering the underbody of a car to be like a platform, then, for a smooth ride, would a round wheel not be the best? The circumference of a circle is at equal radii from the center, but we said it must be at even *distance* between road and platform, or equal *diameters*. Lateral thinking will now ask: "Are there other shapes than wheels which have equal diameters?" (Figure 7-05). Yes, there is one:

The particular knowledge we gained may not be applicable to cars, but it could be an advantage when moving heavy loads such as transformers for example.

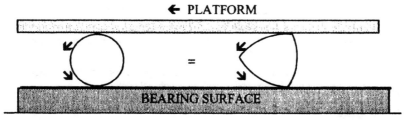

Figure 7-05
Square wheel?

The following will give a more detailed look at and tips for the generation of new ideas:

━━━━━━━━━━━━| TIPS FOR IDEA FINDING |━━━━━━━━━━━━

1. Be optimistic in your approach. Remember that for most things, somehow, somewhere, there is a better way.

2. Consider yourself a thinker - as well as a learner and doer - an idea man or woman as well as a man or woman of action.

3. Develop a honeybee mind. Gather your ideas everywhere. Do not be afraid to associate ideas fully. Let your mind buzz freely from one idea or source to another.

4. Sharpen your nose for problems. Be curious about things that seem wrong or inadequate. Listen to the complaints of others. Jot down your own dissatisfaction with things and situations. Develop an attitude of *constructive* discontent - welcome problems as opportunities not only to accomplish something but to sharpen your creative abilities.

5. Learn to play with ideas At times, you must "regress" - back off from the problem and try to think about it with the naiveté and freshness of a child.

6. Learn to recognize the inhibiting factors or blocks to the free play of the imagination - whether they be perceptive, cultural, emotional or otherwise.

7. Look for the "elegant" answer - don't be satisfied with just any solution to a problem.

8. Be alert for welcome hunches. When you get them, do something about them.

9. Be courageous and independent in your thinking and persistent in the face of frustration and difficulty, but employ an alternating type of persistence so as to invite incubation and insight.

10. Continue to acquire a growing body of knowledge about your field, but don't hesitate to challenge "sacred cows" - long-standing but possibly outmoded or erroneous concepts.

11. Be alert for the unexpected. Serendipity or the "happy discovery" happens only when you are actually seeking something. As Pasteur said, "Inspiration is the impact of a fact on a prepared mind".

12. Organize your approach. Find or devise a methodology that fits your problem and your personality. Break the process up into small step-by-step pieces. If you don't do it in an organized way, you probably won't do it at all.

13. State your problem carefully. Don't let the statement suggest the answer. For instance, if you ask somebody to think up a new way to toast bread, you have already suggested a toaster. What you really want is a new way to dehydrate and crisp the surface of the bread. State it this way and you open up new idea opportunities.

14. Schedule practice sessions - with yourself, that is. Conduct your own private brainstorming session each day. Come up with ideas - good, bad, mediocre. Never mind the duds. Accept all ideas from yourself; don't reject any. Write them down. Unless you drill your mind regularly to produce a bag full of ideas, you haven't really decided to be creative.

15. Carry your idea trap around, a pad and pencil, that is. Keep them with you all the time. Why? Because ideas are elusive. They will drift out of your grasp as readily as they drift in. Better trap them on paper, in black and white.

16. Use idea banks and idea museums. Ideas seldom fall out of the blue. Keep a dream file of clippings, notes from your idea trap, pamphlets, etc., even if you can't work on them right now. Idea museum? This is your reference library. Keep scanning it for ideas. Store them up to solve future as well as present problems.

17. Incubate; relax your mind. After a hard day's work, let it wander. By all means daydream while you take a long walk, meditate. Take a ride, play golf, go fishing. Use the two-day formula: Set your problems aside for a day, then hit it hard after a day's rest.

18. Be enthusiastic, confident. Your will power controls your imagination and is affected by your emotions. So build faith in yourself by scoring successes on little problems before you tackle big ones.

19. Find the right time of day, the time when you are most creative. You know the time when you are full of drive. That's the time to build up a stockpile of ideas. The time for "red light" thinking comes when your mind isn't running creatively.

20. Set a quota - and a deadline too. Force yourself to do a little better each time. Strive for a set number of workable solutions to every problem. A deadline keeps you from putting things off from day to day.

21. Don't kid yourself with vague ideas. Force yourself to reduce them to specific propositions, thus firm up the problems your mind must solve.

22. Expand this list by adding your own tips as you learn more about yourself. CARRY ON FROM HERE.

Generating a new idea is only half the creative process. Transforming the new idea into a practical application that is effective and can be used, is the other half of the process which is:

♦ Extract the principle
♦ Tailor the idea
♦ Transform the idea
♦ Modify the idea
♦ Spell out the conditions for acceptance

Cost optimization may be used to evaluate the conditions for acceptance. Even the best ideas may not be accepted, unless they are cost effective. The next stage, after acceptance, is the implementation. To *implement* approved method study results, the following is recommended:

Implementation, in general, is the essence of management.

Implementing recommendations, in particular, is a powerful medium of management. Effective transfer of information, via report, sets the stage for effective implementation. Having been engaged in a certain phase of work for some time, the transmitter of information (the team), is in a very good position to round up all factors it came across and which have bearing on the next phase.

Plans by the study team aid the manager in implementation. This includes:

♦ Tabulation of recommendations (check list).
♦ A suggested delegation to act and the personnel involved; task groups, meetings, dates.
♦ Flow charts of suggested sequence of action.
♦ Responsibility and authority of participants.
♦ The role of the method study leader. Assistance in implementation by the method study leader formalizes implementation and enlarges the method study function.

A realistic program must consider current work and commitments.

The success of implementation depends largely on the co-operation of those affected.

Control of implementation must be provided to ensure that the program is followed.

Valid measurement and evaluating *techniques* are essential to assess effectiveness of implementation. The purpose of measuring implementation progress is to *motivate* all participants for the success of implementation and to provide a base for evaluating their effectiveness.

7.2. APPLIED PROBLEM SOLVING

This section deals with problem solving for management as applied to both physical operations and management information systems. There is a systematic approach to problem solving. This approach includes systems/methods analysis and work measurement within the framework of work study. Work study was popularized in the UK and stems originally from the work of the Americans Taylor and Gilbreth who called it "time and motion study."

Work study has expanded to include modern aspects of work methods and systems. The term *method study* is often used to include not only work in manufacturing, but also cost optimization, human satisfaction and safety in resource and service industries, government and private enterprises, etc. There are other terms used for similar studies.

Most requests for method studies come from management when things go wrong. Solutions to problems are not always obvious, they are hidden, difficult to define, and easily overlooked.

As mentioned in the previous chapter (thinking processes), there is a tendency that deficiencies are not acknowledged or even become part of a normal operation. The method study expert may be like the doctor who tells the patient: "You should have come earlier, treatment would have been more effective."

It is advisable to review procedures and methods on a regular basis and not wait for problems to become arduous.

Method study could be defined as the critical examination of existing and proposed methods, resulting in their improvement. Its aim is the elimination of unnecessary effort and material waste, or shortly:

```
FINDING BETTER WAYS OF DOING THINGS
```

Method study aims to do the following (RED):

- ♦ Reveal and analyze facts concerning existing conditions.
- ♦ Examine the facts critically.
- ♦ Develop the best possible answer for improvement.

Those aims are relatively easy to achieve. It is much more difficult to *implement* methods improvement than *recommending* changes to "ways of doing things."
It should be recognized, that method study is a dynamic, not static field. Improvement is nearly always possible by changing, combining or eliminating materials, manpower or processes. Whatever the goal, *it is essentially the optimization of costs* which result in a successful implementation of methods improvement.

Operating costs will usually be the final test of any method change. When costs are running higher than normal or suddenly increase, this indicates a need for a study. Where cost changes cannot easily be identified, other *indicators* are:

- ♦ Bottlenecks in production
- ♦ Too much or too little work in progress
- ♦ Excessive movement and backtracking
- ♦ Low utilization of material, labor, space, services resulting in waste

- ♦ High incidence of overtime
- ♦ Heavy rest allowances because of bad working conditions
- ♦ Complaints about the product, increased rejects
- ♦ Poor safety record
- ♦ Follow-up on suggestions made

Before a formal study is initiated, applicable indicators will have to be further explored. The strategy for a study consists of *six basic steps*:

Select:
> This is the definition stage whereby the objectives, techniques, and a timeframe are proposed.

Record:
> The second stage is information gathering and the manner in which data is processed.

Examine:
> This is the most important but least understood part of the study whereby the collected information is critically examined.

Develop:
> The result of the examination is evaluated in regard to cost effectiveness, constructability, acceptability, etc. Recommendations are made and formally presented to management.

Install:
> This is the implementation stage where actions are taken to install a new system or make revisions to existing conditions subject to management approval.

Maintain:
> The new improved version is now operating. Its effectiveness and compliance with new procedures etc. should be watched.

7.2.1 The Select Stage

In selecting the work for study it is usual to give priority to work which offers the greatest margin for improvement. However, there will be those cases where the urgency of the situation demands immediate action (safety considerations). Depending on the organization, the method study analyst may be an individual extensively trained in this field of work and belonging to a full time method study group (staff function) or be assigned temporarily to do the study (line function). If it is temporary work, some training should be given to the people involved before the work commences.

It is suggested, that a formal proposal be given to management outlining the need for and the scope of the study, considerations of personnel, approach and techniques suggested, and economic considerations (cost/benefit).

The second document is the "Terms-of-Reference." This commits management to give full support to the study including the release of funds that are necessary for a successful conclusion.

In conjunction with and following the formal approval should be the *notification* to individuals involved. They must understand the reason for the study in order to cooperate with the investigative team.

Preparing a Proposal:

a) <u>Cover Sheet</u>:	What - Title Where - Organizational Unit When - Date
b) <u>Table of Contents</u>:	A Summary of Headings.
c) <u>Content:</u>	The body of the report, divided into

1.0 INTRODUCTION A brief description of the origin of study request, including outline of what is to be studied, who requested the study and why.

2.0 THE SITUATION A description of the current situation as revealed in a possible "explore phase" and a brief description of how this information was obtained. The former should highlight any deficiencies observed during the explore phase.

3.0 RECOMMENDED ACTION
   3.1 THE STUDY: Type of study (Individual, team, consultant, internal assistance, technical expertise). Scope of study (objectives, criteria to measure effectiveness, goals).
   3.2 STUDY PHASES: This includes required resources, tentative schedule and periodic reports (the use of a barchart may be helpful).

4.0 COSTS AND BENEFITS EXPECTED: Estimated costs in $ for client, study analyst, consultant (if applicable). Anticipated benefits, those that can be costed and those that are intangible.

5.0 APPENDICES: Graphs, charts, tables, pictures, etc.

Preparing a "Terms-of-Reference"

SUBJECT OF STUDY: Brief heading for ready reference and filing purposes.

DATE STARTED: Scheduled begin of study.

THE PROBLEM: Present situation: Brief factual description of the context of the apparent problem, brief history of events leading up to present situation, reference to other pertinent documents. Reason for dissatisfaction with present conditions: What symptoms suggest that some problem exists? What makes the present situation appear to be unsatisfactory? (Relevant opinions as well as facts with identified sources, reference to appropriate documents).

ACTION REQUESTED: What has the investigator initially been asked to do?

OBJECT AND SCOPE:

Criteria of completion: What is to be the end-product of the formal investigation? How will the investigator know, that his work is finished? How far is the investigation to be taken? Does management want

       a) a compilation of facts /data?
       b) a list of questions to elicit inf., opinion?
       c) a list of proposals for further consideration?
       d) fully evaluated proposals for decision and action?
       e) agreed course of action, fully implemented?

Criteria of evaluation: What will be the basis for a management decision? On what bases will alternative courses of action be evaluated, compared, contrasted [financial, technical, safety, acceptability, convenience, etc.)? What is the relative importance of the various factors? If the requirements of some factors conflict, how are they optimized ? Will evaluation depend on measurement or judgment?

Constraints: Time: When must the report be submitted, are there deadlines for specific stages in the investigation? Technical, financial, organizational limits: What may not be changed? What prior decisions must be respected? What areas may not be explored?

TECHNIQUES: What particular techniques are likely to be used, and why? Recording, analysis, evaluation.

NAME OF STUDY MANAGER: Responsible for progress and final report.

NAME OF INVESTIGATORS: Study team.

IMPLEMENTING AUTHORITY: Which person or group will be responsible for implementing accepted conclusions?

7.2.2 The Recording Stage

There is a considerable overlap with other stages since some of the information was already needed at the *select* stage, while other facts need not be recorded until later in the study. This section will deal mainly with the manner in which facts are recorded rather than the extent of the recording necessary.

Objectives: The principal aims of recording are:
- ♦ Obtain adequate and accurate information.
- ♦ Present the facts in a concise, comprehensible form for analysis.
- ♦ Submit study results to management in a way that is easily understood.
- ♦ Provide detailed operating instructions for the use of supervisors and operators.

The Relevant Facts:
    This is the information generally required:
- ♦ The *background* - the history, future prospects, available resources, etc., of the situation under review.
- ♦ The *process* - an account of the activities involved in each of the jobs being done.
- ♦ The *methods* - an account of the various ways, manual or mechanical, in which the activities are carried out.
- ♦ The *movements* - a measure of the nature and amount of movement involved.

Scale:
- ♦ The pursuit of information and the degree of detail in which it is recorded will be determined by the economics of the job and the anticipated requirements of the subsequent analysis. If a radical alteration is made to a process, extensive recording of the existing method may not be needed, since new methods will be devised to suit the new process.

Methods of Obtaining Information:
The collection of data can take many
forms:
- ♦ Group Meetings
- ♦ Interviews
- ♦ Independent Observations
- ♦ The Delphi Method
- ♦ Questionnaires
- ♦ Internal and External Records
- ♦ Transactional Analysis

The Delphi Method

The Delphi Method is a technique that attempts to arrive at the most reliable opinion of a group of experts or very knowledgeable people. The technique consists of repeated questioning of the experts by either questionnaires or interviews, or both, interspersed with controlled opinion feedback. Direct confrontation of the experts with one another is avoided. The goal is to reach a consensus on which to build study recommendations. Below is a typical information flow:

1.	Management:	Recognize the problem
2.	Analyst:	Define problem and prepare primary questions
3.	Management and Analyst:	Select appropriate experts
4.	Analyst:	Prepare the first questionnaire
5.	Analyst:	Send to experts for validity check and review
6.	Experts:	Independently answer questions and give opinion
7.	Experts:	Return completed second questionnaire
8.	Analyst:	Analyze all questionnaires and synthesize
9.	Analyst:	Prepare revised third questionnaire
10.	Experts:	Review revised questions and give opinion
11.	Analyst:	Include feedback in next questionnaire, summarize and prepare argument
12.	Experts:	Make counter arguments or agree
13.	Analyst:	Prepare final draft report with recommendations
14.	Analyst and Experts:	Reach consensus and approve draft
15.	Analyst:	Submit study results to management
16.	Management:	Approve and implement

The above information flow is based on questionnaires. Consensus can also be reached by interviewing individual experts.

Questionnaires

Great caution must be observed when designing a *questionnaire*. Many mistakes have often been made with the use of the technique. The big difference from the interview situation is that the face-to-face interaction which can ensure a more willing and fruitful co-operation is missing. This can seriously affect the validity of the output unless the questionnaire is very carefully designed.

The technique can fail if the recipient has no interest or time to respond; if questions appear silly, spurious or ambiguous; or if a yes/no or multiple check-off answer is to be given to questions which the respondent feels can only be answered by further explanation. Careful phrasing of the questions is vital. The meaning must be clear and if responses are predetermined, the respondent must feel able to choose a valid response from those in front of him.

Transactional Analysis (TA)

This is a theory of personality initially developed for clinical use. It provides a method to
help people understand day-to-day behavior.
Its application to management has gained
popularity in many countries of the world. In
simple terms, TA helps managers or analysts
to deal with people who don't share what
they are thinking and who are not openly expressing opinions, thereby limiting the input
they could be making in meetings and decision-making.

TA tries to overcome covert resistance to organizational objectives, lack of
adjustment to changing situations, incl. Stress, turnover, interviews, difficult
communication etc. Understanding its technique helps the analyst to obtain information which is normally hidden and generally not available. Training in TA
will give managers more interpersonal skills in the resolution of conflicts. *(For
more detail read "Scripts People Live" by Claude Steiner, published by Grove
Press.)*

Recording Techniques

Descriptive Recording:
> This is not always the best way. Facts can sometimes better be pre-
> sented as figures in tabular or graphic form, rather than writing. Below
> are other effective techniques used in method study:

Tabular or Graphic Form:
> The number of operators, machines, components, etc. can be entered in
> tables. Variances in output, quality, resource requirements etc. can be
> plotted on graphs.

Process Charting:
> This is a very effective tool for recording processes, methods and ac-
> tivities. It is extensively used by analysts.

Movement Diagram:
> Movement of people and materials (also information) can be drawn in
> as lines showing the location of various activities and interconnections.
> (Chapter 5.2 describes graphic presentations in detail).

The *process chart* is a pictorial representation of the activities of a process in
which standard symbols are used. Only six symbols are used in making process
charts:

SYMBOL	ACTIVITY	PREDOMINANT RESULTS
○	**Operation**	Produces, accomplishes, changes furthers the process, dismantles, arranges, alters shape, size or shape of an object, etc.
□	**Inspection**	Verifies quantity and/or quality, measuring, grading, etc.

SYMBOL	ACTIVITY	PREDOMINANT RESULTS
⇨	**Transport**	Moves or carries, lifting, trucking, handling, pipe flow, man walking, etc.
D	**Delay**	Interferes or delays, letter in basket, waiting for elevator, queuing, etc.
▽	**Storage**	Holds, keeps or retains, document filed, goods awaiting dispatch, buffer stocks within a process.
◇	**Decision**	Conclusion, implementation.

These basic symbols can be somewhat modified and specifically defined for any particular application, see Figure 7-06 for example.

The process chart is based on the open system concept (Figure 7-07). Furthermore, symbols can be described verbally, by cross-reference numbers or both (Figure 7-08). The symbols are placed vertically, one below the other in sequence, and joined to each other by a vertical line. To the right of each symbol, the activity is described concisely.

Numbering: Activities are numbered from top to bottom and each type is given its own sequence. Numbering sequences start at the top right hand of the chart (main activities) for multiple inputs. When an intersection point is reached, the sequence is resumed at the top of the left-hand branch and brought down to the point of departure from which the sequence continues until another intersection is reached.

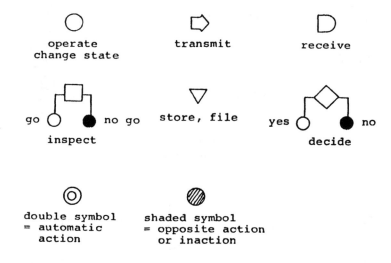

Figure 7-06
Process chart symbols.

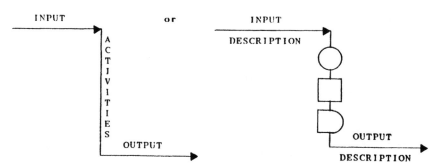

Figure 7-07
Process chart - open system.

Each activity symbol has its own numbering sequence. This convention will help with a tabulation of activity descriptions. Form sheets are designed for this purpose.

Symbols can also be combined such that the outer number is written first, the inner number follows in sequence. The outer symbol will represent the major activity. The example shows a minor operation #8 taking place simultaneously with a more important inspection #5.

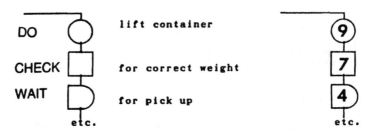

Figure 7-08
Process chart descriptions.

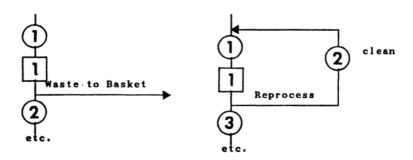

Figure 7-09
Process chart loop.

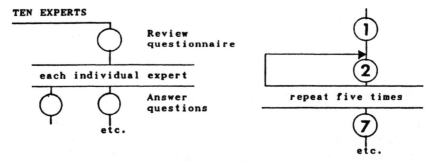

Figure 7-10
Repetitive activities.

Where a number of materials or sub-assemblies are involved, a chart may contain more than one line of vertical symbols. Materials or components discarded during a process, or being transferred to another process is indicated by a line with an arrow head. If it is not transferred to another process, but returned to an earlier stage in the process, a loop is shown (Figure 7-09).

A significant stage in a process can be shown between two double lines. These double lines can also be used to indicate repetitive activities (Figure 7-10).

When alternate routes are used in a process chart, the major flow is on the right hand side (Figure 7-11).

Other activities may occur parallel to the major flow (Figure 7-12).

Additional input is entered from the left at the proper place (Figure 7-13)

The level and scope of activities can be subdivided into first, second and third order of details (17) as shown in Figure 7-14 below (from "Scale Charting" by Ken Brookfield, Journal of Systems Management, August 1972).

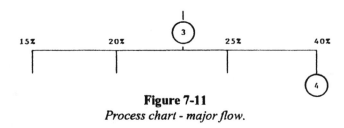

Figure 7-11
Process chart - major flow.

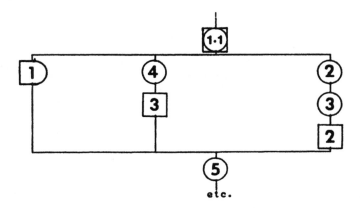

Figure 7-12
Process charting - parallel routes.

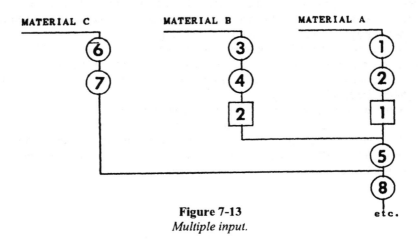

Figure 7-13
Multiple input.

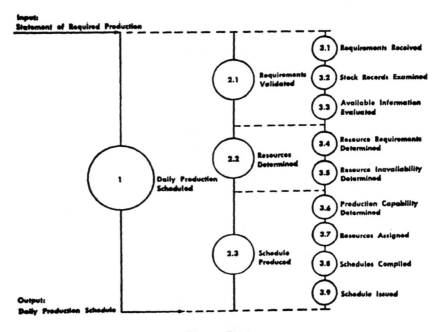

Figure 7-14
Multilevel and scope.

Example: Draw a process chart.

a) go to the warehouse

b) wait your turn

c) obtain cutting tool

d) lift up and inspect it

e) return to job site

f) check material stock

g) pick up and inspect
    metal sheet from pile

h) discard wrong size

i) cut metal sheet

j) put it on pile

k) repeat 20 times (f-g)

h) take a break

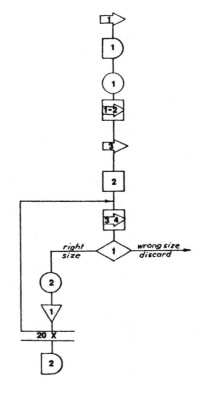

Figure 7-15
Process chart - example.

Very useful is the *Multiple Activity Chart* during the recording of observations. The activities of a number of operators and/or machines can be recorded against a common time scale. Since this chart presents a chronological record of several activities which occur more or less simultaneously, the correct sequence and duration of each must be known (Figure 7-17). Several other means of recording study observations are:

♦ The Flow Diagram
♦ The String Diagram
♦ Models and Templates
♦ Photographs and Movies

In addition are samples from a study showing the delivery of a large transformer (Figure 7-16 and 7-18):

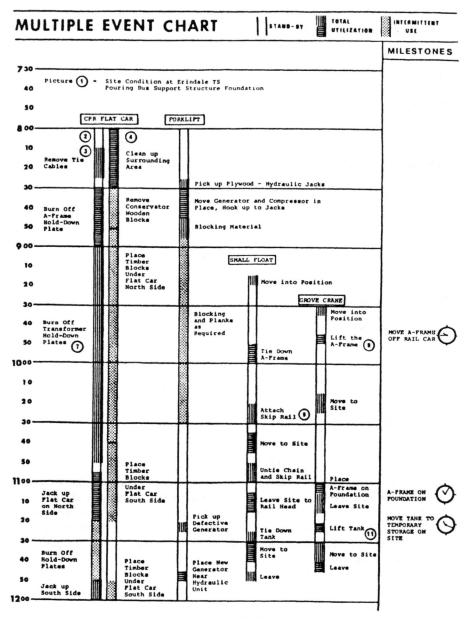

Figure 7-16
Multiple event chart.

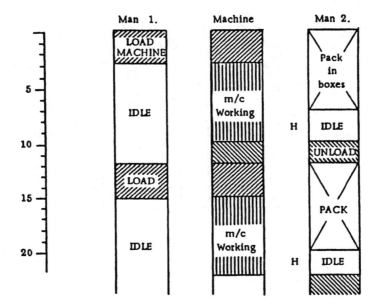

Figure 7-17
Multiple activity chart.

7.2.3 Critical Examination

A detailed procedure for critical examination of collected data has been developed by the B.C. Work Study School operated from the U.B.C. campus. This is a structured approach to problem solving. It uses the essential thought processes of logical and lateral thinking (see 7.1.2-Thinking Processes). Problem examination requires both a clear understanding of present facts and a vision of what might be (17).

It is beyond the scope here to list the detailed procedures of a critical examination as practiced by the trained method study analyst, which includes brainstorming. In general, this is how the processes may be related:

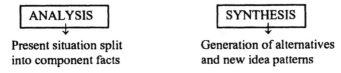

ANALYSIS	SYNTHESIS
Present situation split into component facts	Generation of alternatives and new idea patterns

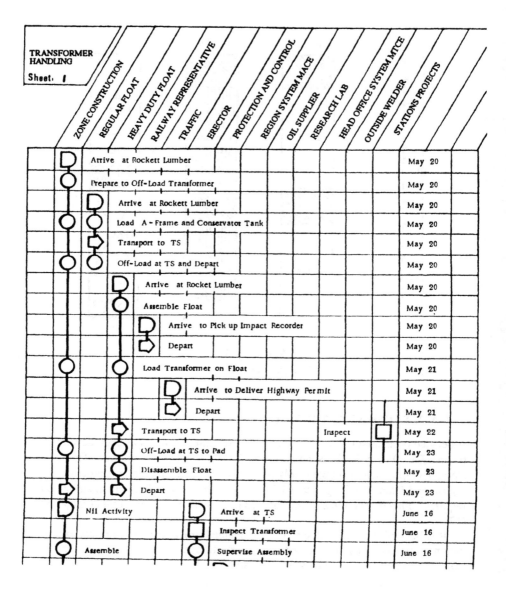

Figure 7-18
Responsibility flow chart.

The facts can be divided into five factors, based on the well-known analysis/syntheses processes:

1. The achievement or result factor   " WHAT "
2. The means applied factor   " HOW "
3. The time and sequence factor   " WHEN "
4. The place or location factor   " WHERE "
5. The person factor   " WHO "

The critical examination can now be structured as follows:

Table 7-01
Critical Examination

| FACTS | ANALYSIS | SYNTHESIS | |
	Reasons	Alternatives	Review
What?	Why that?	What else? consider alternatives	Look for new patterns on the
How?	That way?	How else? eliminate, modify	memory surface (new ideas).
When?	Why then?	When else? numbers, principles	Assess ideas vs. objective.
Where?	Why there?	Where else? substitute, simplify	How can we make it work?
Who?	Why them?	Who else? combine, separate	

Below are random words related to analysis and synthesis that will help to facilitate thought processes:

Analysis: Breakdown, abstract, dissect, resolve, demarcate, distinguish, differentiate, discriminate, separate, detach, divide, divorce, sever, element, component, constituent, factor, ingredient, part, detail, division, fraction, fragment, member, parcel, piece, portion, section, sector, segment, subsystem.

Creativity: Change, alter, decrease, increase, modify, defer, rearrange, delay, incubate, intermit, postpone, suspend, disorder, derange, disturb, unsettle, erase, blot out, cancel, delete, eliminate, exclude, obliterate, omit, humor, joke, act silly, invent, concoct, contrive, create, devise, reverse, capsize, invert, overturn, transpose, etc.

Synthesis: Include, comprehend, embrace, involve, subsume, join, associate, combine, connect, link, relate, unite, order, arrange, marshal, methodize, organize, systematize, plan, design, plot, project, scheme, summarize, aggregate.

7.2.4 Evaluate and Develop

The end result of the initial examination process is a series of notes on the relevance, value and implications of alternatives. Looking at a number of possible ways to improve a system, the analyst must compare them to decide which ones are worthy of further development. He should look at them not only individually, but also in combination. The Development Stage could be summarized as follows:

<u>*Evaluation*</u>
   Economic Evaluation
      Capital Cost (total installed cost)
      Expense Costs (net annual expenses)
   Evaluation of Intangibles
      Organizational
      ♦   Supervision improvement
      ♦   Safety
      ♦   Morale
      Legal
      ♦   Labor Relations
      ♦   Workman's Compensation
      ♦   Copyrights, Patents
      Temporal (timing)
      ♦   Market Factor
      ♦   Union Contract
      ♦   Implementation Period
      Physical
      ♦   Mechanized systems (flexibility)
      ♦   People-systems (adaptability)
      ♦   Housekeeping & maintenance (advantages)
      Technical
      ♦   State of the Art
      ♦   Obsolescence

<u>*Develop*</u> (Develop preferred solutions to yield a comprehensive system):
   Sequential Development
      Feasible Combinations
      Supporting Operations
      New Additions
   Materials Handling
      Travel Distance, Ease of Flow
      Double Handling, Unit loads

    Equipment Utilization
    Stock Locations
   Plant Layout
    Location
    Obstructions, Access
    Capacity
    Uniformity
    Personnel Services
    Regulations
   Office Layout
    Space Utilization
    File Storage and Access
    Future Expansion
    Environment (light, noise, air, decoration)
    Cleanliness
   Human Element
    Body Motion
    Arrangement of Workplace
    Design of Tools and Equipment
    Sense of achievement, responsibility
    Variation in Workpattern
    Opportunity for Recognition
    Sense of Belonging
   Cost/Benefit Analysis

Presentation Of Study Results

   Written Report
    Table of Contents
    Summary including List of Recommendations
    Introduction
    Terms of Reference
    Analysis of Existing Situation
    Discussion of Alternatives
    Costs and Benefits
    Conclusions and Recommendations for Implementation
    Appendices
   Verbal Presentation
    Location, Facilities, Layout
    Audience
    Visual and Audio Aid
    Style

Presenting study results with recommendations to management is the climax of any study. A good write-up and interesting presentation is the best way to get the system approved.

If you are the analyst, a few words of wisdom: Never be disappointed if the "big boss" just says THANK YOU after all your hard work and places your study results in his desk drawer never to be heard of again. There are many reasons why this happens, some of which may be

- ♦ He can not get reasoning across to *his* boss
- ♦ The system is right, but the timing is wrong
- ♦ He cannot afford the implementation costs in spite of total overall savings
- ♦ You did not tell him what he wanted to hear (closed mind)
- ♦ He is playing politics
- ♦ etc. etc.

7.2.5 System Installation

When approved, managers and analysts must work close together to implement the changes to the existing situation. They must consider the nature of the changes and the possible technical and human implications. A decision on the timing of the change will depend on the availability of the necessary material and human resources and on the activity pattern of the group involved. There are two choices, based on capability and an assessment of the risks involved:

Gradual changeover in terms of jobs affected or in terms of the size of change, or *complete changeover,* in which all the changes are rapidly installed and a clean break with the former system is made.

Working closely with management, the analyst has an important part to play. He will assist with

- ♦ Planning the change
- ♦ Preparing operating instructions
- ♦ Training supervisors and operators
- ♦ Observing and reporting the impact of the new system
- ♦ Recommending and installing any unforeseen changes to the system design that have become apparent during the implementation stage

One can have the best system in the world, but if the implementation is poorly planned and executed, the system will not work.

7.2.6 Maintain the System

Maintenance is an important function after the installation of the new system. This applies not only to physical systems but also to information systems. Policies and procedures for maintenance must be worked out and applied so that uncontrolled drift does not occur.

The analyst should continue to be interested in the results resulting from the new processes. That interest should be encouraged and supported by management. After all, the measure of the analyst's work is not in recommendations alone, but also in a successfully operating system.

Similar to handling changes in cost control, changes resulting from method study work should also be subjected to controls. This includes the following "control cycle":

1. *A standard* against performance can be measured
2. *A sensor* to monitor variations from the standard
3. *An analyzer* to explain reasons for deviations
4. *A decision* maker to take action to reduce deviations
5. *An actuator* to apply corrective action to the system
6. *Monitor* to measure effect of action compared to 1) above
7. *Return* to 2) above if necessary.

The above assumes that all is well with the new system in the first place. Since nobody is perfect, situations can arise after the introduction of the new system which require a review of some recommendations that were made originally. This flexibility should not be construed as a failure. After all, the analyst has worked closely with the experts to design and implement the new system. The problem is magnified when the analyst is no longer involved. Misunderstanding the intent of the new system may lead to condemnation and rejection instead of review and updating.

In summary, a great deal depends on the closeness of the relations between the operating departments and the analyst. In a good collaborative atmosphere the analyst will not "police" those who run the new system, he will be working in the management team and assist when difficulties arise or changes occur.

Problem Solving Summary

This will briefly summarize from the point of view of management the nature of Method Study, its techniques, and the effect of its application:

Method Study may be described as the study of the ways of doing things, or as the critical examination of existing and proposed methods with a view to their improvement. Its aim is the elimination of unnecessary work and waste, whether materials, capital equipment or time. This can be achieved

♦ by improving the overall process, and within it the layout and design of buildings, equipment or the workplace,
♦ by selecting the most economical working procedures,
♦ by improving the working environment and
♦ by making more efficient use of human and material resources.

Problem solving is essentially a service to management as well as a part of management itself, providing an assessment of a situation based on all facts available and presented fully and in a balanced way. To ensure that all the aspects of any situation are systematically, objectively and critically examined and assessed, problem solving uses two main interdependent techniques - Method Study and Work (or Activity) Measurement.

Method Study provides a picture of events which must be supplemented by quantitative data if a balanced assessment is to be reached. Quantitative data may include costs, distances, weights, efficiencies and time. It is time that links Work Measurement closely to Method Study.

Work Measurement is defined as the systematic determination of the proper time to be allowed for the effective performance of a defined task carried out by a specified method. It is appropriate to mention also that while it is well known that the techniques of Work Measurement can be used to provide a basis for financial incentive schemes, it is not so often realized that they have other equally if not more important wider uses. Amongst these may be the comparison of alternative methods, economy of workforce, effective planning both of processes and maintenance and realistic labor costing.

The need for improvement within a process is not always apparent. It is often the case that defects in the existing method are accepted simply because they have existed unchecked for a considerable time. An intelligent study of costs and trends can provide useful leads for applying Method Study.

APPENDIX A

COST MANAGEMENT TERMINOLOGY

This is an excerpt of standard cost engineering terminology, based on a draft review (September 1994) by the author of 1994/95 revisions to the American National Standards Institute standard number ANSI - Z94.2. - *Index of Industrial Engineering.*

Accountability. Answerable, but not necessarily charged personally with doing the work. Accountability cannot be delegated but it can be shared.

Accounts Payable. The value of goods and services available for use on which payment has not yet been made.

Accounts Receivable. The value of goods shipped or services rendered to a customer on which payment has not yet been received. Usually includes an allowance for bad debts.

Activity. A basic element of work. An activity occurs over a given period of time [see also Work Item].

Activity Description. Any combination of letters, numbers or blanks which describes and identifies an activity on a schedule.

Activity Duration. The estimate of time units necessary for the accomplishment of the work described by an activity, considering the nature of the work and the resources needed for it.

Activity Splitting. Dividing an activity of stated scope, description and schedule into two or more activities which are rescoped and rescheduled. The sum of the split activities is normally the total of the original.

Actual Cost of Work Performed (ACWP). The direct costs actually incurred and the indirect costs applied in accomplishing the work performed within a given time period. These costs should reconcile with the contractor's incurred cost ledgers that are regularly audited by the client.

Actual Finish Date. The calendar date on which the activity was actually completed. It must be prior to or equal to the data date. The remaining duration of this activity is zero.

Actual Start Date. The calendar date work actually began on activity. It must be prior to or equal to the status date (also called data date).

ADM. [See Arrow Diagramming Method]

Administrative Expense. The overhead cost due to general direction of the company. Generally includes top management salaries at the head office, and the costs of legal, central purchasing, traffic, accounting and other staff functions.

Amortization. (1) As applied to a capitalized asset, the distribution of the initial cost by periodic charges to operations as in capital recovery. Most properly applies to assets with indefinite life.
(2) The reduction of a debt by either periodic or irregular payments.

(3) A plan to pay off a financial obligation according to some prearranged schedule.

Annual Equivalent. (1) In time value of money, a uniform annual amount for a prescribed number of years that is equivalent in value to the present worth of any sequence of financial events for a given interest rate.
(2) One of a sequence of equal end-of-year payments which would have the same financial effect when interest is considered as another payment or sequence of payments which are not necessarily equal in amount or equally spaced in time (see Average Annual Cost).

Annuity. (1) An amount of money payable to a beneficiary at regular intervals for a prescribed period of time out of a fund reserved for that purpose.
(2) a series of equal payments occurring at equal periods of time.

Arrow Diagramming Method (ADM). A method of construction a logical network of activities using arrows to represent the activities and connecting them head to tail. This diagramming method shows the sequence, predecessor, and successor relationships of the activities.

As-built Documentation. (1) The final project schedule which depicts actual scope, actual completion (finish) dates, actual durations, and start dates.
(2) The revised drawings which depict the final configuration of the facility.

Average Annual Cost. The conversion of all capital and operating costs to a series of equivalent equal annual costs.

Backward Pass. Calculation of late finish times (dates) for all uncompleted network activities. Determined by working from the final activity and subtracting durations from uncompleted activities.

Bar Chart. A graphic presentation of work activities shown by a time-scaled bar line (sometimes referred to as a Gantt chart).

Base Period (of a given price index). Period for which prices serve as a reference for current period prices. In other words, the period for which an index is defined as 100 (if expressed in percentage form) or as 1 (if expressed in ratio form).

Base Point (1) For escalation: cost index value for a specific month or an average of several months that is used as a basis for calculating escalation.
(2) For scheduling: beginning mode of network/start (ADM). A node at which no activities end, but one or more activities begin.

Battery Limits. Expression used by the processing industry for on-site facilities.

Break-even Chart. A graphic representation of the relation between total income and total costs for various levels of production, indicating areas of profit and loss.

Break-even Point(s). (1) In business operations, the rate of operations, output, or sales at which income is sufficient to equal operating costs, or operating cost plus additional obligations that may be specified.

(2) The operating condition, such as output, at which two alternatives are equal in economy.

(3) The percentage of capacity operation of a manufacturing plant at which income will just cover expenses.

Breakout Schedule. (Also called Production Schedule). This job site schedule, generally in bar chart form, is used to communicate the day-to-day activities to all working levels on the project as directed by the construction manager. Detail information with regard to equipment use, bulk material requirements, and craft skills distribution, as well as the work to be accomplished, forms the content of this schedule. The schedule is issued on a weekly basis with a two to three-week look ahead from the issue date. This schedule contains from 25 to 100 activities and is called an executing schedule, field schedule, construction schedule, work schedule, checklist, or punchlist.

Budgeted Cost of Work Performed (BCWP). The sum of the budgets for completed portions of in- process work, plus the appropriate portion of the budgets for level of effort and apportioned effort for the relevant time period. BCWP is commonly referred to as "earned value."

Budgeted Cost of Work Scheduled (BCWS). The sum of the budgets for work scheduled to be accomplished (including in-process work), plus the appropriate portion of the budgets for level of effort and apportioned effort for the relevant time period.

Bulk Material. Material bought in lots; generally, no specific item is distinguishable from any other in the lot. These items can be purchased from a standard catalog description and are bought in quantity for distribution as required. Examples are pipe (nonspooled), conduit, fittings, and wire.

Burden. In construction, the cost of maintaining an office with staff other than operating personnel. Includes also federal, state and local taxes, fringe benefits and other union contract obligations. In manufacturing, burden sometimes denotes overhead.

Capital, Budgeting. A systematic procedure for classifying, evaluating, and ranking proposed capital expenditures for the purpose of comparison and selection, combined with the analysis of the financing requirements.

Capital, Cost of. The weighted average of (1) the after-tax cost of long term debt; (2) the cost of common equity capital. Usually expressed as a percent.

Capital, Direct. Cost of all material and labor involved in the fabrication, installation and erection of facilities.

Capital, Indirect. Costs associated with construction but not directly related to fabrication, installation, and erection of facilities. Can be broken down into field costs (temporary structures, utilities, fuels and other consumables, field supervision) and office costs (engineering, drafting, purchasing and office overhead expenses).

Capitalized Cost. (1) The present worth of a uniform series of periodic costs that continue for an indefinitely long time (hypothetically infinite).

(2) The value at the purchase date of the first life of the asset of all expenditures to be made in reference to this asset over an indefinite period of time. This cost can also be regarded as the sum of capital which, if invested in a fund earning a stipulated interest rate, will be sufficient to provide for all payments required to maintain the asset in perpetual service.

Capital Recovery. (1) Charging periodically to operations amounts that will ultimately equal the amount of capital expenditure (see Amortization, Depletion, and Depreciation).

(2) the replacement of the original cost of an asset plus interest.

(3) the process of regaining the net investment in a project by means of revenue in excess of the costs from the project. (Usually implies amortization of principal plus interest on the diminishing unrecovered balance.)

Capital Recovery Factor. A factor used to calculate the sum of money required at the end of each of a series of periods to regain the net investment of a project plus the compounded interest on the unrecovered balance.

Capital, Total. The sum of fixed and working capital.

Capital, Venture. Capital invested in technology or markets new at least to the particular organization.

Capital, Working. The funds in addition to fixed capital and land investment which a company must contribute to the project (excluding startup expense) to get the project started and meet subsequent obligations as they come due. Includes inventories, cash and accounts receivable minus accounts payable. Characteristically, these funds can be converted readily into cash. Working capital is normally assumed recovered at the end of the project.

Cash Flow. The net flow of dollars into or out of the proposed project. The algebraic sum, in any time period, of all cash receipts and disbursements, expenses paid, and investments made. Also called cash proceeds or cash generated. It is needed to plan project funding, to set fiscal budgets, to calculate interest during construction and to evaluate tenders.

Change Order. A document requesting a scope change or correction. It must be approved by both the client and project management before a change can be made to the contract.

Code of Accounts. A numeric or alphanumeric identification to classify goods and services.

Composite Price Index. An index which globally measures the price change of a range of commodities.

Compound Amount. The future worth of a sum invested (or loaned) at compound interest.

Compound Amount Factor. (1) The function of interest rate and time that determines the compound amount from a stated initial sum.
(2) A factor which when multiplied by the single sum or uniform series of payments will give the future worth at compound interest of such single sum or series.

Compounding, Continuous. (1) A compound interest situation in which the compounding period is perceived as zero and the number of periods infinitely great. A mathematical concept that is practical for dealing with frequent compounding and small interest rates.
(2) a mathematical procedure for evaluating compound interest factors based on a continuous interest function rather than discrete interest periods.

Compounding Period. The time interval between dates at which interest is paid and added to the amount of an investment or loan. Designates frequency of compounding.

Compound Interest. (1) The type of interest that is periodically added to the amount of investment (or loan) so that subsequent interest is based on the cumulative amount.
(2) The interest charges under the condition that interest is charged on any previous interest earned in any time period, as well as on the principal.

Conceptual Schedule. A conceptual schedule is similar to a proposal schedule except it is usually time-scaled and is developed from the abstract design of the project. The schedule is used primarily to give the client a general idea of the project scope and an overview of activities. Most conceptual schedules contain between 30 and 200 activities.

Constant Basket Price Index. A price index which measures price changes by comparing the expenditures necessary to provide the same set of goods and services at different points in time.

Constant Dollars. Dollars tied to a reference year by discounting or compounding in order to compare cost for different projects built or scheduled to be built at different time periods.

Constant Utility Price Index. A composite price index which measures price changes by comparing the expenditures necessary to provide substantially equivalent sets of goods and services at different points in time.

Constraint. Any factor which affects when an activity can be scheduled. (See also Restraint.)

Construction Cost. Sum of all costs, direct and indirect, inherent in converting a design plan for material and equipment into a project ready for operation, i.e., sum of field labor, supervision, administration, tools, field office expense and field purchased material costs.

Consumers Price Index (CPI). A measure of time-to-time fluctuations in the price of a quantitatively constant market basket of goods and services, selected as representative of a specific quantitative level of living.

Contingencies. Specific provision for *unforeseen* elements of cost within defined project scope; particularly important where previous experience relating estimates and actual costs has shown that unforeseen events which will increase costs are likely to occur. If an allowance for escalation is included in the contingency it should be as a separate item, determined to fit expected escalation conditions of the project. It does not include *unforseeable* major events such as earthquakes, acts of war.

Contract Dates. The start, intermediate, or final dates specified in the contract that impact the project schedule.

Contracts. Legal agreements between two or more parties, which may be of the type enumerated below:

(1) In *Cost Plus* contracts the Contractor agrees to furnish the Client services and material at actual cost, plus an agreed upon fee for this services. This type of contract is employed most often when the scope of services to be provided is not well defined.

a). Cost Plus *Percentage Burden and Fee* - the client will pay all costs as defined in the terms of the contract, plus "burden and fee" at a specified percent of the labor costs which he is paying for directly. This type of contract generally is used for engineering services. In contracts with some Governmental agencies, Burden items are included in indirect cost.

b). Cost Plus *Fixed Fee* - the client pays costs as defined in the contract document. Burden on Reimbursable Technical Labor Cost is considered part of cost. In addition to the costs and Burden, the client also pays a fixed amount as the contractor's fee.

c). Cost Plus *Fixed Sum* - the client will pay costs defined by contract plus a fixed sum which will cover "non-reimbursable" costs and provide for a fee. This type of contract is used in lieu of a Cost Plus Fixed Fee contract where the client wishes to have the contractor assume some of the risk for items which would be Reimbursable under a Cost Plus Fixed Fee type of contract.

d). Cost Plus *Percentage Fee* - the client pays all costs, plus a percentage for the use of the contractor's organization.

(2) *Fixed Price* types of contract are ones wherein a Contractor agrees to furnish services and material at a specified price, possibly with a mutually agreed upon escalation clause. This type of contract is most often employed when the scope of the services to be provided is well defined.

a). *Lump Sum* - Contractor agrees to perform all services as specified by the contract for a fixed amount. A variation of this type may include a Turn-Key arrangement where the contractor guarantees quality, quantity and yield on a process plant or other installation.

b). *Unit Price* - Contractor will be paid at an agreed upon unit rate for services performed. For example, technical work-hours will be paid for at the unit price agreed upon. Often field work is assigned to a Subcontractor by the Prime Contractor on a unit price basis.

c). *Guaranteed Maximum* (Target Price) - a Contractor agrees to perform all services as defined in the contract document guaranteeing that the total cost to the client will not exceed a stipulated maximum figure. Quite often, these types of contracts will contain special share-of-the-saving arrangements to provide incentive to the Contractor to minimize costs below the stipulated maximum.

d). *Bonus-Penalty* - a special contractual arrangement usually between a Client and a Contractor wherein the Contractor is guaranteed a bonus, usually a fixed sum of money, for each day the project is completed ahead of a specified schedule and/or below a specified cost, and agrees to pay a similar penalty for each day of completion after the schedule date or over a specified cost up to a specified maximum either way. The penalty situation is sometimes referred to as liquidated damages.

Controllable Cost. The type of cost a project manager is able to control. It excludes escalation, interest charges, Government regulation changes.

Cost. The amount measured in money, cash expended, or liability incurred, in consideration of goods and/or services received.

*Cost Accounting...*The historical recording of cash disbursements and accrued cost on a project.

Cost Control. The application of procedures to monitor cost incurredand performance against progress of projects and manufacturing operations with projected completion to measure variances from the authorized plan (budget and schedule) and allow effective action to be taken to achieve minimal costs.

Cost Distribution. The allocation of costs to the proper end account.

Cost Engineer or *Cost Consultant.* An engineer or cost expert whose judgment and experience are utilized in the application of scientific principles and techniques to problems of cost estimation; cost control; business planning and management science; profitability analysis; and project management, planning and scheduling.

Cost Estimation. The determination of quantity and the predicting or forecasting, within a defined scope, of the costs required to construct and equip a facility, to manufacture goods, or to furnish a service. Costs are determined utilizing past experience and calculating and forecasting the cost of resources methods, and management within a scheduled time frame. Included in these costs are assessments and an evaluation of risks and uncertainties. Cost estimation provides the basis for project management, business planning, budget preparation and schedule and cost control.

Cost Flow. The spread of incurred cost over time. It compares cost performance with the approved time-phased plan and is used for cost control. It must not be confused with cash flow.

Cost Index (Price Index). A number which relates the cost of an item at a specific time to the corresponding cost at some arbitrarily specified time in the past.

Critical Activity. Any activity on a critical path.

Critical Path. Sequence of jobs or activities in a network analysis project with the least total slack such that the total duration equals the sum of the durations of the individual jobs in the sequence. There is no time leeway or slack (float) in activity along critical path (i.e., if the time to complete one or more jobs in the critical path increases, the total production time increases) unless the latest allowable time for the terminal event is set by a predetermined duration (see imposed date) for the completion of the project. In this case the slack on the critical path can also be positive or negative.

Critical Path Scheduling or Critical Path Method (CPM). A network planning technique used for planning and controlling elements in a project. By showing each of these elements and associated lead time, the "critical path" can be determined. The critical path identifies those elements that actually control the lead time of the project.

Data Date (DD). The calendar date that separates actual (historical) data from scheduled data, i.e. the date on which the schedule has been updated. Also called Status Date.

Decisions Under Certainty. Simple decisions that assume complete information and no uncertainty connected with the analysis of decisions.

Decisions Under Risk. A decision problem in which the analyst elects to consider several possible futures, the probabilities of which can be estimated. Also called Decisions Under Uncertainty.

Declining Balance Depreciation. Also known as Percent on Diminishing Value. A method of computing depreciation in which the annual charge is a fixed percentage of the depreciated book value at the beginning of the year to which the depreciation applies.

De-escalate. A method to convert present-day costs or costs of any point in time to costs at some previous date via applicable indexes.

Deflation. An absolute price decline for a commodity; also, an operation by means of which a current dollar value series is transformed into a constant dollar value series (i.e., is expressed in "real" terms using appropriate price indexes as defaulters).

Deliverable. A report or product of one or more tasks that satisfy one or more objectives and must be delivered to satisfy contractual requirements.

Demographic Cost Index. Cost indexes developed to deal with geographic cost differences.

Depletion. A form of capital recovery applicable to limited natural resources. It is the reduction in value of an asset that is consumed and thereby depreciated.

Depreciation. (1) Decline in value of a capitalized asset.
(2) A form of capital recovery applicable to a property with two or more years' life span, in which an appropriate portion of the asset's value is periodically charged to current operations.

Deterministic Model. Deterministic model, as opposed to a Stochastic Model, is one which contains no random elements and for which, therefore, the future course of the system is determined by its state at present (and/or in the past).

Development Costs. Also called preliminary cost, they are specific to a project, either capital or expense items, which occur prior to commercial sales found necessary to involve the potentialities of that project for consideration and eventual promotion. Major cost areas include process, product, and market research and development.

Direct Cost. (1) In construction, cost of installed equipment, material and labor directly involved in the physical construction of the permanent facility.
(2) in manufacturing, service and other non- construction industries, the portion of operating costs that is generally assignable to a specific product or process area.

Discounted Cash Flow. (1) The present worth of a sequence in time of sums and money when the sequence is considered as a flow of cash into and/or out of an economic unit.
(2) An investment analysis which compares the present worth of projected receipts and disbursements occurring at designated future times in order to estimate the rate of return from the investment of project. Also called Discounted Cash Flow of Return, Interest Rate of Return, Internal Rate of Return, Investor's Method of Profitability Index.

Distribution. The broad range of activities concerned with efficient movement of finished products from the end of the production line to the consumer; in some cases it may include the movement of raw materials from the source of supply to the beginning of the production line.

Dummy Activity. A constraint dummy has an activity of zero duration and is used to show logical dependency when an activity cannot start before another is complete, but which does not lie on the same path through the network. Normally, the dummy is graphically represented as a dashed line headed by an arrow and inserted between two nodes to indicate a precedence relationship or to maintain the numbering of concurrent activities unique. A procedural dummy in an ADM can have a duration assigned to it.

Early Finish Date (EF). The earliest time an activity may be completed equal to the early start of the activity plus its remaining duration.

Early Start Date (ES). The earliest time any activity may begin as logically constrained by the network for a given date.

Earned Value. The periodic, consistent measurement of work performed in terms of the budget planned for that work. In criteria terminology, earned value is the budgeted cost of work performed. It is compared to the budgeted cost of work scheduled (planned) to obtain schedule performance and it is compared to the actual cost of work performed to obtain cost performance.

Earned Value Concept. The measurement at any time of work accomplished (performed) in terms of budgets planned for that work, and the use of these data to indicate contract cost and schedule performance. The earned value of work done is quantified as the budgeted cost for work performed (BCWP) compared to the budgeted cost for work scheduled (BCWS) to show schedule performance and compared to the actual cost of work performed (ACWP) to indicate cost performance.

Earned Value Reports. Cost and schedule performance reports that are part of the performance measurement system. These reports make use of the earned value concept of measuring work accomplishment.

Ending Node of Network. A node where no activities begin, but one or more activities end.

Escalation. The provision in actual or estimated costs for an increase in the cost of equipment, material, labor, etc., over those specified in the contract, due to continuing price level changes over time.

Escalator Clause. Clause contained in collective agreements, providing for an automatic price adjustment based on changes in specified indices.

Estimate, Cost. An evaluation of all the costs of the elements of a project or effort as defined by an agreed-upon scope. Three specific types (based on degree of definition) of a Process Industry Plant are:

(1) Order of Magnitude Estimate - an estimate made without detailed engineering data. Some examples would be: an estimate from cost capacity curves, an estimate using scale up or down factors, and an approximate ratio estimate. It is normally expected that an estimate of this type would be accurate within plus 50 percent and minus 30 percent.

(2) Budget Estimate - budget in this case applies to the owner's budget and not to the budget as a project control document. A budget estimate is prepared with the use of flow sheets, layouts and equipment details. It is normally expected that an estimate of this type would be accurate within plus 30 percent and minus 15 percent.

(3) Definitive Estimate - as the name implies, this is an estimate prepared from very defined engineering data. The engineering data includes as a minimum nearly complete plot plans and elevations, piping and instrument diagrams, one line electrical diagrams, equipment data sheets and quotations, structural sketches, soil data and sketches of major foundations, building sketches, and a complete set of specifications. It is normally expected that an estimate of this type would be accurate within plus 15 percent and minus 5 percent.

Event. An event is an identifiable single point in time on a project. Graphically, it is represented by a node.

Expense. Expenditures of short-term value, charged to current operating cost, including depreciation, as opposed to land and other fixed capital.

Extra Work. Additional work done within the scope of a project. Those are minor changes, affecting cost negotiable between owner and contractor.

Fee. The charge for the use of the contractor's organization for the period and to the extent specified in the contract.

Feedback. Information (data) extracted from a process or situation used in controlling (directly) or in planning or modifying immediate or future inputs (actions or decisions) into the process or situation.

Feedback Loop. The part of a closed-loop system which allows the comparison of response with command.

Field Cost or *Site Cost.* Engineering and construction costs associated with the construction site rather than with the home office.

FIFO (first in, first out). A method of determining the cost of inventory used in a product. In this method, the costs of materials are transferred to the product in a chronological order. Also used to describe the movement of materials (See LIFO).

First Cost. The initial cost of a capitalized property, including transportation, installation, preparation for service, and other related initial expenditures.

Float. (1) In manufacturing, the amount of material in a system or process, at a given point in time, that is not being directly employed or worked upon.
(2) In construction, the total cushion or slack in a network planning system.

Forecast. An estimate and prediction of future conditions and events based on information and knowledge available at the time of the forecast.

Forward Pass. (1) In construction, network calculations which determine the earliest start/earliest finish time (date) of each activity.
(2) A technique where the scheduler proceeds from a known start date and computes the completion date for an order usually proceeding from the first operation to the last.

Free Float (FF). The amount of time (in work units) an activity may be delayed without affecting the start of any other activity immediately following.

Fringe Benefits. Employee welfare benefits; expenses of employment not paid to the employee, such as holidays, sick leave, S.U.B., social security, insurance, etc.

Function. An expression of conceptual relationships useful in model formulations (e.g., productivity is a function of hours worked).

Future Worth. The equivalent value at a designated future date based on the time value of money.

Gantt Chart. The earliest and best known type of planning chart, also known as the bar chart, it is specially designed to show graphically the relationship between planned performance and actual performance. Named after its originator, H.L. Gantt, a scientific management pioneer.

Gross Domestic Product (GDP). The total domestic output of goods and services at current market prices.

Gross National Product (GNP). The total national output of goods and services at current market prices.

Guideline. A document that recommends methods to be used to accomplish an objective.

Hammock. An aggregate or summary activity between the nodes of two or more activities. All activities between the two or more nodes are tied as one summary activity and reported at the summary level.

Hanger. A beginning or ending node not intended in the network (a break in a network path).

Hedge. In master production scheduling, a quantity of stock used to protect against uncertainty in demand. The hedge is similar to safety stock, except that a hedge has the dimension of timing as well as amount.

Histogram. A chart depicting vertical bars to portray a frequency distribution.

Imposed Date. A date externally assigned to an activity that establishes the earliest or latest date in which the activity is allowed to start or finish.

Imposed Finish Date. A predetermined calendar date set without regard to logical considerations of the network, fixing the end of an activity and all other activities preceding that ending node.

Incremental Costs. The cost and revenue differences between two alternate courses of action.

Incurred Cost. Cost and liability recorded in step with performane, i.e. as the work progresses, regardless when cash payments are made (dollar value of work).

Independent Event. An event which in no way affects the probability of the occurrence of another event.

Indirect Costs. (1) In construction, all costs which do not become a final part of the installation, but which are required for the orderly completion, administration, direct supervision, capital tools, startup costs, contractor's fees, insurance, taxes, etc.
(2) In manufacturing, costs not directly assignable to the end product or process, such as overhead and general purpose labor, or costs of outside operations, such as transportation and distribution. Indirect manufacturing cost sometimes includes insurance, property taxes, maintenance, depreciation, packaging, warehousing, and loading.

Inflation. A rise in the general level of prices.

I-node-(ADM). The node signifying the start of the activity (the tail of the arrow).

In-progress Activity. An activity that has been started but is not completed on the reporting date.

Interest. The cost for the use of capital. Sometimes referred to as the time value of money.

Interest Rate. The ratio of the interest payment to the principal for a given unit of time and is usually expressed as a percentage of the principal.

Interest Rate, Compound. The rate earned by money expressed as a constant percentage of the unpaid balance at the end of the previous accounting period.

Interest Rate, Effective. An interest rate for a stated period (per year unless otherwise specified) that is the equivalent of a smaller rate of interest that is more frequently compounded.

Interest Rate, Nominal. The customary type of interest rate designation on an annual basis without consideration of compounding periods. A frequent basis for computing periodic interest payments.

Interface Activity. An activity connection from one subnet to an activity in another subnet, representing logical interdependence.

Interface Node. A common node for two or more subnets representing logical interdependence.

Intermediate Events. Detailed events and activities, the completion of which are necessary for and lead to the completion of a major milestone.

Intermediate Node. A node where at least one activity begins and one activity ends.

Internal Rate of Return. See Profitability Index.

Inventory. Raw materials, products in process, and finished products required for construction and plant operation or the value of such materials. Also includes other supplies, i.e., chemicals, spare parts.

Investment. The laying out of money or capital in business for the purpose of gaining an income or profit.

J-node, (ADM). The node signifying the finish of the activity (the head of the arrow).

Joint Venture. Two or more companies combine resources to meet the project objective.

Just-in-time (JIT). A delivery system to optimize availability of goods and services. This usually results in lowering inventory cost

Key Activity. An activity that is considered of major significance. Sometimes referred to as a milestone activity.

Labor Cost, (1) Manual - the salary plus all fringe benefits of construction craftsmen and general labor on construction projects and labor crews in manufacturing or processing areas which can be definitely assigned to one product or process area or cost center.

(2) Non-manual - in construction, normally refers to field personnel other than craftsmen and includes field administration and field engineering.

Labor Factor. The ratio between the work hours required to perform an identical task under standard conditions and actual performance.

Lag Relationship. The four basic types of lag relationships between the start and/or finish of a work item and the start and/or finish of another work item are:

(1) Finish to Start

(2) Start to Finish

(3) Finish to Finish

(4) Start to Start

Late Finish (LF). The latest time an activity may be completed without delaying the project finish date.

Late Start (LS). The latest time an activity may begin without delaying the project finish date of the network. This date is calculated as the late finish minus the duration of the activity.

Learning Curve. A graphic representation of the progress in production effectiveness as time passes. The basis for the learning curve calculation is the fact that workers will be able to produce the product more quickly after they get used to making it.

Letter of Credit. A vehicle that is used in lieu of "retention" and is purchased by the contractor from a bank for a predetermined amount of credit that the owner may draw against in the event of default in acceptance criteria by the contractor.

Level Finish/Schedule/ (SF). The date when the activity is scheduled to be completed using the resource allocation process. Level finish is equal to the level start plus duration except when split.

Level Float. The difference between the level finish and the imposed finish date.

Level Start/Schedule (SS). The date the activity is scheduled to begin using the resource allocation process. This date is equal to or later in time than early start.

Life. (1) Economic: that period of time after which a machine or facility should be discarded or replaced because of its excessive costs or reduced profitability. The economic impairment may be absolute or relative.
(2) Physical: that period of time after which a machine or facility can no longer be repaired in order to perform its design function properly.
(3) Service: the period of time that a machine or facility will satisfactorily perform its function without a major overhaul. See also venture life.

LIFO (last in, first out). A method of determining the cost of inventory used in a product. In this method, the costs of material are transferred to the product in reverse chronological order. Also used to describe the movement of goods. (See FIFO.)

Loop. A path in a network closed on itself passing through any node more than once on any given path.

Maintenance. The expense, both for labor and materials, required to keep equipment or other installations in suitably operable condition. Maintenance does not usually include those items which cannot be expended within the year purchased (they must be considered fixed capital).

Major Milestones. The most significant milestones in the project cycle, representing major accomplishments or decision points.

Management Control Systems. The systems (e.g., planning, scheduling, budgeting, estimating, work authorization, cost accumulation, performance measurement, etc.) used to plan and to control the cost and scheduling of work.

Management Science. The application of methods and procedures including sophisticated mathematical techniques to facilitate decision making in handling, direction, and control of projects and manufacturing operations (operations research).

Materiel Cost. The cost of everything of a substantial nature that is essential to the construction or operation of a facility, both of a direct or indirect nature.

Mechanical Completion. A fixed asset becomes mechanically complete upon its in-service date. Mechanical completion is an event.

Milestone. An important or critical event and/or activity that must occur in the project cycle in order to achieve the project objective(s). Also called key event.

Milestone Schedule. A schedule comprised of key events or milestones selected as a result of coordination between the client's and the contractor's project management. These events are generally critical accomplishments planned at time intervals throughout the project and used as a basis to monitor overall project performance. The format may be either network or bar chart and may contain minimal detail at a highly summarized level.

Monitoring. Periodic gathering, validating, and analyzing various data on project status..

Moving Average. Smoothing a time series by replacing a specific value with the mean of itself and adjacent values.

Near-critical Activity. An activity that has low total float.

Net Present Value. Earnings after all operating expenses (cash or accrued non-cash) have been deducted from net operating revenues for a given time period.

Net Profit. Earnings after all operating expenses (cash or accrued non-cash) have been deducted from net operating revenues for a given time period.

Network. A graphic presentation of a project depicting the logic of all selected activities or subnets and their physical interdependence and independence.

Network Analysis. Technique used in planning a project consisting of a sequence of activities and their interrelationship within a network of activities making up a project. (See critical path.)

Network Planning. A broad generic term for techniques used to plan complex projects. Two of the most popular techniques are PERT and CPM.

Node. (1) One of the defining points of a network in planning and scheduling; a junction point joined to some or all of the others by arcs. More specifically, the symbol on a logic diagram at the intersection of arrows (activities). Nodes identify completion and/or start of activities.

(2) A point on a "fork" of a decision tree, where decisions must be made or decisions are evaluated.

Non-work Unit. A calendar unit during which work may not be performed on an activity, such as weekends and holidays.

Offsites. General facilities outside the battery limits of process units, such as field storage, utilities, and administrative buildings.

Onsite. (See Battery limits).

Operating Costs. Used interchangeably with Manufacturing Costs but preferred by the non- manufacturing industries, such as mining or computer services.

Operations Research. Quantitative analysis of industrial and administrative operations with intent to derive an integrated understanding of the factors controlling operational systems and in view of supplying management with an objective basis to make decisions. Frequently involves representing the operation or the system with a mathematical model (see also Management Science).

Opportunity Cost. The profits from alternative ventures that are foregone by using limited facilities for a particular purpose.

Original Duration. The first estimation of work time needed to complete an activity.

Overhead. A cost or expense inherent in the performing of an operation, i.e., engineering, construction, operating or manufacturing, which cannot be charged to or identified with a part of the work, product or asset and therefore, must be allocated on some arbitrary base believed to be equitable, or handled as a business expense independent of the volume of production. Plant overhead is also called factory expense.

Path. The physically continuous, linear series of connected activities throughout a network.

Payoff Period. (1) Regarding an investment, the number of years (or months) required for the related profit or saving in operating cost to equal the amount of said investment.

(2) The period of time at which a machine, facility, or other investment has produced sufficient net revenue to recover its investment costs.

Payroll Burden. Employee's benefits paid by employer such as vacation, sick leave, insurance.

PDM. See Precedence Diagram Method.

PDM Arrow. A geographical symbol in PDM networks used to represent the lag describing the relationship between work items.

PDM Relationships. (1) Finish to Finish-This relationship restricts the finish of the work item until some specified duration following the finish of another work item.

(2) Finish to Start-The standard node type of relationship as used in ADM where the activity of work item may start just as soon as another work item is finished.

(3) Start to Finish-The relationship restricts the finish of the work item until some duration following the start of another work item.

(4) Start to Start-This relationship restricts the start of the work item until some specified duration following the start of the preceding work item.

Percent Complete. A comparison of work completed to the current projection of total work.

Performance Measurement Baseline. (1) The time-phased budget plan against which contract performance is measured. It is formed by the budgets assigned to scheduled work elements and the applicable indirect budgets.

(2) A standard productivity ratio whose level of performance can be expectedat a location under normal or average conditions.

PERT. An acronym for Project Evaluation Review Technique which is a probabilistic technique, used mostly by government agencies, for calculating the "most likely" durations for network activities.

Plan. A predetermined course of action over a specified period of time which represents a projected response to an anticipated environment in order to accomplish a specific set of adaptive objectives.

Planning. The determination of a project's objectives with identification of the activities to be performed, methods and resources to be used for accomplishing the tasks, assignment of responsibility and accountability, and establishment of an integrated plan to achieve completion as required.

Planning Package. A logical aggregation of work within a cost account, normally the far term effort that can be identified and budgeted in early baseline planning, but which will be further defined into work packages, level of effort (LOE), or apportioned effort.

Population. Statistically, all conceivable or hypothetically possible instances or observations of the selected phenomenon.

Precedence Diagram Method (PDM). A method of constructing a logic network using nodes to represent the activities and connecting them by lines that show logic relationships.

Predecessor Activity. Any activity that exists on a common path with the activity in question and occurs before the activity in question.

Present Value (present worth). The discounted value of a series of cash flows at any arbitrary point in time. Also, the system of comparing proposed investments which involves discounting at a known interest rate (representing a cost of capital or a minimum acceptable rate of return) in order to choose the alternative having the highest present value per unit of investment. This technique eliminates the occasional difficulty with profitability index of multiple solutions, but has the troublesome problem of choosing or calculating a "cost of capital" or minimum rate of return. Also called Net Present Value but different from Venture Worth.

Present Worth Factor. (1) A mathematical expression also known as the present value of an annuity of one.

(2) One of a set of mathematical formulas used to facilitate calculation of present worth in economic analyses involving compound interest.

Price. The amount of money asked or given for a product (e.g., exchange value). The chief function of price is rationing the existing supply among prospective buyers.

Price Index. The representation of price changes, which is usually derived by dividing the current price for a specific good by some base period price.

Price Relatives. The ratio of the commodity price in a given period to its price in the base period.

Pricing. The observation and recording (collecting) of prices of commodities.

Probability Distribution. A distribution giving the probability of a value x as a function of x; or more generally, the probability of joint occurrence of a set of variants $X_1...Xp$ as a function of those quantities.

Procurement. The acquisition (and directly related matters) of personal property and nonpersonal services (including construction) by such means as purchasing, renting, leasing (including real property), contracting, or bartering, but not by seizure, condemnation, donation, or requisition.

Production Planning. The function of setting the overall level of a quantative product output. Its prime purpose is to establish production rates that will achieve management's objective in terms of raising or lowering inventories or backlogs, while usually attempting to keep the production force relatively stable.

Productivity. (1) Labor Productivity: Relative measure of labor efficiency (output over input), either good or bad, when compared to an established base or norm.

Profit. (1) Gross Profit - earnings from an on-going business after direct costs of goods sold have been deducted from sales revenue for a given period.
(2) Net Profit - earnings or income after subtracting miscellaneous income and expenses (patent royalties, interest, capital gains) and federal income tax from operating profit.
(3) Operating Profit - earnings or income after all expenses (selling, administrative, depreciation) have been deducted from gross profit.

Profitability. A measure of the excess income over expenditure during a given period of time.

Profitability Analysis. The evaluation of the economics of a project, manufactured product, or service within a specific time frame.

Profitability Index (PI). The rate of compound interest at which the company's outstanding investment is repaid by proceeds for the project. All proceeds from the project, beyond that required for interest, are credited, by the method of solution, toward repayment of investment by this calculation. Also called discounted cash flow, interest rate of return, investor's method, internal rate of return.

Program. An endeavor of considerable scope and enduring in nature as opposed to a project; usually representing some definable portion of the basic agency mission and defined as a item in the agency budget.

Progress. Development to a more advanced stage on a time scale. Progress relates to a progression of development and, therefore, shows relationships between current conditions and past conditions. (See Status.)

Progress Trend. An indication on whether the progress rate of an activity or of a project is increasing, decreasing, or remaining the same (steady) over a period of time.

Project. An endeavor with a specific objective to be met within the prescribed time and dollar limitations and which has been assigned for definition or execution.

Project Duration. The elapsed duration from project start date through project finish date.

Projected Finish Date. The current forecast of the calendar date when an activity will be completed.

Projected Start Date. The current forecast of the calendar date when an activity will begin.

Projection. A series or any set of values extended beyond the range of the observed data (see also Forecast).

Project Life (economic life). Total years of operation for any facility. Sometimes, but not necessarily, equal to depreciable life.

Project Management. The utilization of skills and knowledge in coordinating the organizing, planning, scheduling, directing, controlling, monitoring, and evaluating of prescribed activities to ensure that the stated objectives of a project, manufactured product, or service, are achieved.

Project Manager. An official who has been assigned responsibility for accomplishing a specifically designated unit of a work effort or group of closely related effort on a schedule for performing the stated work funded as a part of the project. He or she is responsible for the planning, controlling, and reporting on his project.

Project Office. The organization responsible for administration of the project management system, maintenance of project files and documents, and staff support for officials throughout the project life cycle.

Project Plan. It is the primary document for project objective and activities. It covers the project from initiation through completion. It includes scope, schedule and cost estimate.

Proposal Schedule. Usually the first schedule issued on a project and accompanies either the client's request or the proposal. It contains key engineering, procurement, and construction milestones historical data, and any client-supplied information. Usually presented in bar chart form or summary level CPM network, this schedule is used for inquiry and contract negotiations.

Range Estimating. The degree of dispersion or variability around the expected or "best" value which is estimated to exist for the economic variable in question; e.g., a quantitative measure of the upper and lower limits which are considered reasonable for the factor being estimated.

Rate of Return on Investment. The efficiency ratio relating profit or cash flow incomes to investments. Several different measures of this ratio are in common use. (See Return on Average Investment; Return on Original Investment; Profitability Index.)

Regression. This term was originally used by Galton to indicate certain relationships in the theory of heredity but it has come to mean the statistical method developed to investigate the behavior of two or more related variables, applied to fit and smooth scattered data.

Remaining Available Resources. The difference between the resource availability pool and the level schedule resource requirements. Computed from the resource allocation process.

Remaining Duration. The estimated work units needed to complete an activity as of the data date.

Remaining Float (RF). The difference between the early finish and the late finish.

Reprogramming. A comprehensive replanning of the efforts remaining in the contract resulting in a revised total allocated budget which exceed the contract budget base.

Required Return. The minimum return or profit necessary to justify an investment.

Resource. Any factors, except time, required or consumed to accomplish an activity.

Resource Allocation Process (RAP). The scheduling of activities in a network with the knowledge of certain resource constraints and requirements. This process adjusts activity level start and finish dates to conform to resource availability and use.

Resource Availability Date. The calendar date when a resource pool becomes available for a given resource code.

Resource Availability Pool. The amount of resource availability for any given allocation period.

Responsibility. Originates when one accepts the assignment to perform assigned duties and activities. The acceptance creates a liability for which the assignee is held answerable for and to the assignor. It constitutes an obligation or account-ability for performance.

Restraint. An externally imposed factor affecting when an activity can be sched-uled. The external factor may be labor, cost, equipment, or other such resource.

Return on Average Investment. The ratio of annual profits to the average book value of fixed capital, with or without working capital.

Return on Original Investment. The ratio of expected average annual after tax profit (during the earning life) to total investment (working capital included). Similar in usefulness and limitations to payoff period.

Risk. The consequence of a penalty or reward when action is considered in an uncertain environment.

Safety Stock. The average amount of stock on hand when a replenishment quan-tity is received. Its purpose is to protect against the uncertainty in demand and in length of the replenishment lead time.

Salvage Value. (1) The cost recovered or which could be recovered from a used property when removed, sold, or scrapped. A factor in appraisal of property value and in computing depreciation.
(2) The market value of a machine or facility at any point in time. Normally, an estimate of an asset's net market value at the end of its estimated life.

Schedule. (1) In construction, applying dates to the time estimate.
(2) In manufacturing, a listing of jobs to be processed through a work center, de-partment, or plant and their respective start dates as well as other related informa-tion.

Schedule Variance. The difference between BCWP and BCWS. At any point in time it represents the difference between the dollar value of work actually per-formed (accomplished) and that scheduled to be accomplished.

Scheduling. The assignment of desired start and finish dates to each activity in the project within the overall time cycle required for completion according to plan.

Scheduling Variance. The difference between projected start and finish dates and actual or revised start and finish dates.

Scope. Defines the work to be done to achieve the project objective. It puts limits and boundaries on the project and is documented by the parameters for a project to which the company is committed.

Scope Change. A deviation from the project scope originally agreed to in the contract. A scope change can consist of an activity either added to or deleted from the original scope. A contract change order is needed to alter the project scope.

Secondary Float (SF). The same as the total float, except that it is calculated from a schedule date set upon an intermediate event.

Sensitivity. The relative magnitude of the change in one or more elements of an engineering economy problem that will reverse a decision among alternatives.

Sensitivity Analysis. An analysis of the effect a change in parameters will have on the solution of a mathematical problem.

Shutdown Point. The production level at which it becomes less expensive to close the plant and pay remaining fixed expenses out-of-pocket rather than continue operations; that it, the plant cannot meet its variable expense.

Simple Interest. (1) Interest that is not compounded--is not added to the income-producing investment or loan.
(2) The interest charges under the condition that interest in any time period is only charged on the principal.

Simulation. The technique of utilizing representative or artificial operating and demand data to reproduce, under test, various conditions that are likely to occur in the actual performance of a system. Frequently used to test the accuracy of a theoretical model or to examine the behavior of a system under different operating policies.

Sinking Fund. (1) A fund accumulated by periodic deposits and reserved exclusively for a specific purpose, such as retirement of a debt or replacement of a property.
(2) A fund created by making periodic deposits (usually equal) at compound interest in order to accumulate a given sum at a given future time for some specific purpose.

Site Preparation. An act involving grading, landscaping, installation of roads and siding, of an area of ground upon which anything previously located had been cleared so as to make the area free of obstructions, entanglements or possible collisions with the positioning or placing of anything new or planned.

Skewness. An expression for nonsymmetrical "tailing" of a distribution.

Slack. See Total Float.

Slack Time. The difference in calendar time between the scheduled due date for a job and the estimated completion date. If a job is to be completed ahead of schedule, it is said to have slack time; if it is likely to be completed behind schedule, it is said to have negative slack time.

Standard Deviation. The most widely used measure of dispersion of a frequency distribution.

Standard Error of Estimate. An expression for the standard deviation of the observed values about a regression line, i.e., an estimate of the variance likely to be encountered in making predictions from the regression equation.

Standard Error of The Mean. The standard deviation of the distribution, divided by the square root of the number of cases.

Start-up. That period after the date of initial operation, during which the unit is brought up to acceptable production capacity and quality within estimated production costs.

Start-up Costs. Extra operating costs to bring the plant on stream incurred between the completion of construction and beginning of normal operations.

Status. The condition of the project at a specified point in time. An instantaneous snapshot of the then current conditions. (See Progress.)

Status Line. A vertical line on a time-scaled schedule indicating the point in time (date) on which the status of the project is reported. Often referred to as the time now line. (See Data date.)

Stochastic. The adjective "stochastic" implies the presence of a random variable.

Straight-line Depreciation. Method of depreciation whereby the amount to be recovered (written off) is spread uniformly over the estimated life of the asset in terms of time periods or units of output.

Subcontract. Any agreement or arrangement between a contractor and any person in which the parties do not stand in the relationship of an employer and an employee.

Subnet. The subdivision of a network into fragments.

Subsystem. An aggregation of component items (hardware and software) performing some distinguishable portion of the function of the total system of which it is a part.

Successor Activity. Any activity that exists on a common path with the activity in question and occurs after the activity in question.

Sum-of-digits Method. Also known as sum-of-the-years-digits method. A method of computing depreciation in which the amount for any year is based on the ratio: (years of remaining life)/(1+2+3...+n), n being the total anticipated life.

Sunk Costs. (1) The unrecovered balance of an investment. It is a cost, already paid, that is not relevant to the decision concerning the future that is being made. Capital already invested that for some reason cannot be retrieved.
(2) A past cost which has no relevance with respect to future receipts and disbursements of a facility undergoing an economical study.

System. A combination of hardware (equipment and facilities) and related software (procedures, etc.) designated to perform a unique and useful function. A system has input, processes and output.

Systems or *Method Studies.* The development and application of methods and techniques for analyzing and assessing programs, activities and projects to review and assess efforts to date and to determine future courses and directions. These studies include cost/benefit analysis, environmental impact analysis, assessment of the likelihood of technical success, forecasts of possible futures resulting from specific actions, and guidance for energy program planning and implementation.

Tangibles. Things that can be quantitatively measured or valued, such as items of cost and physical assets.

Target Date. The date an activity is desired to be started or completed; either externally imposed on the system by project management or client, or accepted as the date generated by the initial CPM schedule operation.

Task. Any definable unit of work. It must have an identifiable start and ending and usually produces some recognizable result.

Terms of Payment. Defines a specific time schedule for payment of goods and services and usually forms the basis for any contract price adjustments on those contracts that are subject to escalation.

Time-limited Schedule. The scheduling of activities so predetermined resource availability pools are not exceeded unless the further delay will cause the project finish to be delayed.

Time Series. Any quantitative entity (Whrs, %, $) projected upon a time scale.

Time Value of Money. (1) The cumulative effect of elapsed time on the money value of an event, based on the earning power of equivalent invested funds. (See future worth and present worth.)
(2) The expected interest rate that capital should or will earn.

Total Float (TF). The amount of time (in work units) that an activity may be delayed from its early start without delaying the project finish date. Total float is equal to the late finish minus the early finish or the late start minus the early start of the activity.

Turnover Ratio. The ratio of annual sales to investment. Inclusion of working capital is preferable, but not always done.

Uncertainty. Unknown future events which cannot be predicted quantitatively within useful limits.

Unforeseeable Events. They are impossible to predict such as earthquakes, acts of war, sabotage.

Unforeseen Events. They are difficult to predict or to quantify in an estimate. They are missing information, equipment breakdown, adverse weather conditions, contract non-performance. Cost impacts are covered by contingencies.

Unit Cost. Dollars per unit of production.

Update. To revise the schedule to reflect the most current information on the project.

Weights. Numerical modifiers used to infer importance of commodities in an aggregative index.

Work Breakdown Structure (WBS). A product-oriented family tree division of hardware, software, facilities and other items which organizes, defines and displays all of the work to be performed in accomplishing the project objectives.

Work Item. It is an itemized breakdown of the work performed and quantities installed at the commodity level. It includes the installation of pipe, the pouring of concrete, meters of cable pulled, etc.

Work Package. A segment of effort required to complete a specific job such as a research or technological study or report, experiment or test, design specification, piece of hardware, element of software, process, construction drawing, site survey, construction phase element, procurement phase element, or service, which is within the responsibility of a single unit within the performing organization. The work package is usually a functional division of an element of the lowest level of the WBS.

Yield. The ratio of return or profit over the associated investment, expressed as a percentage or decimal usually on an annual basis. See rate of return on investment.

Zero Base Budget. Usually applicable on programs of long duration. No residues of the previous period budget are transferred to the next period. Very strict cost control applies.

APPENDIX B

METRIC (SI) UNITS IN COST MANAGEMENT

Based on an article *"The International System of Units (SI)"* AACE Cost Engineering Journal, November 1986 and publications by the Canadian Metric Association (43).

The International System of Units

Recognizing the value of having one common worldwide language, managers have now become quite familiar with metric conversion in North America. There is plenty of literature available on all aspects of the SI system (43). Because the use of metric terms is not extensive in cost management, Appendix B will only give a cursory overview of the "Système International d'Unités (SI)".

Background

Our customary measuring system is nearly the same as that brought by the colonists from England, which in turn had its origin in a variety cultures such as the Babylonian sexagesimal counting system (360° circle), Egyptian (common fractions, i.e. repeated halving), Roman (base number 12 as in dozen, hours, months), Celts and Vikings (base number 20).

The Chinese were the first to introduce the decimal notation (ten fingers and the abacus) which later appeared in Arabia in the 15th century.

Through colonization and dominance of world commerce the English system of weights and measures spread to many parts of the world, including the American colonies.

The French Academy of Science proposed using the decimal system in 1790, where weights and measures were based on the number 10. In 1866 the metric system was made legal in the United States.

The SI is not the European metric system.

The British Association for the Advancement of Science introduced the CGS or centimeter/gram/second system in 1873, which was adopted by 35 nations, but not by Britain. Later on the more practical units meter and kilogram (MKS system) came into being. This was the *European metric system*. It had a number of conversion factors which made it almost as unwieldy as the British system, e.g.

1 Liter = 1000.028 cm = 1 kg water at 4° Centigrade at sea level or
1 gram-force = 980.7 dyne   or   1 atmosphere = 0.76 cm Hg
1 horsepower = 735.5 m-kg/sec = 175.8 Cal/sec or 1 torr = 0.133 dynes/cm^2.

In a rare demonstration of unity among nations, the General Conference on Weights and Measures (CGPM) adopted an extensive revision and simplification of the metric system in 1960. The word "metric" was officially changed to SI.

The European old metric system is *definitely not* what we are converting to in North America.

SI is coherent, absolute, unique

The SI consists of seven fundamental base units, two supplementary units, and several derived units with special names. The supplementary units were changed to derived units in October 1995, resulting in only two major categories.

It also uses prefixes that relate to base units in intervals of one thousand. In the lower range, for popular use, the intervals are ten (centi-, hecto-).Thanks to those prefixes, most measurements can be expressed with numerical values of two to four digits. The relationship to the base unit is immediately evident from the prefix. There is no need for any more than *one basic unit for any given quantity.* Length is always expressed in meters [m] with prefix if applicable [mm, cm, km]. This eliminates different names for different lengths such as miles, fathoms, chains, feet, knots, etc. with many conversion factors. SI quantities relate by simply moving the decimal point.

The SI is coherent:

There are no factors relating different units; they are all related to each other by unity (Figure B-01). *One* newton is the force required to give a mass of *one* kilogram an acceleration of *one* meter per second-squared. If this force is exerted though a length of *one* meter, it produces the energy of *one* joule. If this takes place in *one* second, the power produced is *one* watt. Furthermore, if this force is distributed over an area of *one* meter, it produces a pressure of *one* pascal. We find the same coherence with the other base units. For example, the derived units for ampere are volt, ohm, siemens, coulomb, henry, weber, tesla and farad.

The SI is absolute:

SI expressions are unqualified, that is force is always expressed in newtons, there is no other name for force; mass is kg and pressure is pascal. The imperial system used "pound" for force (psi) and also for mass. "Ounces" could denote weight or volume.

The SI is unique:

The same units are used wether we do thermal, mechanical or electrical calculations. This is a tremendous advantage because all engineering disciplines talk the same language. *One* unit of power, the watt, is used now where *nineteen* units were used previously (Btu, cal, HP, etc.).The pascal is replacing some thirty different units including obsolete metric units.

Conversion tables do not teach us the SI

The advantage of the SI over the customary and the old "metric" system is obvious but not necessarily appreciated by the general public in its day-to-day living.

Many still seem to be deprived of facts and may not really understand the system. It is not the learning of the new, but the unlearning of the old which is so painful.

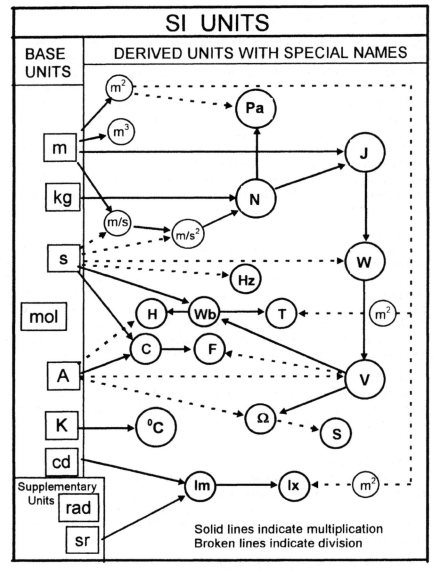

Figure B-01
The SI is coherent.

We can make it easy for ourselves by using computers and conversion tables which instantly give us SI measurements.

Our customary system worked in the past, but it is standing on its own. The SI is a different improved system which also stands on its own. The two systems are incompatible. Instant "converters" do not convert conventional units into SI units; instead they translate the existing into the "metric" equivalent of the existing.

Soft conversion:

This subjective approach is called "soft" conversion. What we are doing when we "soft" convert is add another expression to the proliferation of expressions already in existence for each quantity of measurement. During the transition period of conversion we cannot avoid changing existing physical quantities into their SI equivalents. There is a danger that we use an overabundance of insignificant digits (see also Chapter 1.1.1). A 20 foot hydro pole is not 6096 mm long. A jar of pickled eggs (Figure B-02) shows a conversion from 7 oz to 198 g. Those are probably fluid ounces and should have been converted into 200 cm^3 (commonly known as 200 mL).

Figure B-02
Soft conversion.

SI changeover (hard conversion):

England's double bed size has been 54 inches wide and 75 inches long (137 cm × 191 cm). Briton's Bedding Federation changed this into a 150 by 200 cm bed. This is a true changeover, taking advantage of the opportunity to increase the size of the bed at the same time. The same holds true with standard door openings and other modular building components. We have the opportunity now to improve the standards we use and update, houseclean, rationalize, remove duplication and otherwise improve upon the standards we use.

SI Changeover is an Investment

As cost concious managers we are well aware that practical action by corporate bodies must be justified by monetary consideration. The SI changeover is a deliberate practical move and must ultimately yield a net gain. Therefore, it is an in-

vestment, not an expense (see also Chapter 2.1.2 - The Project Life Cycle).
Keeping track of costs and benefits is, therefore, an important part of "metric conversion" activity.

The SI changeover has an infinite benefit duration, similar to the invention of the
wheel. Even if we let the cost lie where they fall, benefits will accrue when conversion is complete (Figure B-03). The SI changeover is only a part of a bigger
change, the standardization of industrial products and processes. Standardization
includes rationalization (reduction in the total number of sizes and ranges) of
products, which is where a great deal of the benefits of conversion will occur. The
benefits of rationalization remain long after the expenses of conversion are paid.

About 20 years ago, a well known Cost Engineer, Professor Frederick C. Jelen
(5) defined metric conversion costs as shown in Figure B-04. His emphasis was
on timing. The main factors which influence timing are the need to restrict the
period of dual working (hybridizing) and the desire to *take advantage of the
benefits as soon as possible.* It has taken us 200 years to implement a worldwide
uniform measuring system, but a completion date has not yet been established in
the United States.

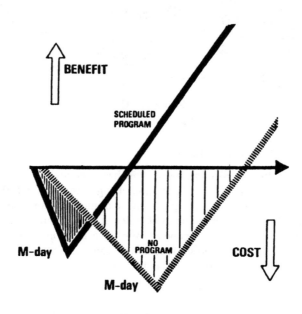

Figure B-03
Cost/benefit graph.

We have a project whose objective has not completely been defined. It has no time or cost plan. Furthermore, it has no project leader. What would we call a project like this ?

FREDERIC C JELEN

——————————— REGISTERED PROFESSIONAL ENGINEER ———————————

PROFESSOR OF
CHEMICAL ENGINEERING

P O BOX 10053
LAMAR UNIVERSITY
BEAUMONT, TX 77710
Tel (713) 838-7021

June 15, 1976

Mr. Kurt Heinze
Ontario Hydro
620 University Avenue
Toronto, Ontario
CANADA   M5G 1X6

Dear Mr. Heinze:

With regard to the cost of metrication, perhaps the real issue is not so much the cost but rather determining the optimum time at which to make the conversion.

If one accepts the fact that metrication must be made, the only decision to be made is how to do it most economically. To be sure a knowledge of the cost is desirable, but it will be greatly shaded by the accounting method used and many subjective evaluations.

One is really interested in the economy of timing the steps in the conversion. The methods in the book COST AND OPTIMIZATION ENGINEERING are mathematically adequate including inclusion of inflation. However, the assignment of benefits and costs will not be easy. Nevertheless, the principles are clear and an in depth study should give the optimum timing for conversion. Obviously this will not be the same for all organizations.

Thank you for writing. If you plan to be at the annual meeting in Boston in July, please look me up.

Sincerely,

F C Jelen

Figure B-04

Conversion costs.

Rules for Style and Usage

(1) Symbols and prefixes are the same in *all* languages.
(2) Symbols are written in lower case, except when the unit is derived from a proper name, i.e. m = meter;   W = watt. The exception is the non-SI but commonly used unit for cm^3 , the liter = L. This is to avoid confusion between l and 1.
(3) Symbols and prefixes are printed in upright type. There is no spacing between prefixes and units, e.g.

	km	cm	ng
not	*km*	*cm*	*ng*
and not	k m	c m	n g

(4) Symbols are not pluralized, e.g. 100 g or 50 km not 100 gs or 50 kms

(5) There is no period after a symbol except at the end of a sentence, e.g.
m (not m.),   mL (not mL.)
(6) A sentence does not start with a symbol and prefix:
The symbol for kiligram is kg (not: kg is the symbol for..)
(7) Preference is given to decimal notation:
Use 3.25 % rather than 3¼ %; 0.75 km rather than ¾ km.
But "I walked 2 km in half an hour" (not in 0.5 h)
(8) For values less than one (1), a zero (0) is used, i.e. 0.56 (not .56)
(9) There is a space between the last digit of a number and the first letter of the symbol except for °C, e.g. 36 kg (not 36kg); but 22°C (not 22 °C)
(10) The multiplication symbol is used instead of a dot or period:
5 × 7 (not 5 · 7 or 5 . 7))
(11) Compound prefixes must never be used:
3 mg (milligram) not 3 µkg (microkilogram)
(12) Spaces are used instead of commas to put large numbers in reading blocks of three, e.g. 23 456 789.24 not 23,456,789.24
(13) Only one unit is used to designate quantities:
5.36 m (not 5 m, 36 cm) or 3.7 kg (not 3 kg, 700 g)
(14) An oblique stroke (/) with no spaces is used with symbols rather than the word "per":
km/h (not km / h) and (not km per h)
When writing units in full, then kilometer per hour (not kilometer/hour)
(15) Time units start with the *largest* unit, e.g.:

Year	Month	Day	Hour	Minute	Second
1996	07	06	18	21	08
or		1996 . 07 . 06 . 18 : 21 : 08			

It is extremely important that we move toward a common international system of YYMMDD.

Europe is committed to change from DDMMYY to the ISO standard. The USA still uses MMDDYY, which can be very confusing and legally costly in international trade. The adverse effect will have great repercussions around the year 2000. If we have a commitment with China for example (they use ISO dating) for 03 - 02 - 01 (ISO date), we will probably make payments earlier than necessary (2 March 2001) instead of 1 February 2003.

(16) SI units are pronounced on the *first* syllable:
centimeter, kilometer, Celsius, micrometer, megahertz; never kilometer.
Instruments are pronounced micrometer, thermometer, speedometer.
(The spelling in Canada is different:, e.g. kilometre vs. thermometer).
(17) Numeral and symbol should not be separated, neither should a number be hyphenated:

Toronto has a population of 4 000 000 people.	The distance from NewYork City to Philadelphia is 160 km when travelling by car.

not

Toronto has a population of 4 - 000 000 people.'	The distance from NewYork City to Philadelphia is 160 km when travelling by car.

(18) The choice of the appropriate multiple of an SI unit in governed by the application, preferably having values between 0.1 and 1000.

        3.94 mm instead of 0.003 94 m

(19) The symbol is placed behind the numeral:

        15.7 km or 350 mL; (not km 15.7 or mL 350)

        (Prefixes may also apply to dollars, e.g. 250 k\$ or 5 M\$)

(20) Temperature will be expressed in "degrees Celsius" or °C without space after the numeral (not centigrade or degees C), e.g.

    40°C and –40°C; (not + 40 °C, neither minus 40°C or 40°C minus).

Please note, that there are no warmer or cooler temperatures, they can only be higher or lower (how do we warm up a temperature?). Furthermore, 20°C is not "half as cold" as 40°C

(21) Number and symbol should not be separated by an adjective.

    Write   You get "200 km free" with your car rental.

    Do not write   You get "200 free km" with your car rental"

(22) If a presentation in dual values is unavoidable, give preference to the SI value: 20 kg (44 lbs.); not 44 lbs. (20 kg)

(23) Former units of area and volumn will be changed as follows:

Square meter (formerly sq.m) now becomes m^2, pronounced square meter, not meter square. Similarly, cubic centimeter is now cm^3, not cu.cm. or c.c. (It is acceptable to use mL instead of the SI unit cm^3 in case of liquids and gases).

(24) In order not to confuse quantity symbols in engineering with SI symbols, the ISO recommends to use italic type lettering where practical, e.g.

m = mass; d = depth or diameter; M = bending moment; A = area of section.

m = meter; d = deci-; M = mega-; A = ampere

Gravity = g = 9.8 m/s^2. *Force* = F = m (kg) × a (m/sec^2).

Sources of SI Information

AACE International, *Cost Engineers' Notebook*, 209 Prairie Avenue, Suite 100, Morgantown, WV 26507.

American National Metric Council (ANMC), Publications Department, 4330 East West Highway, Suite 1117, Bethesda, MD, 20814 - 4408.

American National Standards Institute (ANSI), *ANSI/IEEE Std. 268-82*, 11 W 42nd St., New York, NY 10036.

American Society of Civil Engineers (ASCE), *Metric Units in Engineering - Going SI*, Revised Edition 1995, 345 East 47th Sreet, New York NY 10017 - 2398.

American Society for Testing and Materials (ASTM), *ASTM E380, Standard Practice For Use Of The International System Of Units* and *Units In Building Design And Construction*, 1916 Race St., Philadelphia, PA 19103.

Canadian Metric Association (CMA), *Metric Fact Sheets* (out of print), 481 Guildwood Pkwy., Scarborough, ON, Canada M1E 1R3.

Canadian Standards Association (CSA), *Metric Practice Guide*, *CSA Z234*, 178 Rexdale Blvd., Rexdale ON, Canada M9W 1R3

National Bureau of Standards (NBS), *Various Publications*, U.S. Department of Commerce, Gaithersburg, MD 20899.

National Research Council of Canada, *Manual on Metric Building Practice*, Otttawa, ON Canada K1A 0R6.

R.S. Means Co. *Building Construction Cost Data, Metric Addition*, R.S. Means Co., Box 800, Kingston, MA 02364.

U.S. Metric Association, 10245 Andasol Ave., Northridge, CA 91325-1504.

REFERENCES

(1) MORE CONSTRUCTION FOR THE MONEY, The Business Roundtable Policy Committee, 200 Park Avenue, New York, NY 10166
AACE contributers: Among many other active AACE members, the following were specifically identified: G. Wallace Bates, Jack F. Enrico James A. Bent, Brisbane H. Brown, John D. Borcherding, Frank Duda

(2) AACE INTERNATIONAL and AACE-Canada, Inc. Reprints are by permission of AACE. The address of their headquarters is: 209 Prairie Ave., Suite 100, Morgantown, WV, USA, 26505, Telephone (304) 296-8444. Fax 304 - 291-5728. Publications include a) Cost Engineers' Notebook, b) Cost Engineering, c) Transactions from annual meetings

(3) PROJECT MANAGEMENT INSTITUTE. Reprinted from "A Guide to the Project Management Body of Knowledge: Exposure Draft - August 1994" with permission of the Project Management Institute, 130 South State Rd., Upper Darby, PA 19082, a worldwide organization of advancing the state-of-the-art in project management. Other publications include a) Project Management Body of Knowledge (PMBOK), b) Project Management Journal, c) PM Network

(4) ELEMENTARY BUSINESS AND ECONOMIC STATISTICS, Alva M. Tuttel, McGraw-Hill Book Company, Inc.,1957

(5) JELEN'S COST AND OPTIMIZATION ENGINEERING, edited by Kenneth K. Humphreys, McGraw-Hill, Inc., Third Edition, 1991

(6) MANAGING CAPITAL EXPENDITURES FOR CONSTRUCTION PROJECTS, K. M. Guthrie, CCE, et al.; Craftsman and also various AACE Transactions

(7) CAPITAL PROJECT LIFE CYCLES, D.K. Clancy & D.W. Finn, Cost & Management, July/August 1985, p.p. 6

(8) RANGE ESTIMATING AND THE USE OF RISK ANALYSIS, Kurt G. Heinze, CCE, Technical Presentation to AACE members, February 25, 1982 Sheraton Centre, Toronto

(9) MANAGEMENT ATTITUDES TOWARD RISK, Henry C. Thorne, AACE 22nd Annual Meeting, San Francisco, 1973 and 1978

(10) THE MECHANISM OF MIND, and LATERAL THINKING, Edward-deBono, Penguin Books Ltd., 1973 and 1978

(11) QUICK AND EFFECTIVE RISK ANALYSIS, G.T. Kreamer, AACE 21st Annual Meeting, Milwaukee, 1977

(12) SUCCESSFUL CONSTRUCTION COST CONTROL, Hira N. Ahuja, John Wiley & Sons, 1980

(13) APPLIED COST ENGINEERING, Second Edition, F.D. Clark and A.B. Lorenzoni, Marcel Dekker, Inc., 1985

(14) INSTALLATION OF A CONSTRUCTION COST CONTROL SYSTEM, Kurt G. Heinze, The Bruce Nuclear Generating Station, Ontario Hydro Library, Sept.1981

(15) A COST ENGINEERING CONTROL SYSTEM OVERVIEW, David J. Peeples, AACE Cost Engineering, April 1985

(16) SYSTEM AND ORDER, The royal Bank of Canada Monthly Letter; Vol.52,No.5, May 1971

(17) SYSTEMS/METHODS ANALYSIS, Lecture notes, Brookfield Management Systems, Consulting and Training Ltd., Vancouver, B.C.

(18) BETTER COST REPORTS THROUGH GRAPHIC PRESENTATION, Kurt G. Heinze; AACE 19th Annual meeting, Orlando, Fl. 1975

(19) INFORMATION REQUIREMENT STUDY, Heinze, contributer to Ontario Hydro Design and Construction Division, 1982

(20) SKILLS AND KNOWLEDGE OF COST ENGINEERING, 2nd Edition, AACE International, 3rd edition, Donald F. McDonald

(21) CASH FLOW AND INCURRED COST, Robin F. deSchulthess; AACE 1978 Transactions, San Francisco, CA

(22) WORK BREAKDOWN STRUCTURES IN CONSTRUCTION, Dr. Gui Ponce-Campos and Paul Ricci, AACE 1978 Transactions, San Francisco, CA

(23) MANAGING THE ENGINEERING AND CONSTRUCTION OF SMALL PROJECTS, Richard E. Westney, Marcel Dekker, Inc.,1985

(24) CONSTRUCTION PRODUCTIVITY IMPROVEMENT, James J.Adrian Elsevier Science Publishing Co., Inc., New York, 1987

(25) PERFORMANCE MEASURES BY MEANS OF ACTIVITY SAMPLING, Kurt G. Heinze, AACE 1984 Transactions, Montreal, Que.

(26) FACTORS WHICH INFLUENCE PRODUCTIVITY ON LARGE PROJECTS John D. Borcherding, AACE 1978 Transactions, San Francisco

(27) THE LAYOUT OF TEMPORARY CONSTRUCTION FACILITIES, P.F.Rad & B.M.James, AACE Cost Engineering, April 1983

(28) SCALE CHARTING, Kenneth L. Brookfield, Lecture notes, Brookfield Management Systems, Consulting and Training Ltd., Vancouver, B.C., August 1972

(29) PROJECT CONTROL FOR ENGINEERING, Construction Industry Institute, Publication 6-1, Austin, Texas 1986

(30) GÖDEL, ESCHER, BACH, Douglas R. Hofstädter, Random House of Canada Ltd., Vintage Books Edition, 1980

(31) AUTHOR'S working experience with Ontario Hydro

(32) QUALIFICATION OF ESTIMATES, J.A. Rebata, AACE Transactions 1978, D-2

(33) CONSTRUCTION COST ESTIMATING FOR PROJECT CONTROL, Dr. James Neil

(34) MURPHY'S LAW, Collections by Arthur Bloch, Price/Stern /Sloan Publishers 1979

(35) PROJECT MANAGEMENT, Third Edition, Harold Kerzner, Ph.D., Van Nostrand Reinhold, New York, 1989

(36) PROJECT COST CONTROL FOR MANAGERS, Bill G. Tompkins, Gulf Publishing Co., Houston, 1985

(37) SCIENTIFIC AMERICAN, June 1981 and October 1985

(38) PROJECT CONTROL STRATEGIES FOR MOTIVATING MANAGEMENT, R.A. Mazzini, 1985 AACE Transactions

(39) IMPROVE SCHEDULE FORECASTING VIA EARNED VALUE, Donald J. Cass, 1994 AACE Transactions

(40) USA GENERAL ACCOUNTING OFFICE, 1987 Document GAO/RCED 88-16, Washington DC

(41) SOFTWARE AND SYSTEMS, Chris Vandersluis, Computing Canada, Nov. 1994

(42) NEURAL NETWORK APPLICATIONS to Cost Engineering, Robert A McKim, Cost Engineering, July 1993

(43) METRIC FACT SHEETS, The Canadian Metric Association, 481 Guildwood Pkwy., Scarborough, ON, M1E 1R3

(44) POWERSHIFT, Alvin Toffler, Bantam Books, 1990

(45) CONSTRUCTION SPECIFICATIONS INSTITUT'S MASTERFORMAT, Paul D. Giammalvo, AACE Cost Engineering, Vol. 36/No.7, July 1994.

INDEX